SEDR 104

PROJECT MERCURY FAMILIARIZATION MANUAL

NASA

MANNED SATELLITE CAPSULE

MCDONNELL AIRCRAFT

©2011 Periscope Film LLC   All Rights Reserved   ISBN #978-1-935700-68-5

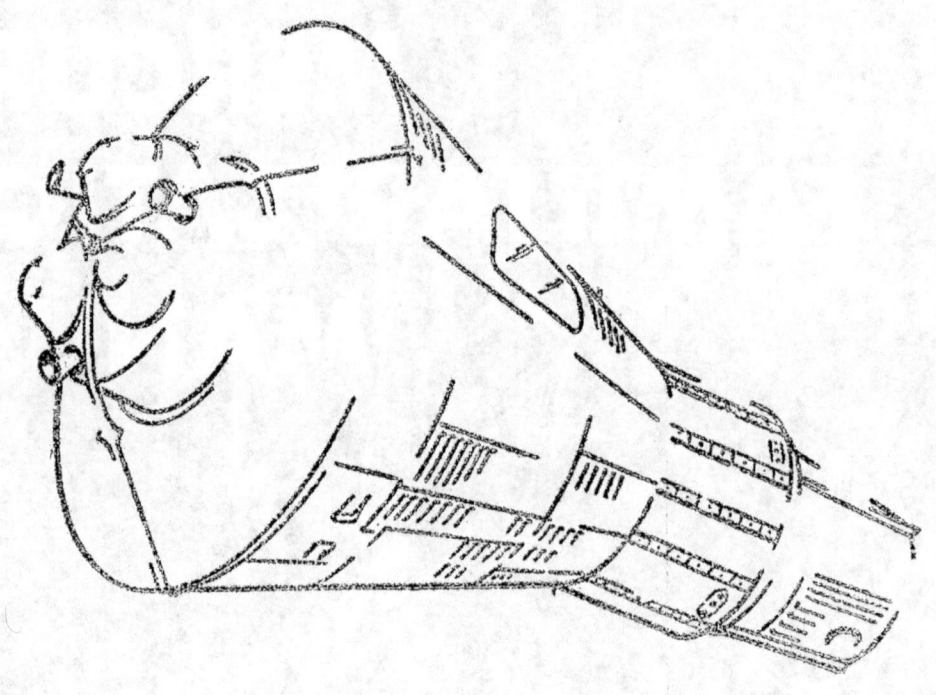

This text has been digitally watermarked to prevent illegal duplication.

PROJECT MERCURY FAMILIARIZATION MANUAL

NASA

MANNED SATELLITE SPACECRAFT
ONE DAY MISSION

1 DECEMBER 1962

MCDONNELL

SEDR 104

FOREWORD

This document is applicable to Spacecraft No. 20 and its one-day orbital mission and supersedes basic SEDR 104-18, dated 1 June 1962, Manned Satellite Spacecraft 18-Orbit Configuration. This document covers Spacecraft No. 20 as delivered but does not include changes generated after delivery.

The purpose of this document is to present a clear operational description of the spacecraft systems and major components. A comparison between systems/components installed in previous spacecraft and those installed in Spacecraft No. 20 can be made by the use of earlier issues of this document. Spacecraft numbered 2, 3, 4, 5, 6, 7, 11, and 14 are covered in the 1 February 1961 issue of SEDR 104 Revised 1 August 1961. Spacecraft numbered 9, 13, 16, 18 and 19 are covered in the 1 November 1961 issue of SEDR 104-3 Revised 1 February 1962.

SECTION INDEX

SECTION I
 INTRODUCTION

SECTION II
 CABIN

SECTION III
 MAJOR STRUCTURAL ASSEMBLIES

SECTION IV
 ENVIRONMENTAL CONTROL SYSTEM

SECTION V
 STABILIZATION CONTROL SYSTEM

SECTION VI
 SEQUENCE SYSTEM, LAUNCH, RETROGRADE OR ABORT

SECTION VII
 SEQUENCE SYSTEM, LANDING THROUGH RECOVERY

SECTION VIII
 ESCAPE AND JETTISON ROCKET SYSTEM

SECTION IX
 POSIGRADE ROCKET SYSTEM

SECTION X
 RETROGRADE ROCKET SYSTEM

SECTION XI
 ELECTRICAL POWER AND INTERIOR LIGHTING SYSTEMS

SECTION XII
 COMMUNICATION SYSTEM

SECTION XIII
 NAVIGATIONAL AIDS

SECTION XIV
 INSTRUMENTATION SYSTEMS

SECTION I

INTRODUCTION

TABLE OF CONTENTS

TITLE	PAGE
Description	1-3
Cabin	1-5
Major Structural Assemblies	1-5
Environmental Control System	1-5
Stabilization Control System	1-5
Sequence System	1-6
Rocket Motors	1-6
Electrical Power And Interior Lighting System	1-6
Communication System	1-6
Navigational Aids	1-8
Instrumentation System	1-8

 MCDONNELL — SEDR 104

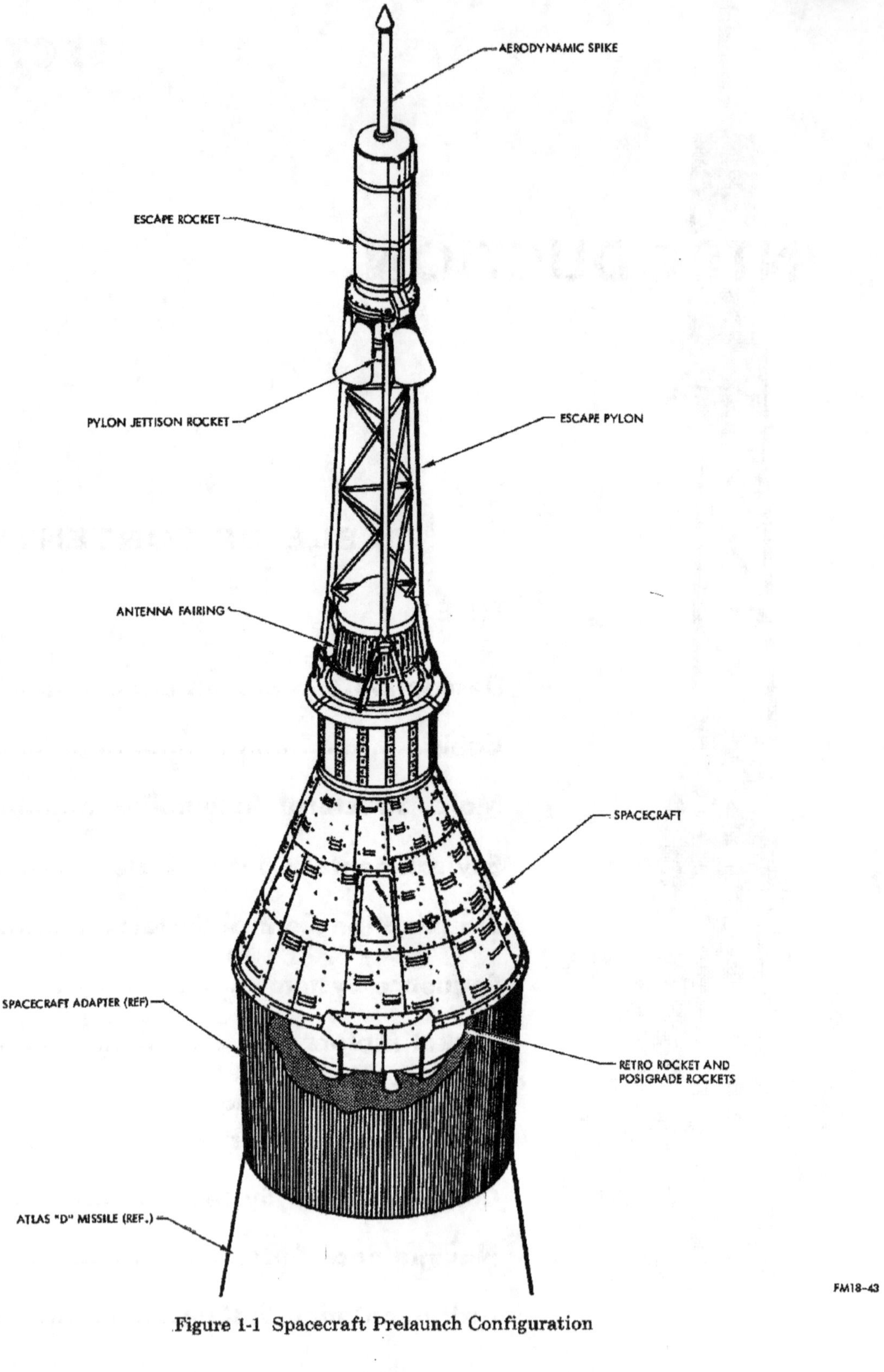

Figure 1-1 Spacecraft Prelaunch Configuration

SEDR 104

I. INTRODUCTION TO PROJECT MERCURY

1-1. **DESCRIPTION**

The Mercury Spacecraft described in this manual is designed for a one-day earth-orbiting mission. This mission is an extension of the three-orbit and six-orbit flights of the Mercury vehicle.

The One-Day Spacecraft is similar to the earlier configurations measuring 74 inches in diameter at the widest part of the spacecraft and 115 inches (See Figure 1-2) from the heat shield to the end of the recovery cylinder. The spacecraft shape resembles a truncated cone topped by a short cylinder and retopped by a shorter truncated cone. The first cone contains the astronaut and his supporting systems, the cylinder contains the recovery aids and parachutes for landing, the second cone is the antenna fairing which contains the bicone antenna and horizon sensing elements plus the drogue chute.

Straddling the antenna fairing is a triangular-shaped trusswork of steel tubing forming the escape tower. Two solid propellant rockets are mounted on top of the tower: one escape, one jettison.

At the base of the spacecraft, a heat shield of ablating glass fiber composition is attached to the impact landing skirt. The complete assembly is held securely to the spacecraft until landing. The retrograde rocket motor pack is attached to the heat shield and is jettisoned after retro-fire.

The first cone is a double-wall structure, the inner wall forming a pressure vessel (the cabin) and the outer shell forming a heat shield.

The following paragraphs briefly describe the material covered in this manual.

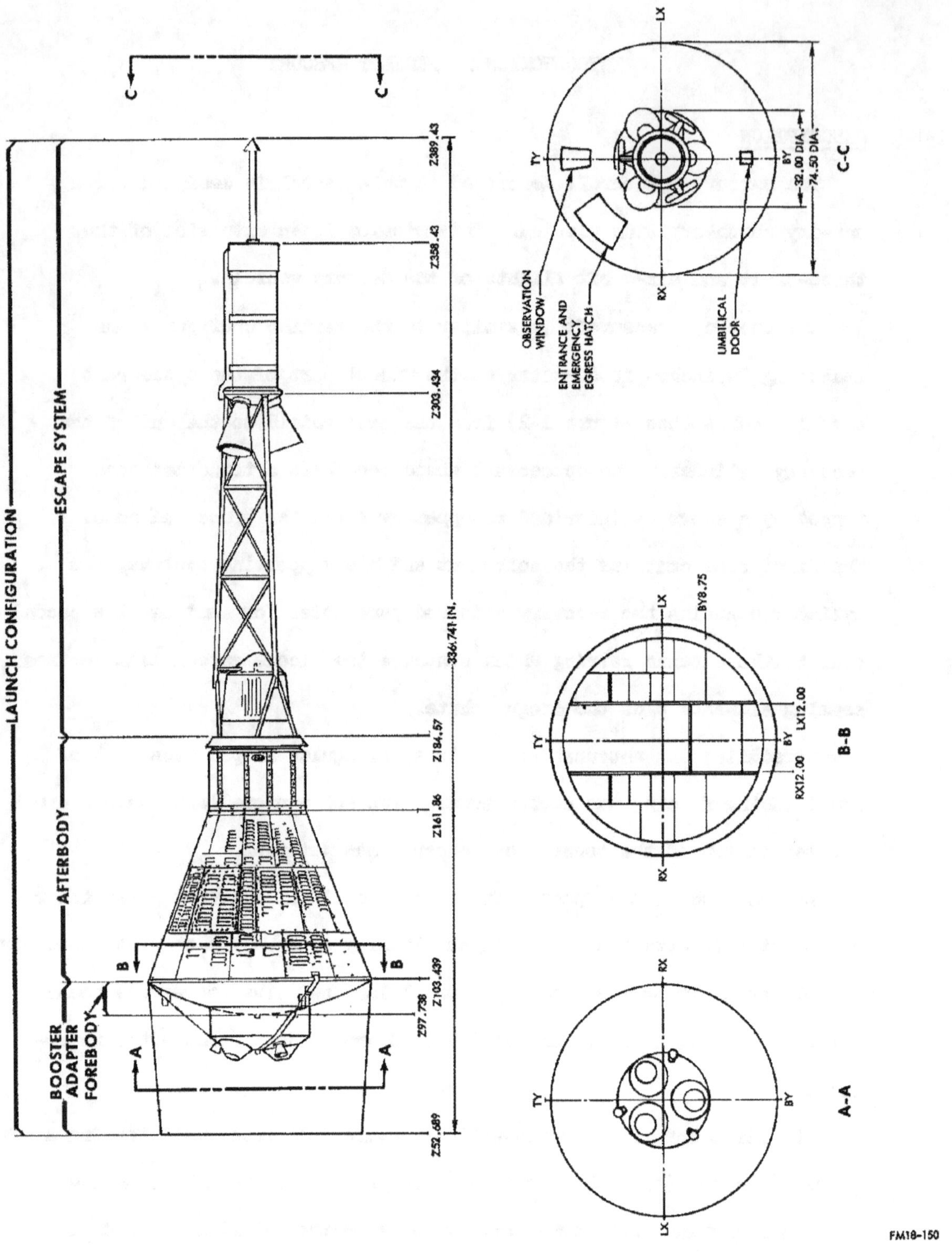

MCDONNELL — SEDR 104

Figure 1-2 Principal Dimensions

1-2. CABIN

The inner wall of the spacecraft forms the cabin structure and houses the astronaut in a pressurized environment. The equipment within the cabin is arranged so that all operating controls and emergency provisions are accessible to the astronaut when in the normal restrained position. See Section II.

1-3. MAJOR STRUCTURAL ASSEMBLIES

The structure is of a conventional semi-monocoque construction utilizing titanium for the primary structure. The outer shell is designed to protect the internal cabin from excessive heating, noise and meteorite penetration. The internal cabin is designed to provide a safe environment for the astronaut. See Section III.

1-4. ENVIRONMENTAL CONTROL SYSTEM

The Environmental Control System provides a livable environment for the astronaut by controlling the gaseous composition, temperature, humidity and pressure as well as cooling the electronic equipment aboard the spacecraft. See Section IV.

1-5. STABILIZATION CONTROL SYSTEM

The Stabilization Control System provides stabilization and orientation of the spacecraft from the time of separation from launch vehicle until antenna fairing separation. The system operates either automatically or manually using hydrogen peroxide as a propellant. The hydrogen peroxide is forced over a catalyst where it decomposes into steam and oxygen producing the exhaust jet controlling the attitude of the spacecraft. See Section V.

SEDR 104

1-6. **SEQUENCE SYSTEM**

The Sequence System is set in operation at liftoff providing the initial signal which is a basic reference point. This reference is used to establish the order of events in relation to time. The system functions from liftoff through the post landing phases of the flight in normal or abort sequences as flight conditions dictate. See Sections VI and VII.

1-7. **ROCKET MOTORS**

The escape rocket is used to carry the spacecraft out of danger in the event of a failure. The jettison rocket separates the tower from the spacecraft under normal and abort conditions. See Section VIII.

The posigrade rocket motors are used to accomplish separation between the spacecraft and the booster. See Section IX.

The retrograde rocket motors slow the spacecraft sufficiently to re-enter the earth's atmosphere. See Section X.

1-8. **ELECTRICAL POWER AND INTERIOR LIGHTING SYSTEM**

The spacecraft is supplied 6, 12, 18 and 24 volt d-c power from six silver-zinc batteries. Battery voltage, transformed to 115 volt, 400 cycle, single-phase a-c power, is supplied by two main and one standby inverters. The interior lighting consists of two fluorescent flood lights mounted on either side of the astronaut and a series of warning telelights on the instrument panels. See Section XI.

1-9. **COMMUNICATION SYSTEM**

During flight and recovery phases, communication equipment and tracking aids are used. The flight communication equipment includes UHF and HF

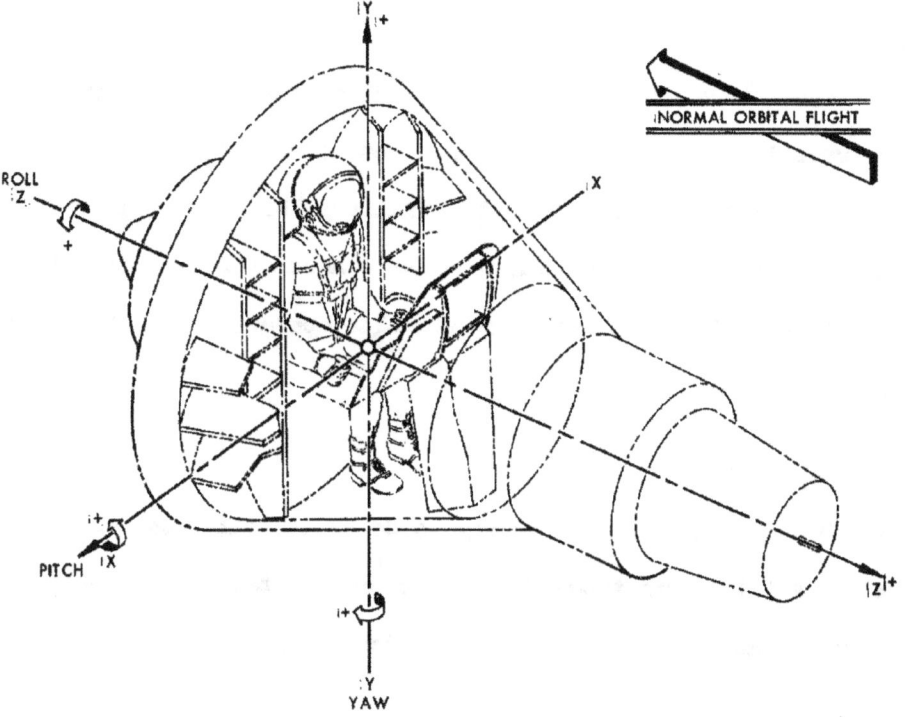

PITCH

PITCH IS DEFINED AS THE ROTATION OF THE SPACECRAFT ABOUT ITS X-AXIS. THE PITCH ANGLE IS ZERO DEGREES (0°) WHEN THE Z-AXIS LIES IN A HORIZONTAL PLANE. USING THE ASTRONAUT'S RIGHT SIDE AS A REFERENCE, POSITIVE PITCH IS ACHIEVED BY COUNTERCLOCKWISE ROTATION FROM THE ZERO DEGREES (0°) PLANE. THE RATE OF THIS ROTATION IS THE SPACECRAFT PITCH RATE AND IS POSITIVE IN THE DIRECTION SHOWN AS ARE THE CONTROL MOVEMENTS WHICH CAUSE IT. THE CONTROL HANDLE MOVES TOWARD THE ASTRONAUT AND THE POSITIVE +, PITCH REACTION JET FIRES.

YAW

YAW IS DEFINED AS THE ROTATION OF THE SPACECRAFT ABOUT IT'S Y-AXIS. CLOCKWISE ROTATION OF THE SPACECRAFT, WHEN VIEWED FROM ABOVE THE ASTRONAUT, IS CALLED RIGHT YAW AND IS DEFINED AS POSITIVE (+).

THIS MOVEMENT IS PRODUCED BY POSITIVE CONTROL MOTION. THE CONTROL HANDLE IS ROTATED CLOCKWISE (AS VIEWED FROM ABOVE THE ASTRONAUT) AND THE POSITIVE (+) YAW REACTION JET FIRES. YAW ANGLE IS CONSIDERED ZERO DEGREES (0°) WHEN THE SPACECRAFT IS IN NORMAL ORBITAL POSITION (BLUNT END OF SPACECRAFT FACING LINE OF FLIGHT). WHEN THE POSITIVE Z-AXIS OF THE SPACECRAFT IS DIRECTED ALONG THE ORBITAL FLIGHT PATH (RECOVERY END OF SPACECRAFT FACING LINE OF FLIGHT), THE YAW ANGLE IS 180°.

ROLL

ROLL IS DEFINED AS THE ROTATION OF THE SPACECRAFT ABOUT ITS Z-AXIS. CLOCKWISE ROTATION OF THE SPACECRAFT, AS VIEWED FROM BEHIND THE ASTRONAUT, IS CALLED RIGHT ROLL AND IS DEFINED AS POSITIVE (+). THIS MOVEMENT IS INITIATED BY MOVING THE CONTROL HANDLE TO THE RIGHT, THEREBY FIRING THE POSITIVE (+) ROLL ACTION JET. WHEN THE X-AXIS OF THE SPACECRAFT LIES IN A HORIZONTAL PLANE, THE ROLL ANGLE IS ZERO DEGREES (0°).

ACCELEROMETER POLARITY WITH RESPECT TO GRAVITY

WITH THE SPACECRAFT IN THE LAUNCH POSITION THE Z-AXIS WILL BE PERPENDICULAR TO THE EARTH'S SURFACE AND THE Z-AXIS ACCELEROMETER WILL READ +1 "G".

Figure 1-3 Spacecraft Polarity Orientation With Respect To Astronaut

SEDR 104

voice transmitters and receivers, command receiver and decoder, telemetry transmitter and "C" and "S" band beacons for radar tracking. The recovery equipment includes the auxiliary UHF Rescue and the HF Rescue Beacons used for DF-homing on the spacecraft. The normal HF and UHF transmitters and receivers are also operative to aid in recovery. See Section XII.

1-10. NAVIGATIONAL AIDS

The Navigational Aids necessary to determine altitude, course, velocity, attitude and time of re-entry are located conveniently for quick reference by the astronaut. See Section XIII.

1-11. INSTRUMENTATION SYSTEM

The Instrumentation System monitors pressures, temperatures, operation of various units throughout the spacecraft and events that occur throughout the flight. The astronaut's respiration rate and volume, temperature and heart action are monitored continuously; blood pressure is monitored periodically by a start-stop measurement cycle which requires active astronaut participation in starting the cycle. Heart action monitoring (EKG) is interrupted during blood pressure measurement. Data is routed to the telemetry equipment and to the tape recorder. See Section XIV.

SECTION II

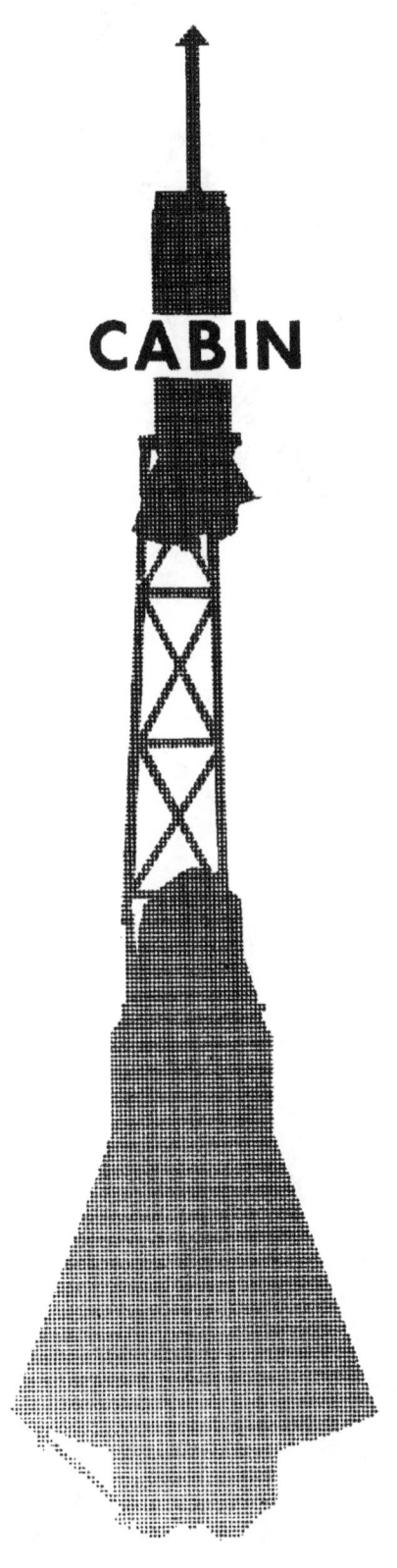

CABIN

TABLE OF CONTENTS

TITLE	PAGE
Arrangement	2-3
Support Couch	2-3
Restraint System	2-6
Controls	2-6
Instrument Panels	2-6
Static System	2-9
Survival Equipment	2-13
Food and Water	2-13
Astronaut's Apparel	2-13
Spacecraft Recovery	2-16

 SEDR 104

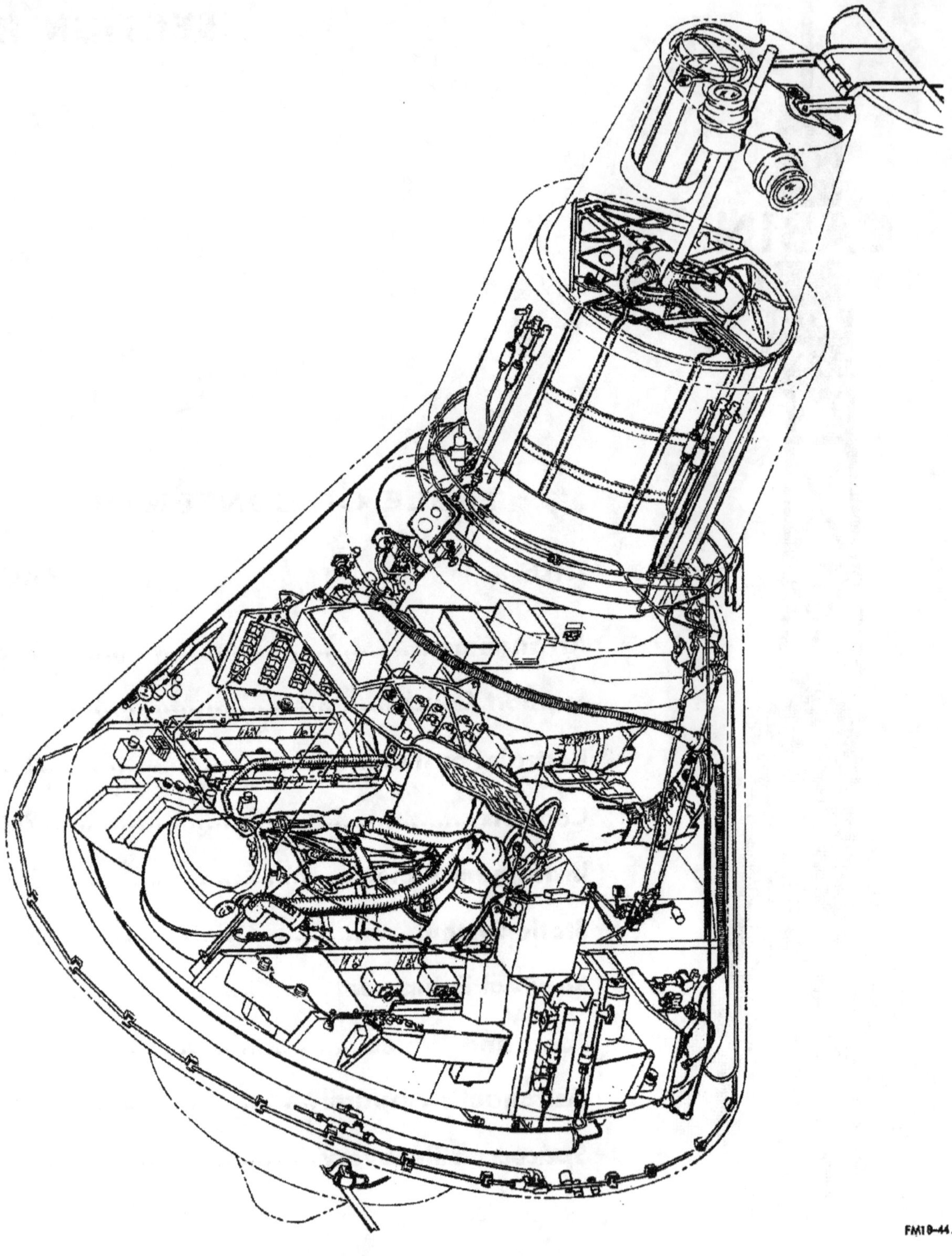

Figure 2-1 Interior Arrangement

II. CABIN

2-1. **ARRANGEMENT**

The equipment within the cabin, Figures 2-1 and 2-2, is arranged so that all normal and emergency controls are accessible to the astronaut when in the normal restrained position. The cabin equipment basically consists of the support couch, and restraint system, environmental system, navigational aids, flight and abort control handles, instrument and display panels, food and water supply, survival kit, communication equipment, camera and other equipment as needed for the mission.

2-2. **SUPPORT COUCH**

The astronaut's support couch, Figure 2-3, is designed to firmly support the astronaut's body during launch, re-entry and landing phases of the mission. The support couch also protects the astronaut from loss of consciousness during peak acceleration periods and from possible injury at impact. The couch is centrally located adjacent to the large pressure bulkhead.

The couches are individually tailored to each astronaut. They consist of a honeycomb material, common to all couches, and an individually tailored glass fiber shell. The shell is padded with a fitted liner for the astronaut's comfort. The honeycomb material is crushable aluminum and the shell is polyester resin-reinforced glass fiber laminate. Honeycomb material is also installed between the large pressure bulkhead and the couch to absorb impact loads. The couch is fabricated in sections to enable installation through the hatch opening.

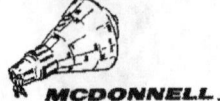

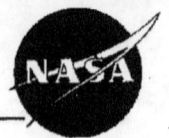

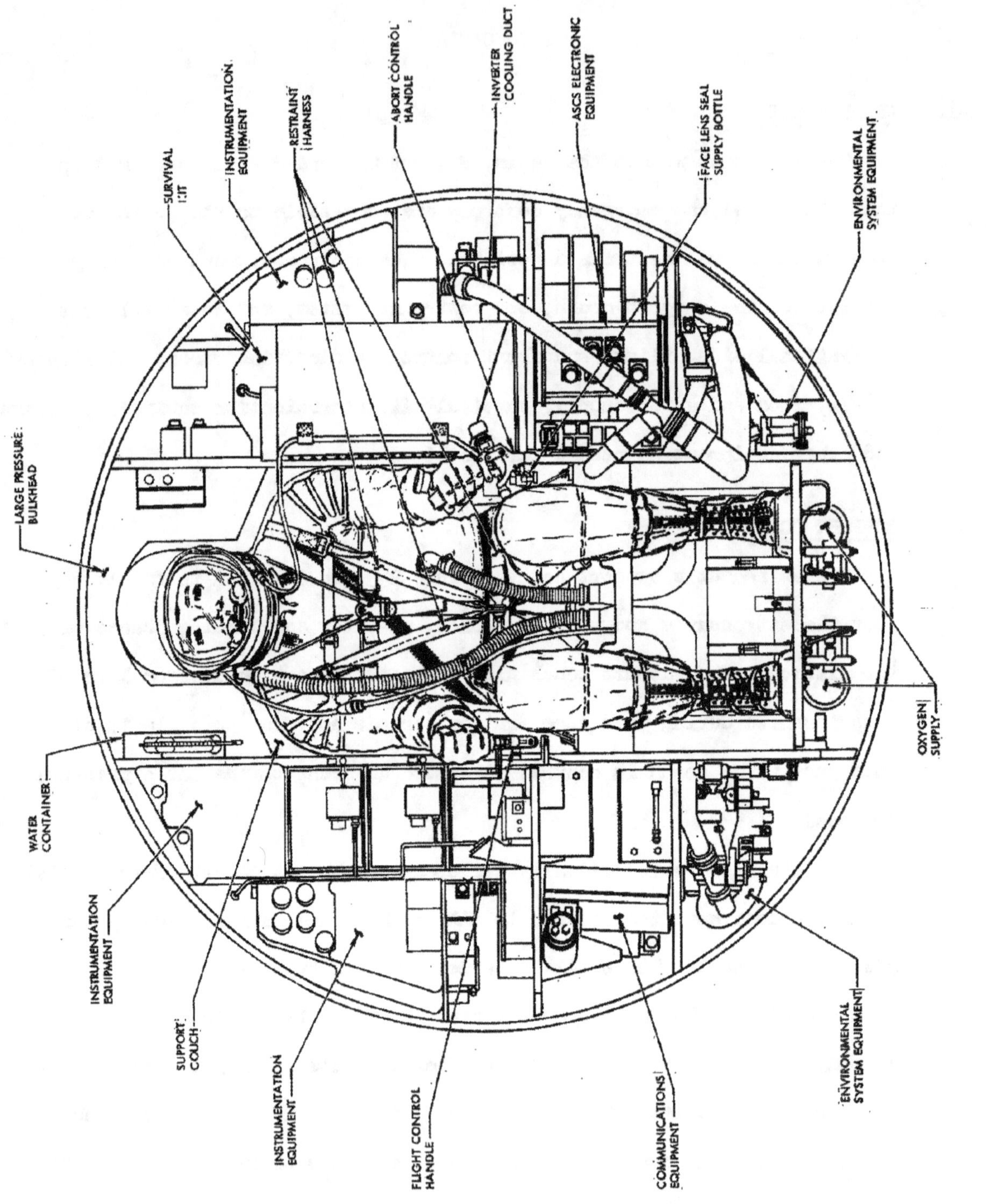

FM18-105-1A

Figure 2-2 Cabin Equipment (Sheet 1 of 2)

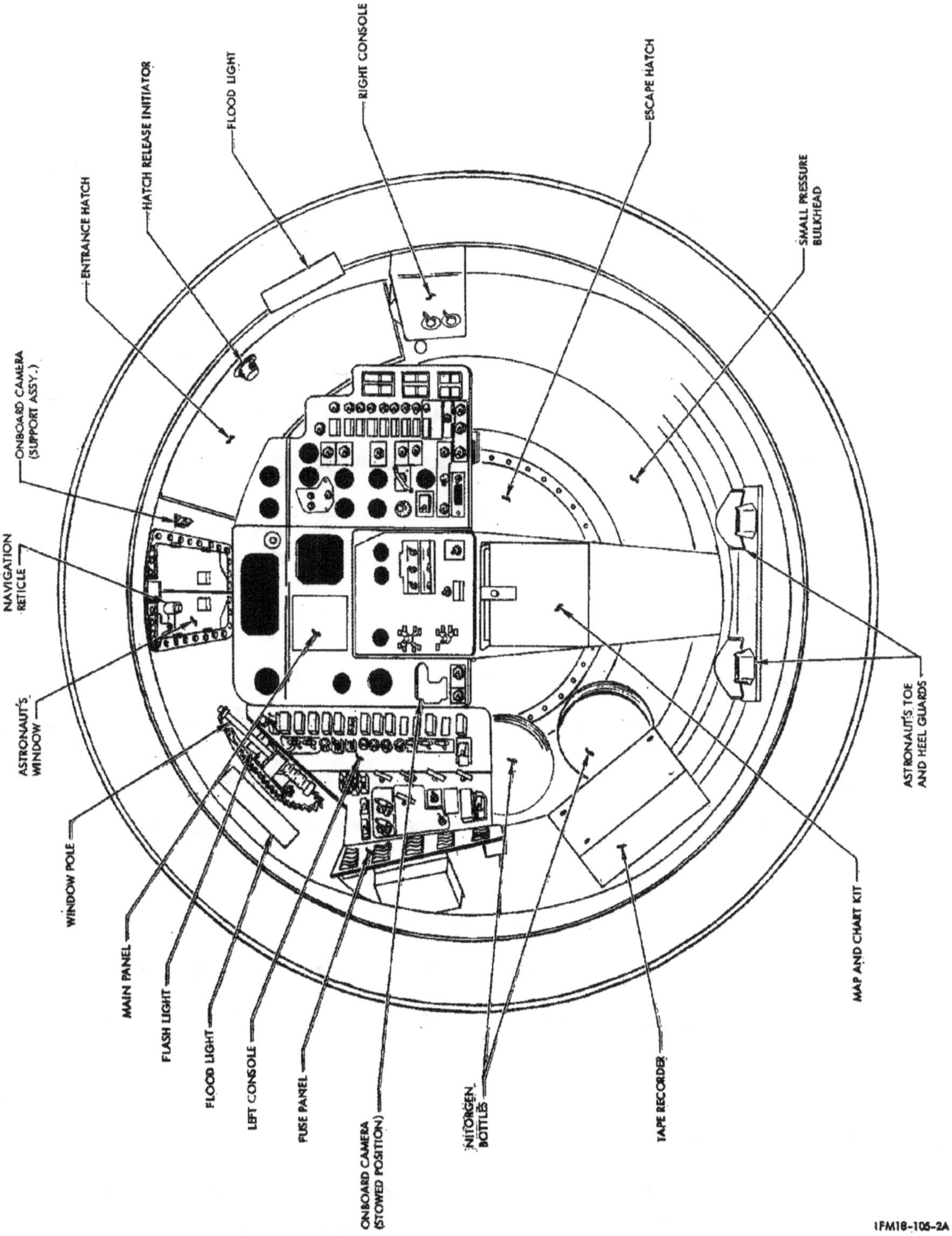

Figure 2-2 Cabin Equipment (Sheet 2 of 2)

SEDR 104

2-3. **RESTRAINT SYSTEM**

The astronaut's restraint system, Figure 2-4, is designed to firmly restrain the astronaut in the support couch during maximum acceleration, deceleration and to aid positioning during weightless conditions. The restraint system consists of shoulder, chest, and crotch straps, lap belt, and toe and heel guards. The shoulder straps may be adjusted to restrain or release the astronaut, by a harness reel control handle, located forward of the upper left side of the support couch. The chest and shoulder straps firmly restrain the upper torso. The lap belt and crotch strap supports the lower torso. The toe and heel guards support the feet. The astronaut's hands and arms are restrained by gripping the abort and flight control handles, located near the ends of the support couch arm rests.

2-4. **CONTROLS**

Spacecraft controls are located forward of each arm rest of the support couch. An emergency escape handle is located forward of the support couch left arm rest. The escape handle is utilized to initiate the abort sequence. To prevent inadvertent actuation of the escape system, the escape handle is provided with a manual lock. The manual control handle, located forward of the support couch right arm rest, is utilized as an override system to control flight attitude of the spacecraft in the event the automatic system fails or for normal manual flight control if desired for a portion of the mission.

2-5. **INSTRUMENT PANELS**

The instruments and controls are located on the main instrument panel, left and right consoles. See Figures 2-5 and 2-6. The main instrument

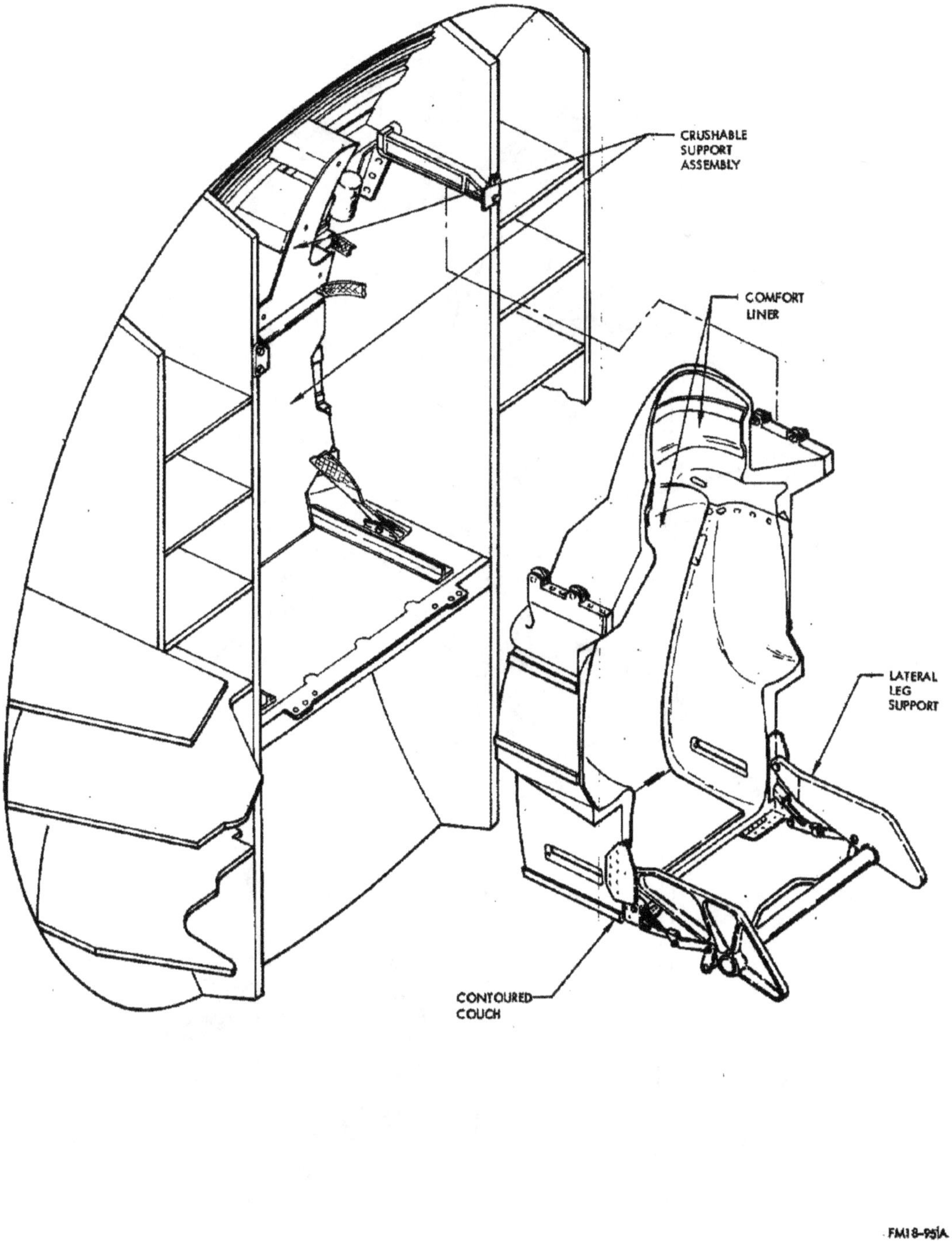

SEDR 104

Figure 2-3 Astronaut's Support Couch

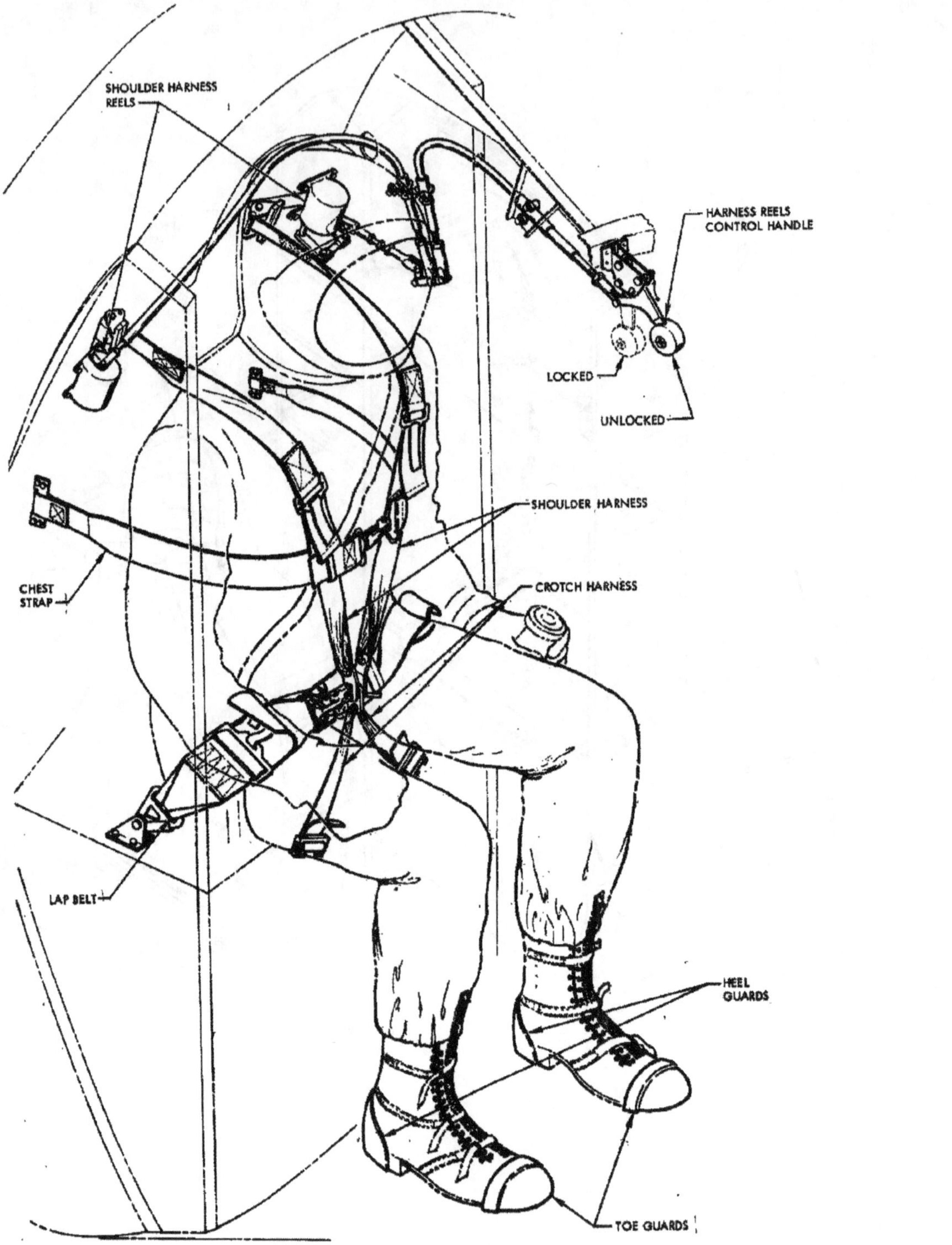

Figure 2-4 Astronaut's Restraint System

panel is located directly in front of the astronaut's support couch as viewed by the astronaut. The satellite clock, attitude indicator, communication controls, warning lights, electrical switches, and indicators are located on the main instrument panel.

The left hand console is located on the left side of the main panel and is accessible and visible to the astronaut when in the fully restrained position. The console includes a telelight sequence, indicators and controls. The right hand console, located below the entrance hatch, includes controls for the environmental control system. A separate coaxial switch is included on the right hand side for HF antenna selection by the astronaut. The Navigational Reticle, which permits optical alignment of the retrofire attitude of the spacecraft, is supported by a stowable bracket. The reticle when stowed is located to the left of the astronaut's window; when in operating position, the illuminated reticle is directly in line with the astronaut's left eye and the window and forms an angle of $20\frac{1}{2}°$ with the longitudinal axis of the spacecraft. A camera support, which causes a camera angle also of $20\frac{1}{2}°$, with respect to the longitudinal axis of the spacecraft, is mounted to the right of the window. Additional equipment will be added to the instrument panels as it is needed. A window pole, located adjacent and to the left of the observation window, enables the astronaut to actuate controls with the suit pressurized.

2-6. **STATIC SYSTEM**

The static system provides for atmospheric pressure measurement necessary for switches and indicators to function in the spacecraft. The altimeter, rate of descent indicator, four baroswitches and a pressure transducer are connected to the static access port in this system. See Figure 2-7.

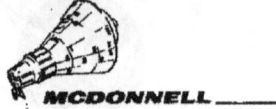

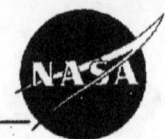

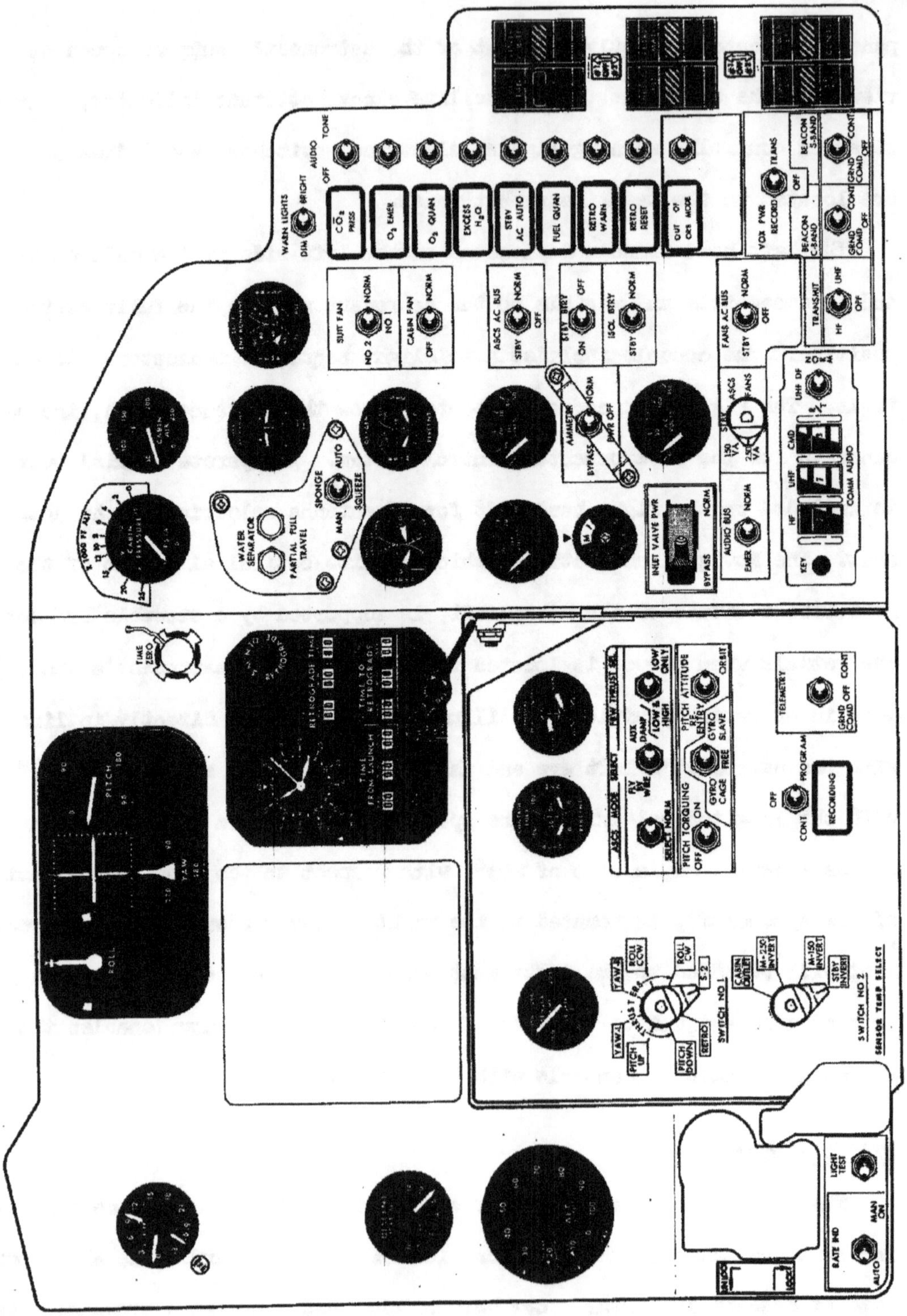

FM98-93A

Figure 2-5 Main Instrument Panel

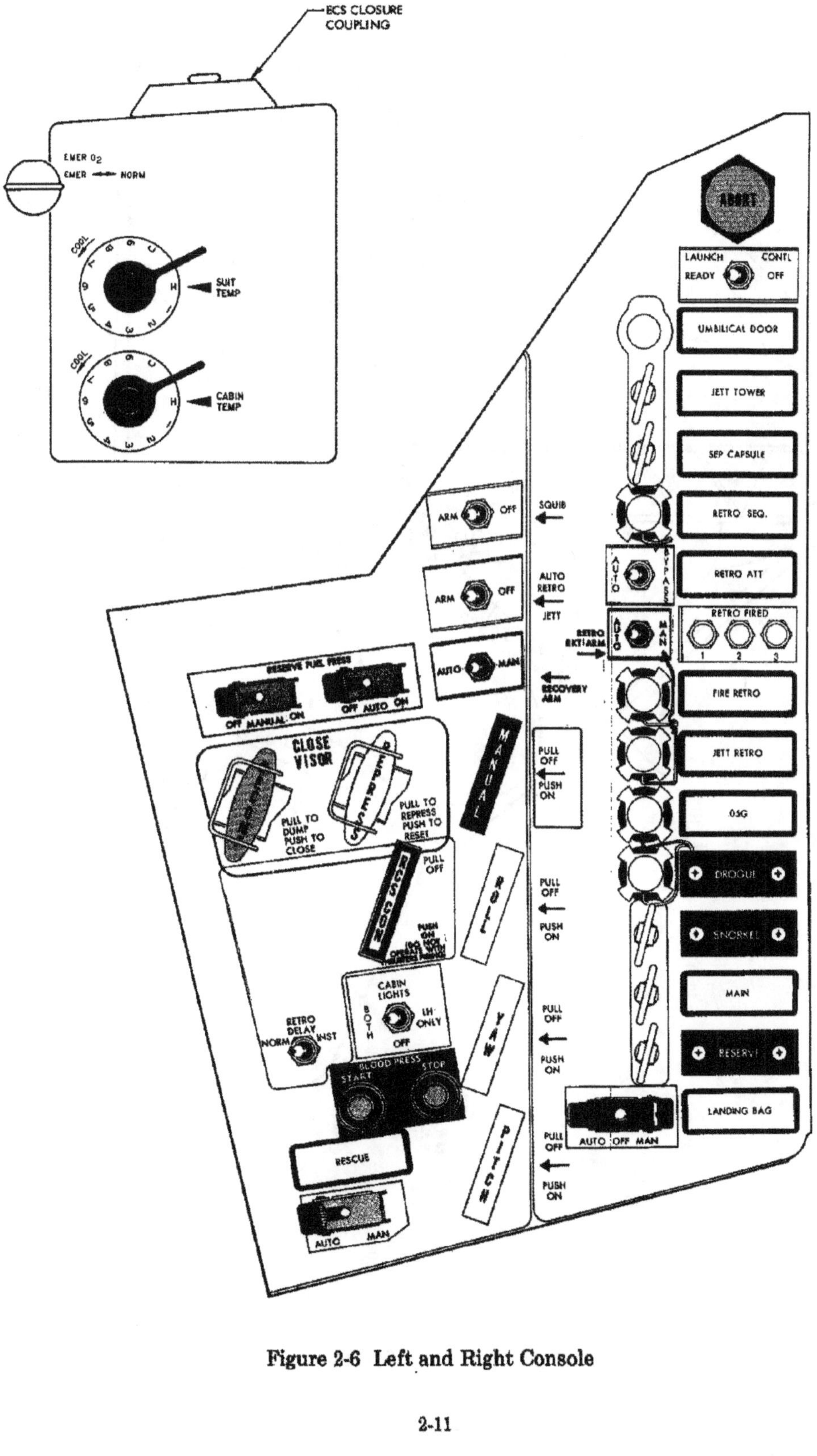

Figure 2-6 Left and Right Console

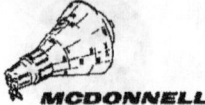

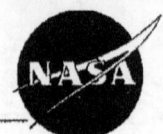

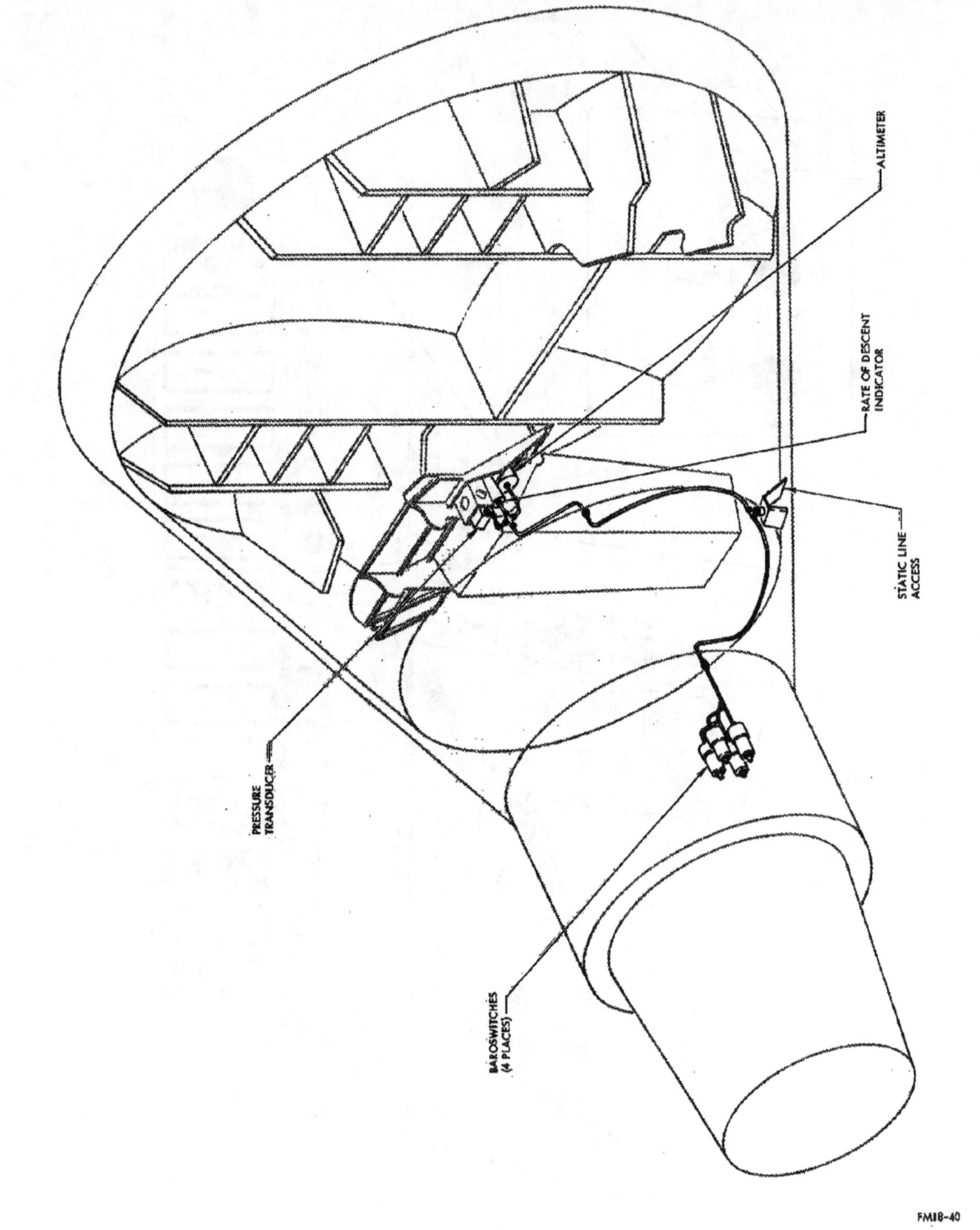

FM18-40

Figure 2-7 Static System

SEDR 104

2-7. SURVIVAL EQUIPMENT

The survival kit, see Figure 2-8, stowed at the left side of the couch, contains the following:

1 Water Container	1 First Aid Kit	1 Food Container
1 Life Raft	1 Bar Soap	1 Container of Matches
1 Desalting Kit (For 8 pts)	3 Morphine and Anti-Seasickness Inj.	1 Whistle
1 Shark Repellant Package		1 Nylon Cord (10 ft).
1 Tube Zinc Oxide	1 Sun Glasses & Case	1 Signal Light
3 Dye Markers	1 SARAH Rescue Beacon	1 Pocket Knife
1 Receiver-Beacon Transceiver	1 Signal Mirror	

Plus additional items deemed necessary for the mission.

A flashlight is located adjacent and to the left of the observation window.

2-8. FOOD AND WATER

Food will be supplied as required by the mission. A six pound water container provides the normal source of drinking water for the astronaut. A secondary water supply containing up to five and one-half pounds of water is located in the survival kit. Tank containing forty-nine pounds of water are provided to supply the Environmental Control System Cabin and suit circuit heat exchangers. The water remaining in these tanks, at the end of the flight provides an additional backup source of drinking water.

2-9. ASTRONAUT'S APPAREL

The astronaut's apparel consists of a completely enveloping pressure suit with helmet, suitable undergarments, and boots. The helmet faceplate can be opened while the spacecraft interior is pressurized although normal

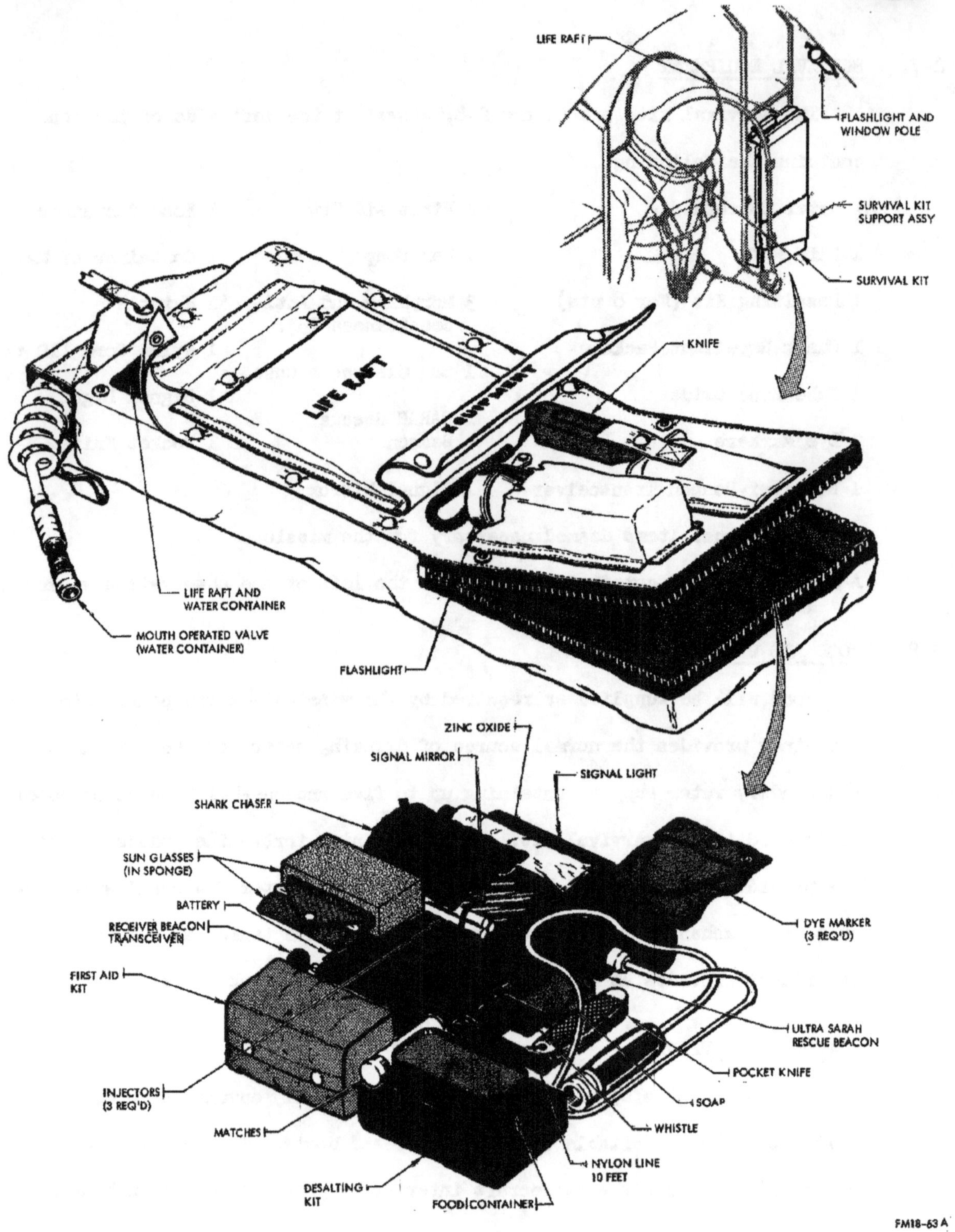

Figure 2-8 Survival Kit

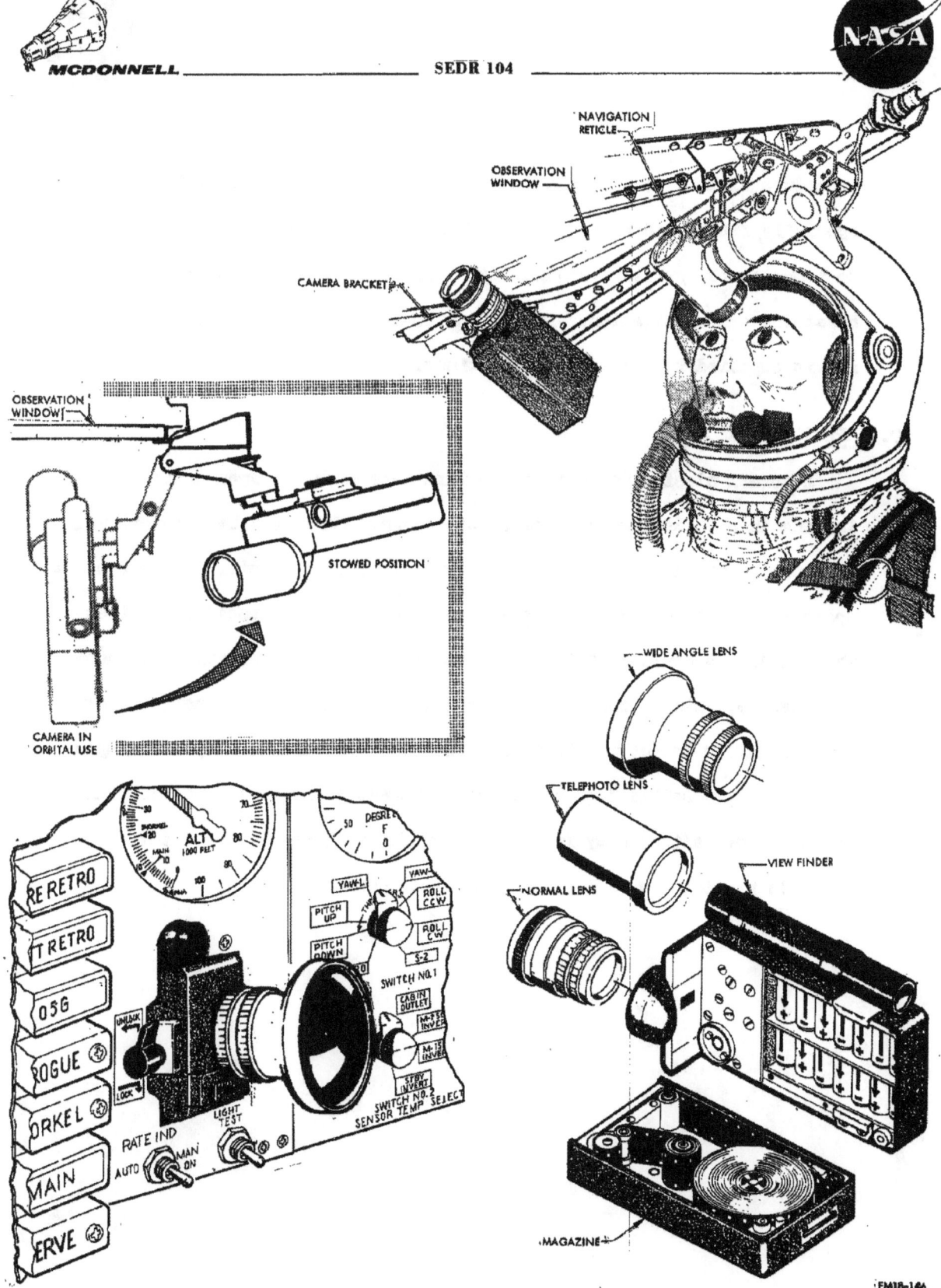

Figure 2-9 Pilots Camera

procedure will be to keep the faceplate closed. Each astronaut is specially fitted and trained to use his suit. Oxygen, regulated as to temperature, pressure and humidity, is supplied to the suit for breathing and ventilation. For the astronaut's comfort, ventilating air is supplied to the suit at all times.

2-10. SPACECRAFT RECOVERY

A normal mission is intended to terminate with the spacecraft landing in a predetermined area of the ocean. Ships and helicopters will be standing by in the recovery area with provisions to pick up the buoyant spacecraft immediately after landing. Considering the possibility that the spacecraft could land in other than the intended recovery area; numerous devices, both electronic and visual, are automatically energized or deployed after impact to aid in locating the spacecraft. Depending upon the weather, possible damage and the astronaut's physical condition, the astronaut may either stay in the spacecraft or egress to the liferaft which is provided as part of the survival equipment.

SECTION III

MAJOR STRUCTURAL ASSEMBLIES

TABLE OF CONTENTS

TITLE	PAGE
Introduction	3-3
Spacecraft Forebody	3-4
Spacecraft Afterbody	3-6
Entrance Hatch	3-7
Observation Window	3-9
Small Pressure Bulkhead	3-11
Large Pressure Bulkhead	3-11
Recovery Compartment	3-13
Antenna Fairing	3-14
Destabilizer Flap	3-16
Impact Landing System	3-16
Escape Tower	3-20
Pylon-Spacecraft Clamp Ring	3-22
Missile Adapter	3-22
Spacecraft-Adapter Clamp Ring	3-25
Retro-Package	3-27

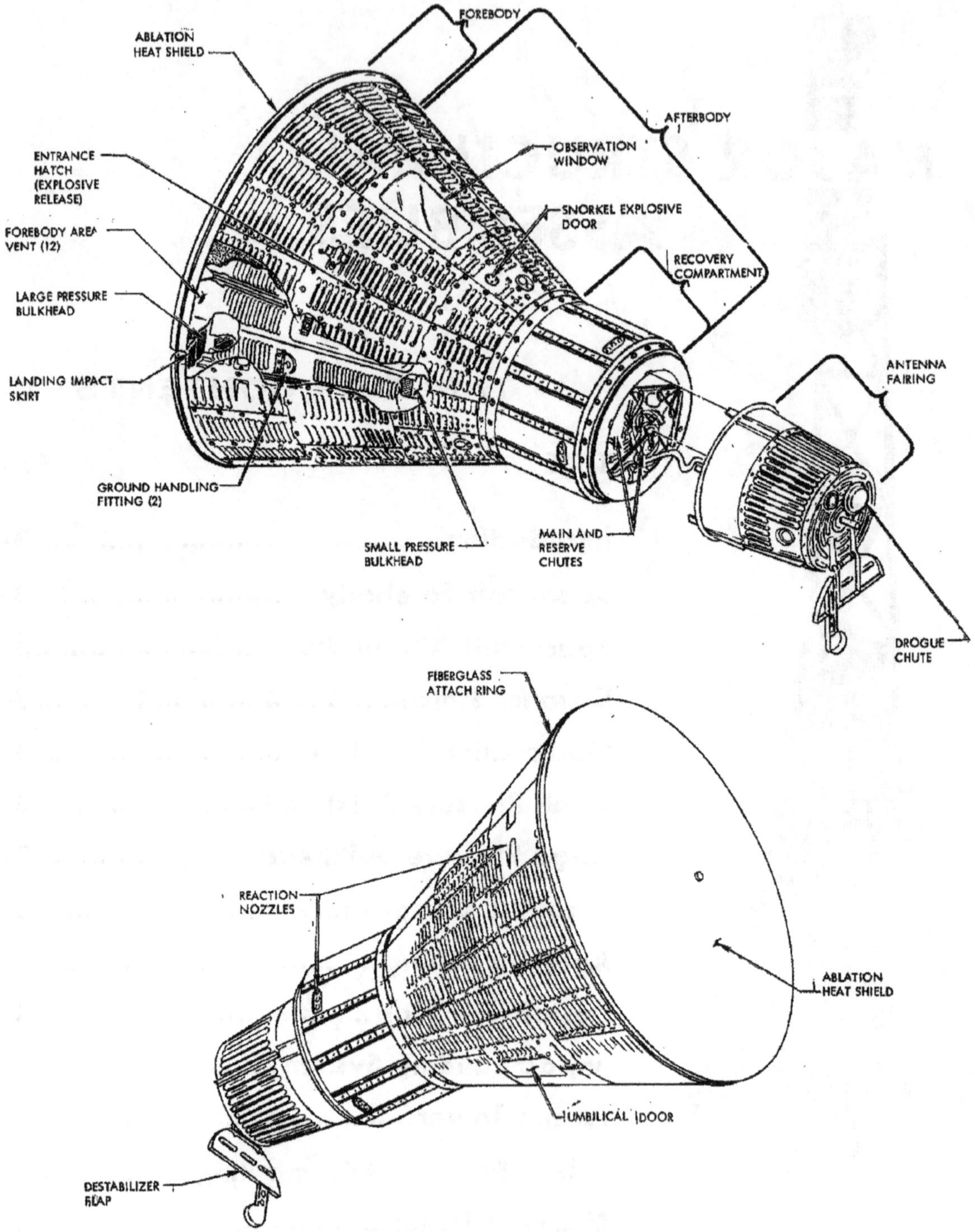

Figure 3-1 Spacecraft Structure

SEDR 104

III. MAJOR STRUCTURAL ASSEMBLIES

3-1. **INTRODUCTION**

The Project Mercury spacecraft, Figure 3-1, is designed to contain an astronaut, during orbital flight. The spacecraft also contains recording, environmental stabilization and other equipment necessary for the flight.

The spacecraft is of a conventional semi-monocoque construction utilizing titanium for the primary structure. The structure is designed to protect the internal cabin from excessive heating, noise and meteorite penetration. The spacecraft is basically a conical configuration consisting of a forebody and afterbody. The forebody is the large dish shape structure called the heat shield. The afterbody consists of a conical mid-section attached to a small cylindrical section. During the orbital flight, the forebody is forward with respect to the flight path. Provisions are incorporated to permit cabin entry, normal and emergency exit and exterior viewing.

Prior to flight, an escape tower and antenna fairing are attached to the afterbody cylindrical section. The escape tower, designed to aid in spacecraft-missile emergency separation, consists of a pylon framework equipped with rockets. The antenna fairing is a cylindrical shaped structure containing the radio main receiving and transmitting antenna. The escape tower is jettisoned during the launch phase or during an escape sequence. During the landing phase the antenna fairing is ejected and serves to deploy the main chute.

3-2. **FOREBODY**

The forebody, Figure 3-1, mainly consists of a large, blunt, dish shaped structure that is supported by the large pressure bulkhead and adjoins the afterbody conical section. The large pressure bulkhead internally separates the forebody from the afterbody. The forebody dish shaped structure is an ablation heat shield that is designed to protect the spacecraft from extreme thermal conditions during re-entry flight. It is also designed to prevent damage upon landing impact. The heat shield is attached to the heat shield attach ring, which in turn is riveted to the conical section inner skin. The heat shield attach ring incorporates elongated holes, for the installation of the heat shield to the spacecraft and to allow for thermal expansion. The ablation shield is designed to ablate under heating conditions and is constructed of fiberglass to form a smooth contour. A retrograde package assembly is attached to the heat shield by means of three straps. The retrograde package is jettisoned from the spacecraft following retrograde rocket firing, which initiates spacecraft re-entry.

The forebody area, between the large pressure bulkhead and the heat shield, is vented to atmosphere through a series of vents located around the periphery of the spacecraft forebody, adjacent to the forebody and afterbody junction. Three toroidal shaped hydrogen peroxide tanks and six reaction control nozzles, each covered with Min-K heat insulation, are located in the forebody area. The forebody area also houses the heat shield release pneumatic system. A landing impact skirt is also stored in the forebody area. The rubber impregnated fiberglass impact skirt, attached to the heat shield attach ring and the heat shield, is designed to absorb high energy shock loads encountered during a landing

on land or water; and also to stabilize the spacecraft during an astronaut's egress, following a landing in the water. During the landing phase, the heat shield is released, and extends the full length of the impact skirt. Upon heat shield contact with land, air within the impact skirt is forced out through a series of holes located in the impact skirt wall which in turn provides a cushion-like effect. To prevent damage to the large pressure bulkhead in the event the heat shield strikes the spacecraft during landing, the large pressure bulkhead incorporates a reinforced laminated fiberglass shield assembly. The fiberglass shield assembly is attached to the torus tank support brackets. Sandwiched between the fiberglass shield and the large pressure bulkhead are sections of honeycomb. Twenty-four straps fabricated of stainless steel are located about the periphery of the impact landing skirt to prevent tearing of the impact landing skirt during high horizontal velocity water landings. Twenty-four stainless steel cables are located inside the impact landing bag and alternately located in relation to the steel straps. The cables retain the heat shield to the spacecraft in the event strap failure should occur. The afterbody conical section exterior shingle arrangement extends beyond the large pressure bulkhead, to the forebody heat shield, and encloses the equipment located between the large pressure bulkhead and the heat shield. Located adjacent to the forebody and afterbody juncture, and bolted to the heat shield attach ring, is a fiberglass attach ring. During spacecraft-adapter installation, the fiberglass attach ring and the adapter attach flange are clamped together with a segmented clamp ring. Receptacles for the retro-package, adapter, and the clamp ring pneumatic

and electrical connectors are located under the forebody shingles adjacent to fiberglass attach ring. Five spring loaded access doors, for the receptacles are incorporated in the shingles.

3-3. **AFTERBODY**

The afterbody conical mid-section mainly consists of a pressurized cabin that is supported between a small pressure bulkhead and the large pressure bulkhead. The cabin interior wall is lined with channeled frames to provide additional structural strength and equipment attach points. The mid-section is constructed of a conically formed inner and outer titanium shell, seam welded together. The outer skin is beaded to form small sealed pressure panels capable of withstanding high pressures and structural loads. The outer conical skin is reinforced with longitudinal hat stringers. A blanket of thermoflex insulation is bonded, in between the hat stringers, to the outer (beaded) conical skin. Min-K insulation is also installed over the hat sections and covered with a shingle arrangement. The shingle arrangement is similar to the shingle installation used on the recovery system compartment. The forward end of the conical section is attached to the forebody heat shield. The combination of the conical section beaded outer skin, the hat section reinforcements, thermoflex insulation and external shingle installation provide the spacecraft with adequate heat, noise and meteorite protection.

Located in the bottom of the conical section, as viewed during normal flight attitude, is a spring loaded door that enclosed the ground checkout umbilical receptacle. The door, hinged to the spacecraft structure, automatically closes when the umbilical is disconnected.

Two auxiliary hoist fittings, attached to left and right side of the spacecraft, provide ground handling attach points. The hoist fittings are removed prior to launch. An explosive door, namely the snorkel door, is provided in the spacecraft shingles, between the small pressure bulkhead and the conical-cylindrical sections juncture. This door is exploded from the spacecraft during landing. Actuation of the door enables cool air to be drawn into the spacecraft through the snorkel valve when the cabin air inlet valve opens.

3-4. ENTRANCE HATCH

An entrance hatch, Figure 3-2, is located on the right side of the afterbody conical section as viewed from the crew member station. Entrance hatch construction, similar to the conical section construction, consists of an inner and outer (beaded) skin seam welded together and reinforced with hat stringers. An explosive charge, moulded in the hatch sill, is provided to quickly release the hatch and enable the astronaut to egress rapidly. An explosive charge initiator, located in the upper aft corner of the hatch, is linked to an internal release control initiator. Prior to launch, the hatch is bolted and sealed into position with bolts, and two corrugated shingles are installed over the hatch. The bolts are inserted through the entrance hatch sill, which incorporates the explosive charge, and threaded into the spacecraft sill. A magnesium gasket, with inlaid rubber, forms the hatch seal when the hatch is bolted into position. Two hatch shingles are attached to the hatch stringers, but in no manner are they attached to spacecraft shingles. (This enables the hatch to separate cleanly, upon ignition of hatch explosive charge.) Following impact, the astronaut

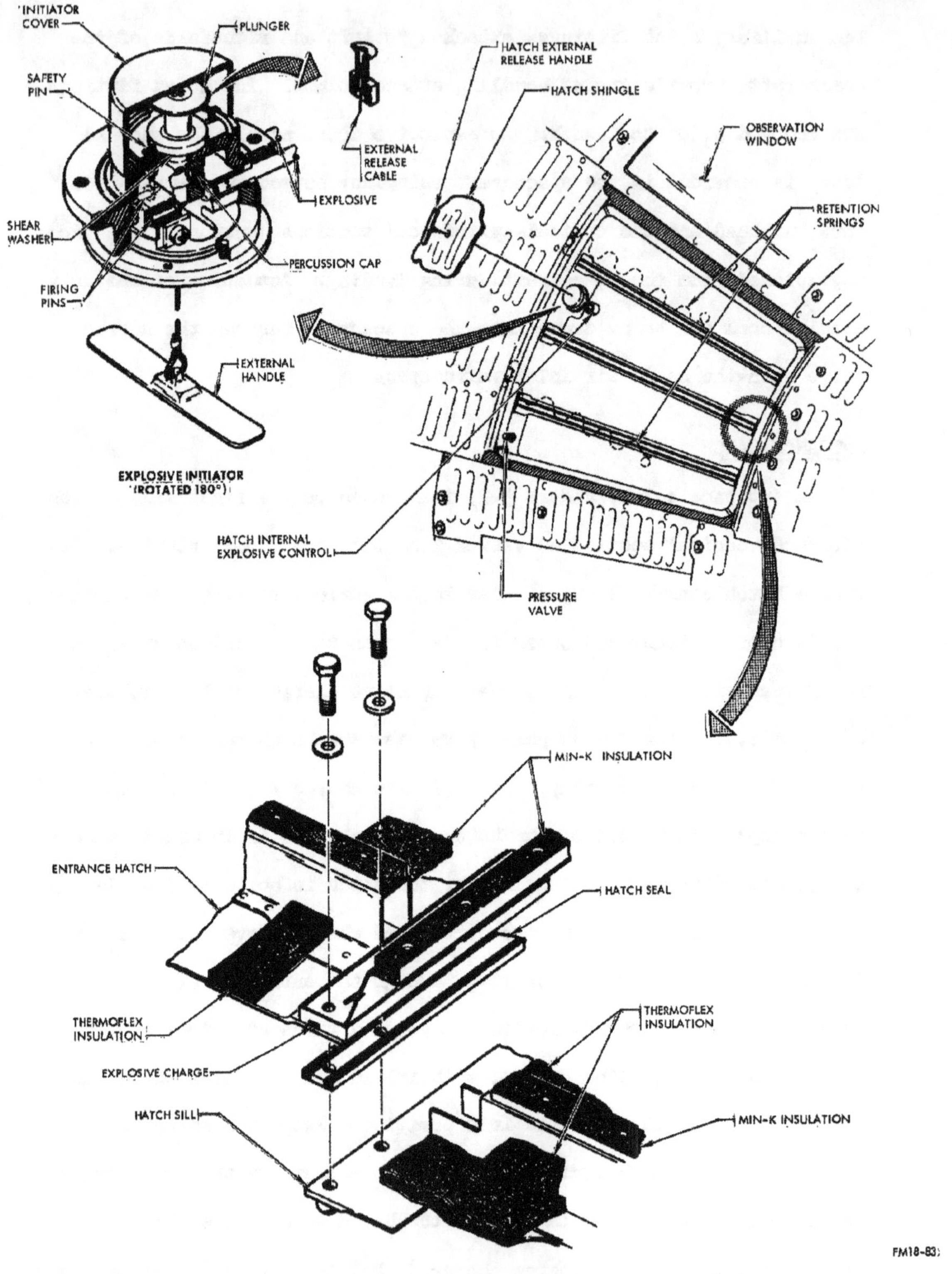

Figure 3-2 Entrance Hatch.

removes the initiator cap from the initiator, and the safety pin from the initiator plunger. By depressing the initiator plunger, the initiator's two spring-loaded firing pins strike the explosive charge percussion caps and detonate the explosive charge. This action explodes the hatch from the spacecraft. An exterior hatch release control is also provided to enable ground personnel to explode the hatch in the event the astronaut is unable to do so. Hatch retention springs, secured by pip pins, are incorporated on the inner side of the entrance hatch to prevent injury to ground personnel in the event the initiator plunger is accidently depressed. Two pressure valves, located in the hatch, enables pressurization and purging of the spacecraft during ground checkout operations.

3-5. OBSERVATION WINDOW

An observation window, Figure 3-3, located on the afterbody conical section, provides the astronaut with external viewing. The window, located above the main instrument panel, consists of an inner and outer assembly. The inner window assembly made of three glass panes is sealed in a titanium frame which is attached to the cabin wall. Each glass pane is independently sealed to provide a pressure seal between the panes. The outer window assembly consists of a glass pane sealed in a titanium frame, that is attached to the spacecraft outer skin. The outer window assembly is sealed separately, from the inner window assembly, to provide a complete seal. The outer window conforms to the curvature of the conical section. The observation window is equipped with filters, enabling the astronaut to regulate external light entering the cabin. A shield provided

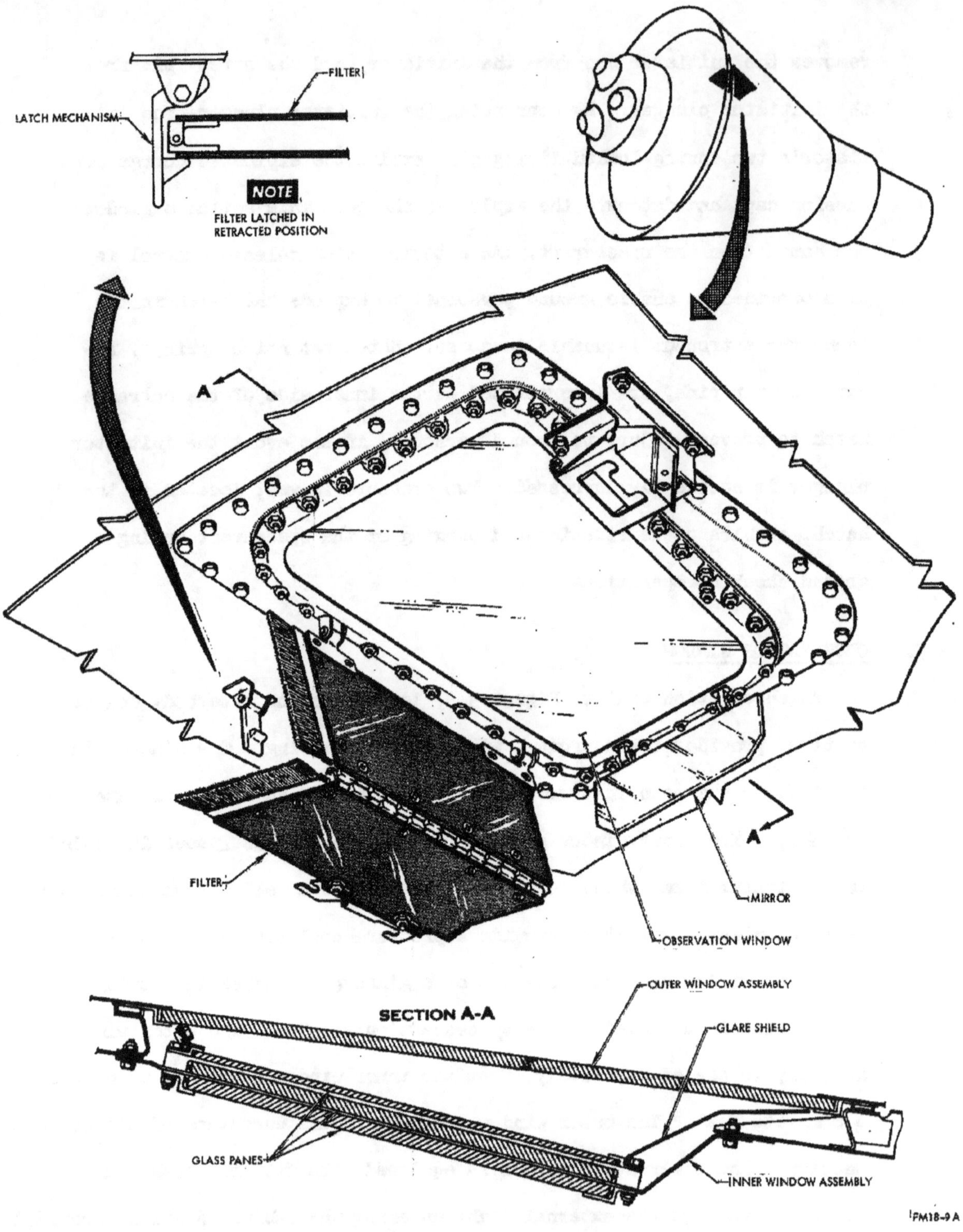

Figure 3-3 Observation Window

between the inner and outer window assemblies eliminates images reflected from the inner spacecraft structure. The observation window includes a two position mirror assembly which increases the astronaut's angle of observation.

3-6. SMALL PRESSURE BULKHEAD

The small pressure bulkhead internally separates the cabin pressurized area from the recovery system compartment and structurally supports the aft conical section. A sealed escape hatch, Figure 3-4, internally actuated, is provided in the small pressure bulkhead to enable the astronaut's exit following spacecraft landing. The dish-shaped escape hatch is constructed of a beaded aluminum skin spotwelded to an inner skin, that is reinforced with structural "Z" shaped members. The hatch outer flanged edge fits into the small pressure bulkhead sill and is held in place with a retaining ring. Expanding the retainer ring by raising the hatch handle, wedges the retainer ring between the bulkhead sill and the hatch flanged edge and forces the hatch flange aft to provide a sealing action. The titanium small pressure bulkhead is seam welded to the conical section inner skin and bolted to the conical hat stringer flanges.

3-7. LARGE PRESSURE BULKHEAD

The large pressure bulkhead supports the forward end of the conical section and internally separates the pressurized cabin from the forebody heat shield. The large pressure bulkhead is constructed of a combined inner and outer titanium skin. The outer skin is beaded and seam welded to the inner skin. The bulkhead is reinforced with horizontal

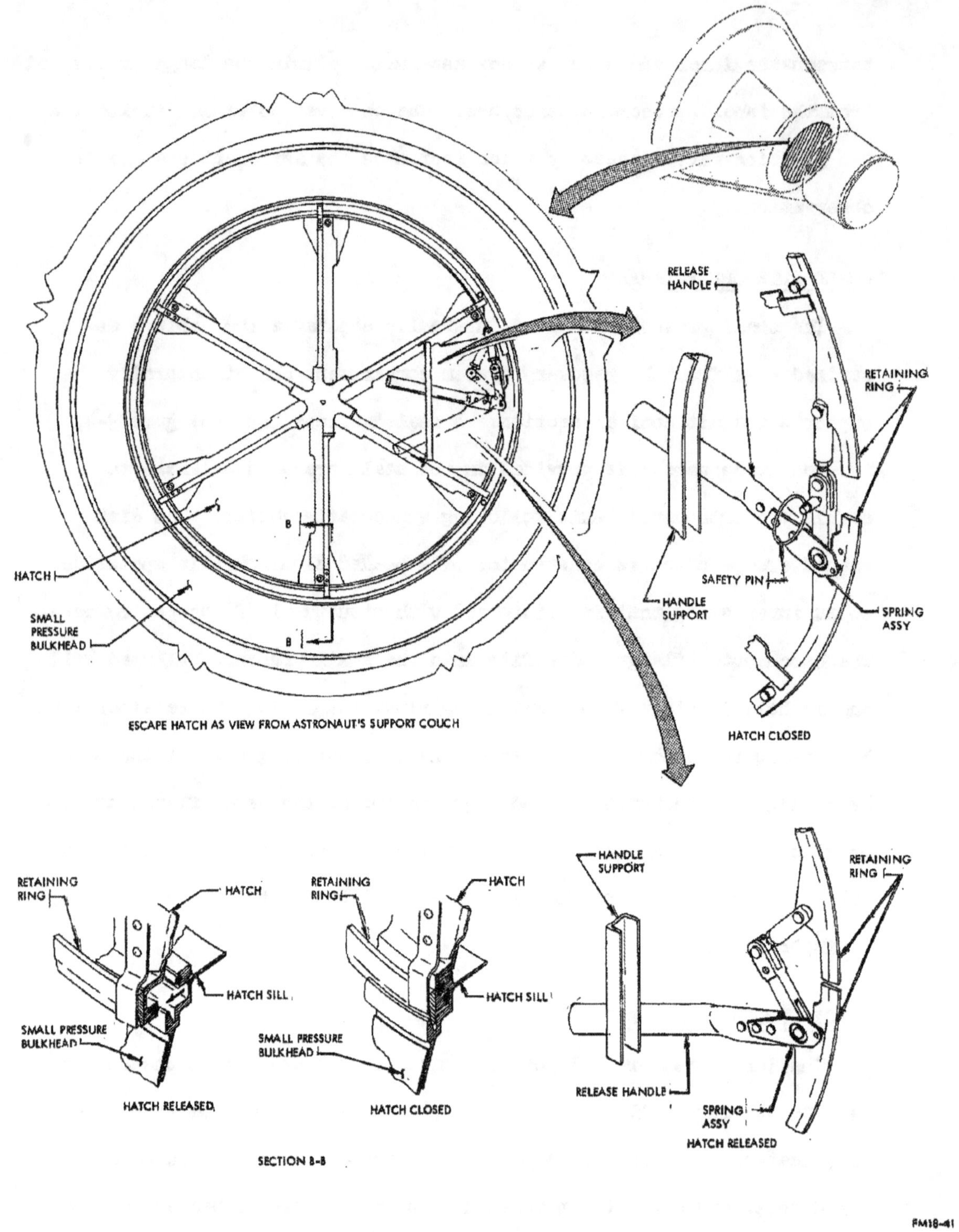

MCDONNELL — SEDR 104

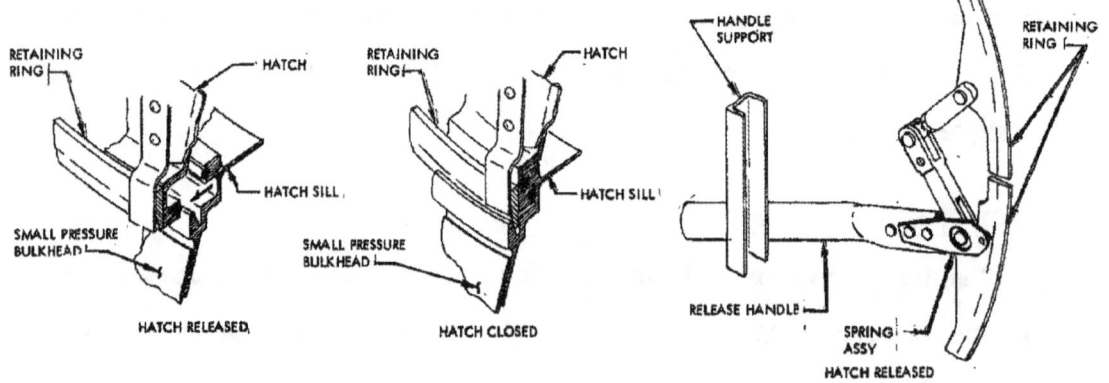

Figure 3-4 Escape Hatch.

channels installed on the outer skin. The bulkhead inner skin is provided with two vertical channels, centrally located and spaced, that furnish structural attach point for the astronaut support couch. Honeycomb shelves are provided on the bulkhead inner skin, outboard of the two vertical channels, for equipment installation. The bulkhead outer flange ring is bolted to the conical section inner skin and the bulkhead is also bolted to the conical section inner attach ring. Vents are provided in the large pressure bulkhead to enable overboard venting of the spacecraft battery vapors and environmental control system exhaust steam.

3-8. RECOVERY COMPARTMENT

The spacecraft afterbody, Figure 3-1, basically consists of the short cylindrical section and the truncated cone shaped structure. The cylindrical section is referred to as the spacecraft recovery system compartment and contains the landing parachutes, recovery aids and the reaction control nozzles. The truncated cone shaped structure, referred to as the afterbody conical section, encloses the pressurized cabin. The recovery compartment is connected to the pressurized cabin by a small pressure bulkhead. The recovery system compartment is a cylindrical formed titanium skin structure, reinforced with longitudinal hat stringers, and covered with a corrugated beryllium shingle arrangement. A layer of thermoflex insulation is installed between the hat stringers and the external shingles to prevent excessive heating within the compartment. The shingles are individual panels bolted to the hat sections with allowances for thermal expansion. A set of reaction control exhaust nozzles are internally located every 90°, between the compartment

inner skin and the external shingle installation. The recovery system compartment interior is structurally divided into a left and right section. The compartment left section houses the recovery aids, electrical wiring and plumbing routed through the compartment. The right section of the compartment houses a fiberglass container, structurally divided into two sections that contain the main and reserve parachutes. The container can be removed by the astronaut from the cabin following landing, to permit egress through the recovery compartment.

3-9. ANTENNA FAIRING

The antenna fairing, Figure 3-5, is a cylindrical shaped structure that houses the pitch and roll horizon scanners, and the main receiving and transmitting antenna. The antenna fairing basic structure is of titanium construction and is covered with "Rene-41" shingles. An eight inch window assembly is located around the outer base of the fairing and acts as a dielectric between the top of the fairing and spacecraft. The window assembly consists of a silicone base, fiberglass insulation, vycor glass and teflon strips. In line with the three teflon strips and attached to the antenna fairing shingles, are three laminated fiberglass guides. The fiberglass guides and teflon strips prevent damage to the antenna fairing when the escape tower is jettisoned. An aluminum bi-conical horn is internally located at the base of the antenna fairing. An electric insulator and lockfoam, located above the bi-conical horn, aid in antenna fairing insulation. A roll horizon scanner is located at the top of the antenna fairing. A pitch horizon scanner is located in the side of the fairing, in line with the roll horizon scanner. The fairing is attached to the spacecraft by a mortar gun located in the

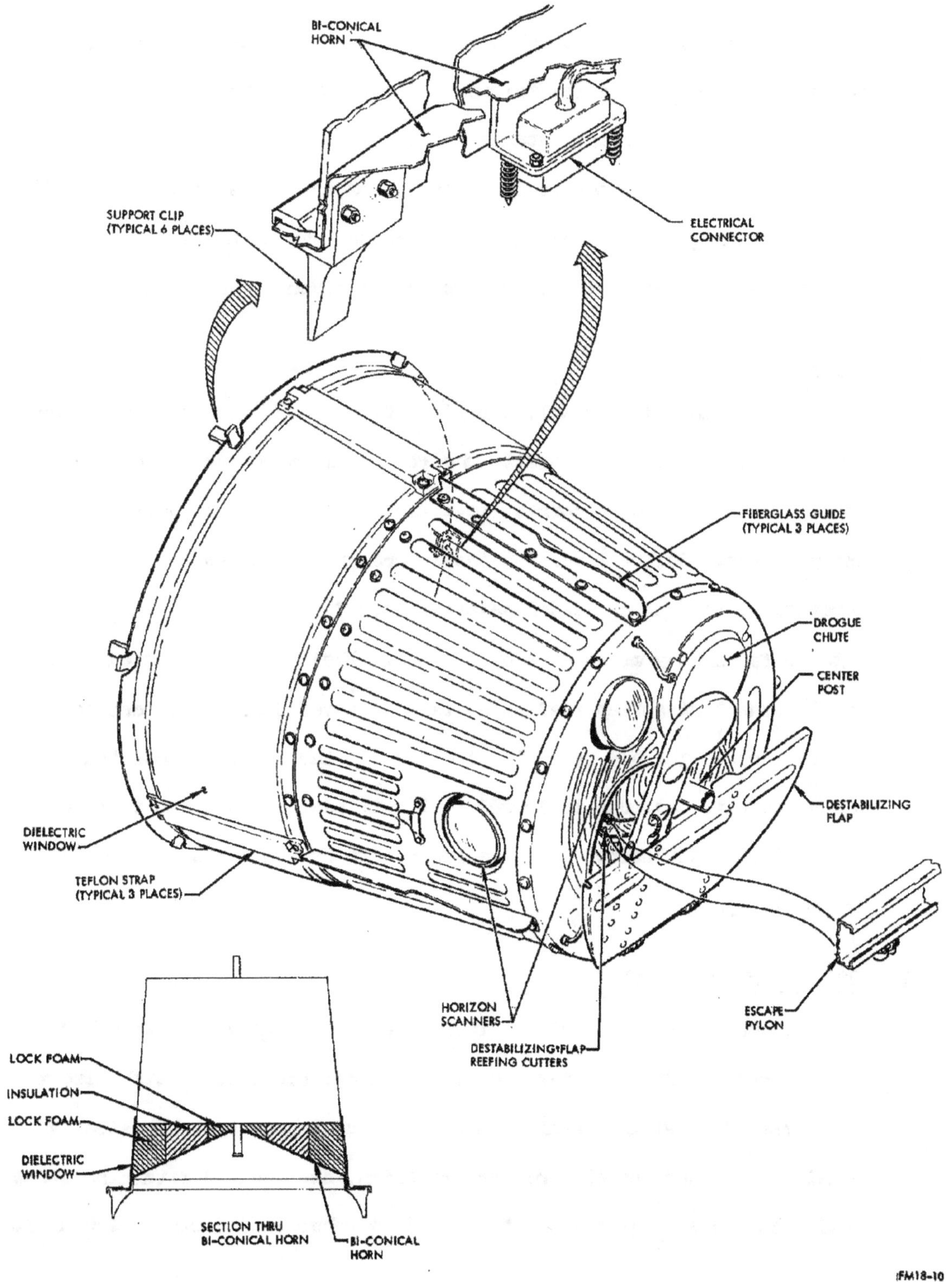

MCDONNELL — SEDR 104

Figure 3-5 Antenna Fairing

3-15

recovery compartment. A steel post located in the center of the fairing is used as a guide when the fairing is jettisoned. Three index pins and six support clips, in the antenna fairing lower mating flange, align with three holes and six brackets in recovery compartment mating flange. The antenna fairing also houses the drogue chute. Three cables retain the drogue chute risers to the fairing when the chute is deployed.

3-10. DE-STABILIZER FLAP

A spring loaded de-stabilizer flap, Figure 3-5, is attached to the top of antenna fairing, opposite the roll horizon scanner. The de-stabilizer flap ensures correct re-entry attitude during abort and re-entry phases. During launching phase, and up to the spacecraft-tower separation, the spring loaded de-stabilizer flap is held against the antenna fairing by means of a nylon cord. The nylon cord routed through two de-stabilizing flap reefing cutters is severed after the escape tower is jettisoned, thereby releasing the flap to the outboard position. When the spacecraft descends to 10,000 feet altitude, the antenna fairing is automatically jettisoned from the spacecraft by the firing of the fairing mortar gun.

3-11. IMPACT LANDING SYSTEM

The impact landing system, Figure 3-6, is designed to absorb high energy shock loads encountered during landing; and also to stabilize the spacecraft following a landing in water. The impact system basically consists of a heat shield release mechanism, heat shield retaining straps (24), heat shield retention cables (24), a rubberized cloth impact skirt

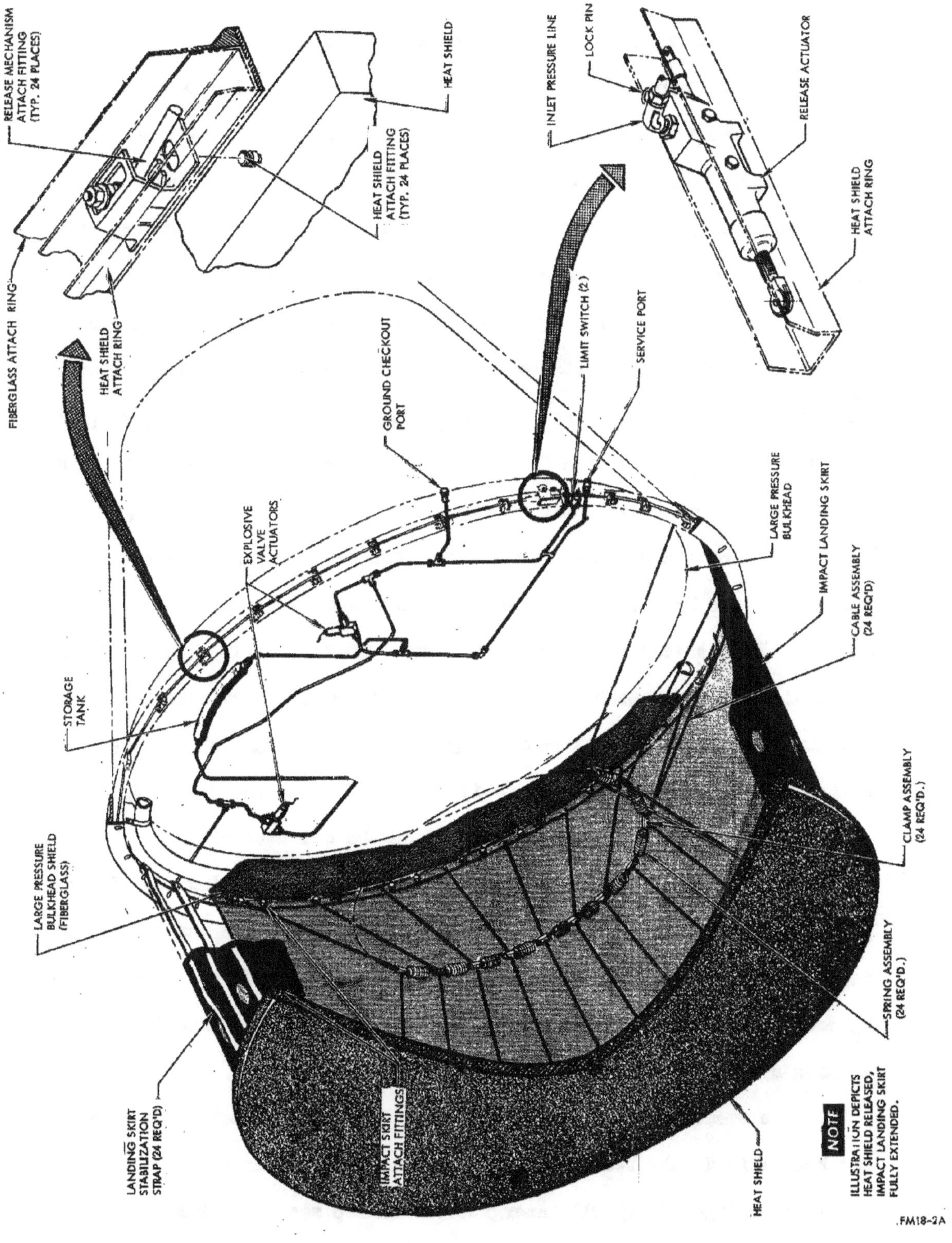

Figure 3-6 Impact Landing System

and a fiberglass shield assembly. The impact skirt is stored in the forebody area. During the normal landing phase, the 10,000 feet barostats energizes the Main Deploy Relay to eject the antenna fairing, which in turn deploys the main parachute. (See Figure 3-7). Ejection of the antenna fairing closes the antenna fairing separation sensing switch, which in turn directs 24 V d-c electrical power to the Main Inertia Switch Relay #1 (time delay relay). Twelve seconds later, the Main Inertia Relay #1 energizes to direct electrical power to the heat shield release system limit switches and also to energize the Landing Bag Relay. Energizing the Landing Bag Relay directs 24 V d-c electrical power to ignite the two heat shield release explosive squib valves.

Ignition of the squib valves allows 3,000 psig nitrogen pressure to flow to the two heat shield release mechanism actuators. This action moves the heat shield from the spacecraft. Simultaneously with the actuation of the release mechanism, the mechanism two limit switches close to energize the Landing Bag Extension Signal Relay. Energizing the signal relay, directs electrical power to illuminate the Landing Bag Telelight (green), indicating a safe condition. When the actuator piston fully travels to the open limit, the actuator is locked by a spring loaded lock pin. The impact landing bag circuit is de-energized while the spacecraft is in orbit. Placing the landing bag switch in the "AUTO" position during spacecraft re-entry will arm the circuit allowing normal operation of the landing system.

In the event the heat shield mechanism failed to actuate, and release the heat shield, the two limit switches will remain open and the Landing Bag Warning Light Relay will energize within two seconds. This in turn directs power to illuminate the Landing Bag Telelight (red), indicating

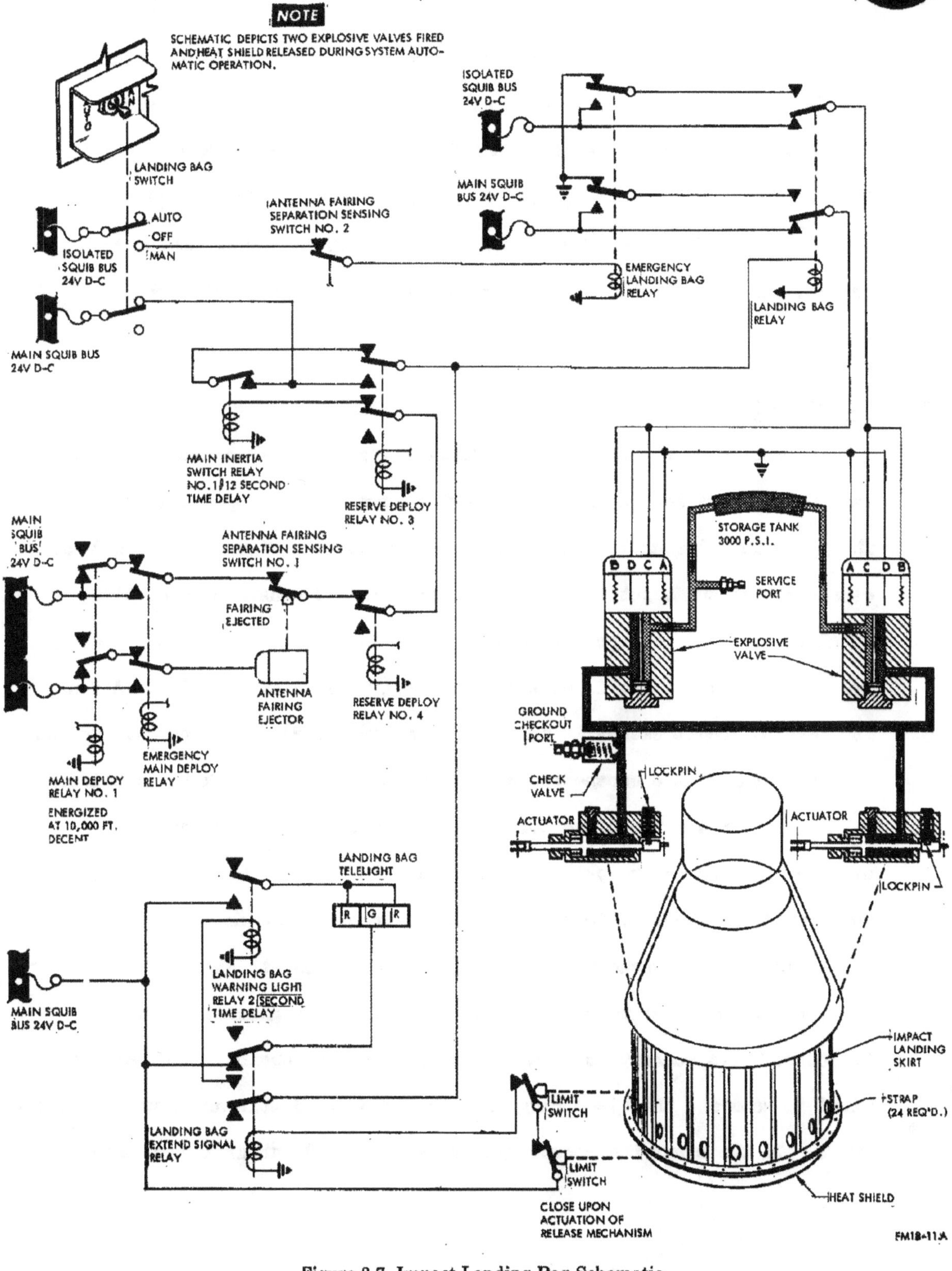

Figure 3-7 Impact Landing Bag Schematic

an unsafe condition to the astronaut. The astronaut should place the LANDING BAG SWITCH to the "MAN" position. Placing the landing switch to the manual position energizes the Emergency Landing Bag Relay, which in turn provides power to ignite the two heat shield release explosive squib valves. Ignition of squib valves will actuate the mechanism to release the heat shield, and in turn the limit switches will close to illuminate the telelight (green).

3-12. ESCAPE TOWER

The escape tower (Figure 3-8), designed to aid in spacecraft-missile emergency separation, consists of a pylon framework equipped with rockets. The pylon is a triangular shaped structure that is designed to support an escape rocket and a jettison rocket. The pylon is constructed of 4130 tubular steel and is approximately ten feet in length. The base of the pylon structural tubing is bolted to a steel flanged attach ring. A four foot escape rocket casing is bolted to the top (apex) of the pylon. Bolted to the bottom of the escape rocket casing is a jettison rocket. Electrical wiring is routed through the structural tubing, from the rockets to connectors, located on the pylon attach ring. The pylon tubular structure is covered with heat protective material. Prior to launch the pylon is installed onto the spacecraft, by clamping the pylon attach ring to the recovery system compartment with a chevron shaped, segmented clamp ring. Explosive bolts connect the clamp ring segments in tension. The bolts are fired to separate the clamp ring when the pylon is jettisoned from the spacecraft. In the event the escape system is activated during launch phase,

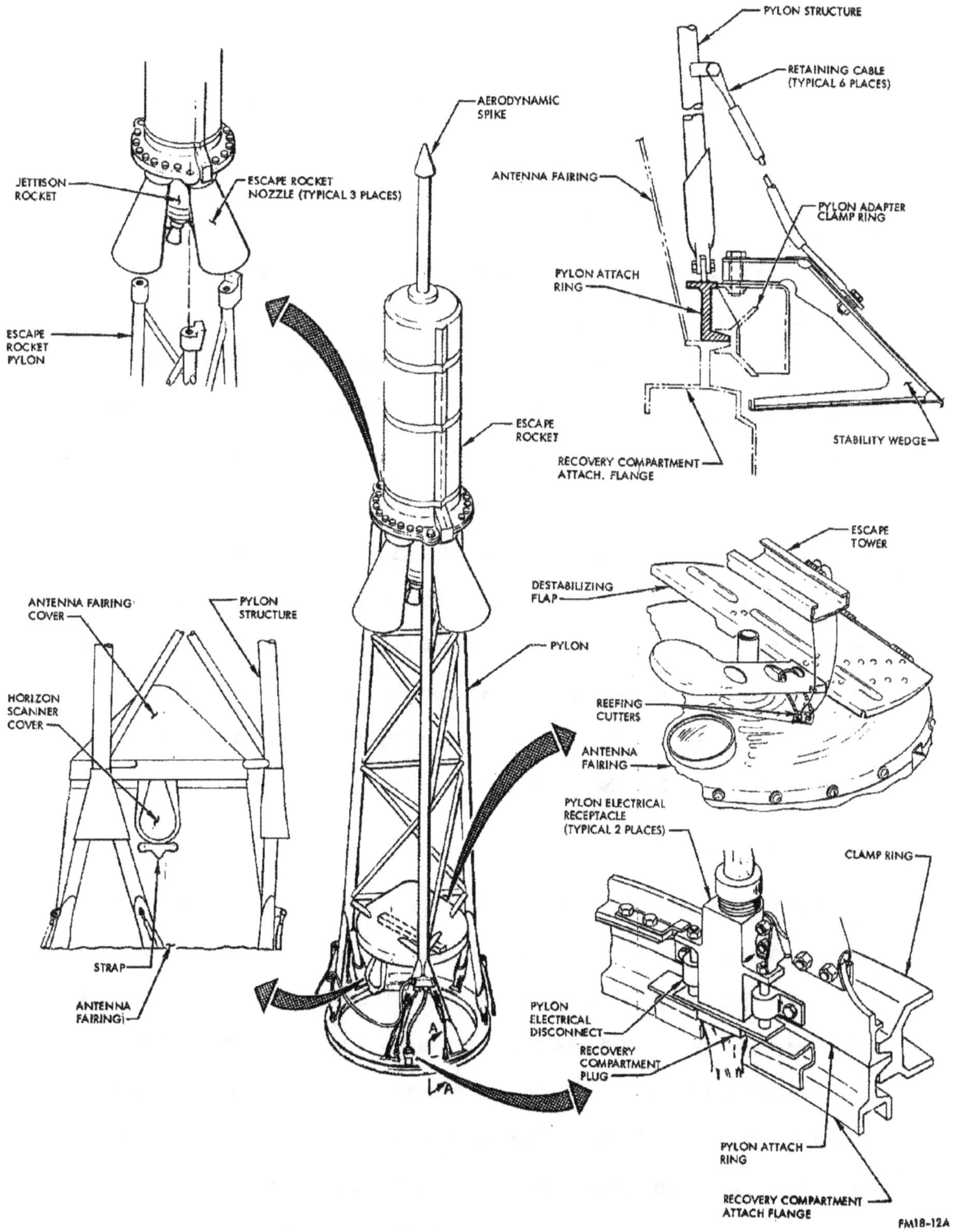

Figure 3-8 Escape Tower

SEDR 104

the escape tower is fired to propel the spacecraft away from the missile and then the jettison rocket is fired to separate the pylon from spacecraft.

3-13. **PYLON-SPACECRAFT CLAMP RING**

The clamp ring consists of three chevron shaped segments that clamp the pylon attach ring to the recovery system compartment flange. Three explosive bolts, with dual ignition provisions, connect the ring segments in tension. The clamp ring is basically the same in design as the spacecraft-adapter clamp ring (Figure 3-10), but considerably smaller in size. The clamp ring retains the pylon to the spacecraft until the clamp ring explosive bolts are fired which in turn separates the clamp ring. An aerodynamic stability wedge attached to pylon clamp ring, aids in stabilizing the spacecraft during the launch phase. Six cable straps, bolted to the pylon structure and clamp ring stability wedge, aid in spacecraft-pylon separation by retaining the clamp ring segments of the pylon when the explosive bolts are fired.

3-14. **MISSILE ADAPTER**

The missile adapter, Figure 3-9, is a slightly tapered, cylindrical shaped structure that is designed to mate the spacecraft with the Atlas missile. Upon adapter and spacecraft installation to the missile, the adapter is bolted to the missile and the spacecraft is attached to the adapter. The adapter is of semi-monocoque construction and is approximately four feet in height. The adapter basically consists of an outer corrugated titanium skin assembly, riveted and seam welded to an inner titanium skin assembly and internally reinforced with two titanium support rings, riveted between the ends of the adapter. A steel flanged

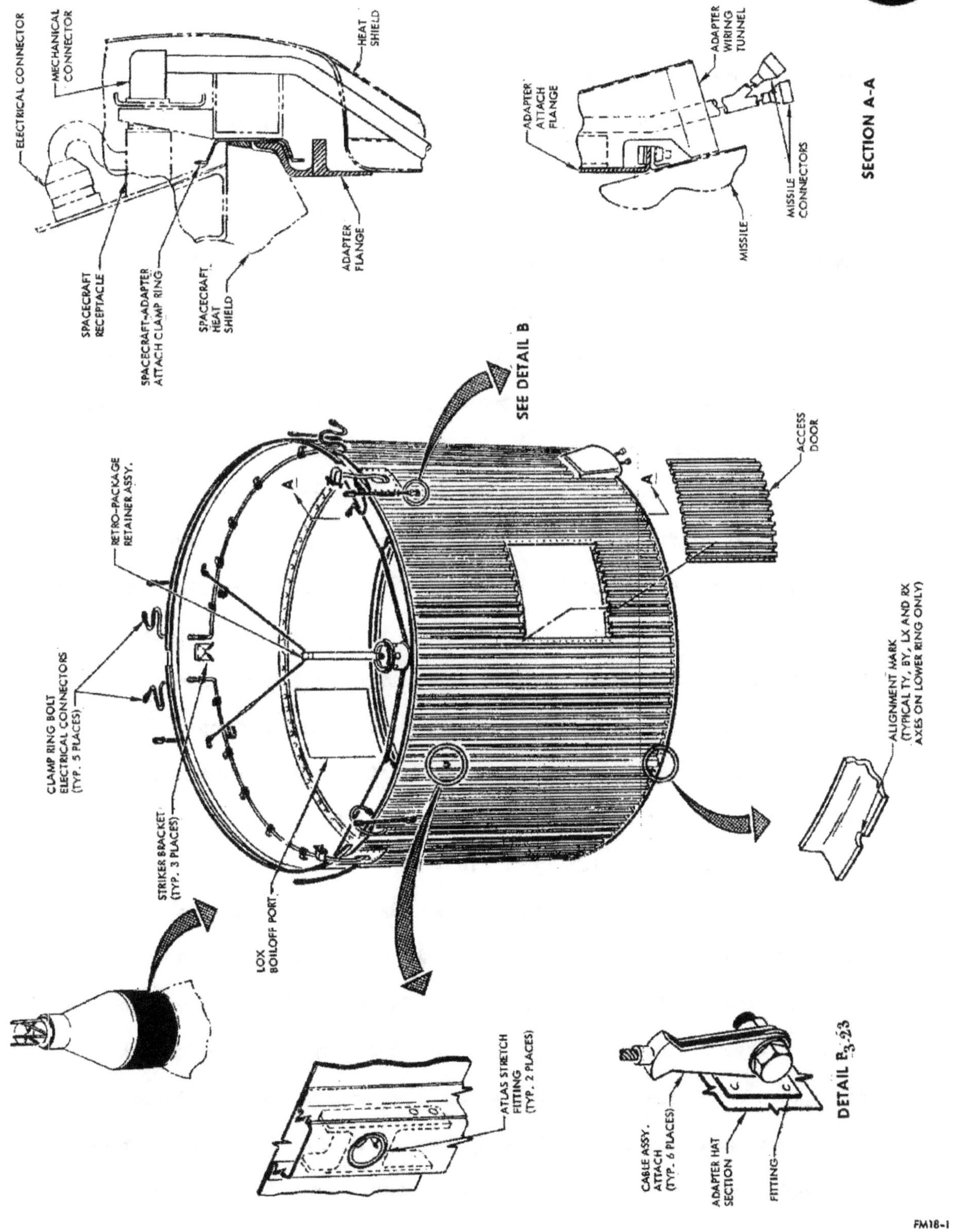

Figure 3-9 Atlas Missile Adapter

ring is riveted to the bottom, inner surface of the adapter. The flanged ring is provided with holes to enable the attachment of the adapter to the missile with bolts. Alignment marks are provided on the ring for proper adapter-missile alignment. Riveted to the top, inner surface of the adapter is an aluminum flanged ring. The adapter aluminum ring mates with the spacecraft forebody fiberglass attach ring, during spacecraft to adapter installation. An alignment mark on the adapter ring BY axis enables proper alignment of spacecraft to adapter. The top of the aluminum ring is slotted at 120° intervals, to provide adequate clearance for the retrograde rocket assembly attach straps, when the spacecraft is attached to the adapter. A metal striker bracket is riveted internally, every 120°, to the adapter skin. When the spacecraft is attached to the adapter, these striker brackets depress (open) the spacecraft adapter separation sensing switches, located on the bottom of the retrograde rocket assembly attach straps. The spacecraft is attached to the adapter by installing a chevron shaped, segmented clamp ring over the mated flanges of the forebody fiberglass attach ring and the adapter upper ring.

A retainer assembly, attached to the adapter interior skin, is provided to prevent the retro-package and the explosive bolt fragments from striking the Atlas missile adapter LOX tank. The retainer assembly is a cup shaped structure, that fits over the retro-package dome, and is supported by three metal straps that are attached to the adapter with cable assemblies. A vent port, located in the adapter skin, receives the missile boil-off valve tube and enables the relieving of liquid oxygen from the missile. Opposite the liquid oxygen boil-off

port, is an adapter door installation. The door installation provides access to the booster and heat shield area while on the pad. A fiberglass shield attached above the vent port opening, streamlines the adapter and shields the boil-off tube. Two stretch fittings, located 180° apart at the upper section of the adapter, provide a means of supporting (stretching) the missile while in the vertical position following adapter installation. Six cable assemblies, attached to fittings spaced around the adapter outer corrugated skin, are attached to the clamp ring that attaches the spacecraft to the adapter. The cables retain the clamp ring to the adapter following spacecraft-adapter separation.

3-15. SPACECRAFT-ADAPTER CLAMP RING

The spacecraft-adapter clamp ring, Figure 3-10, is provided to attach the spacecraft to the adapter. The clamp ring secures the spacecraft to the adapter throughout the launching phase until the clamp ring is separated by means of explosive bolts, allowing the separation from the adapter. The clamp ring consists of three chevron shaped segments, that when installed, mate with the forebody fiberglass attach ring and the adapter upper support ring. Three explosive bolts, with dual ignition provisions, connect the three clamp ring segments in tension. A metal striker bracket is bolted, every 120°, to the inside of the clamp ring. When the clamp ring is installed, the striker brackets depress the spacecraft ring separation sensing switches, located in outer periphery of the spacecraft forebody.

The exterior of the clamp ring is covered with a heat shield that protects the explosive bolts from excessive heating. The heat shield

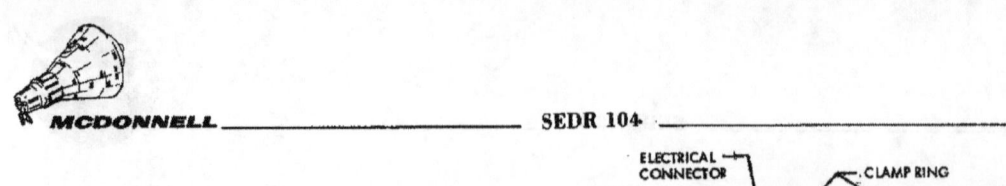

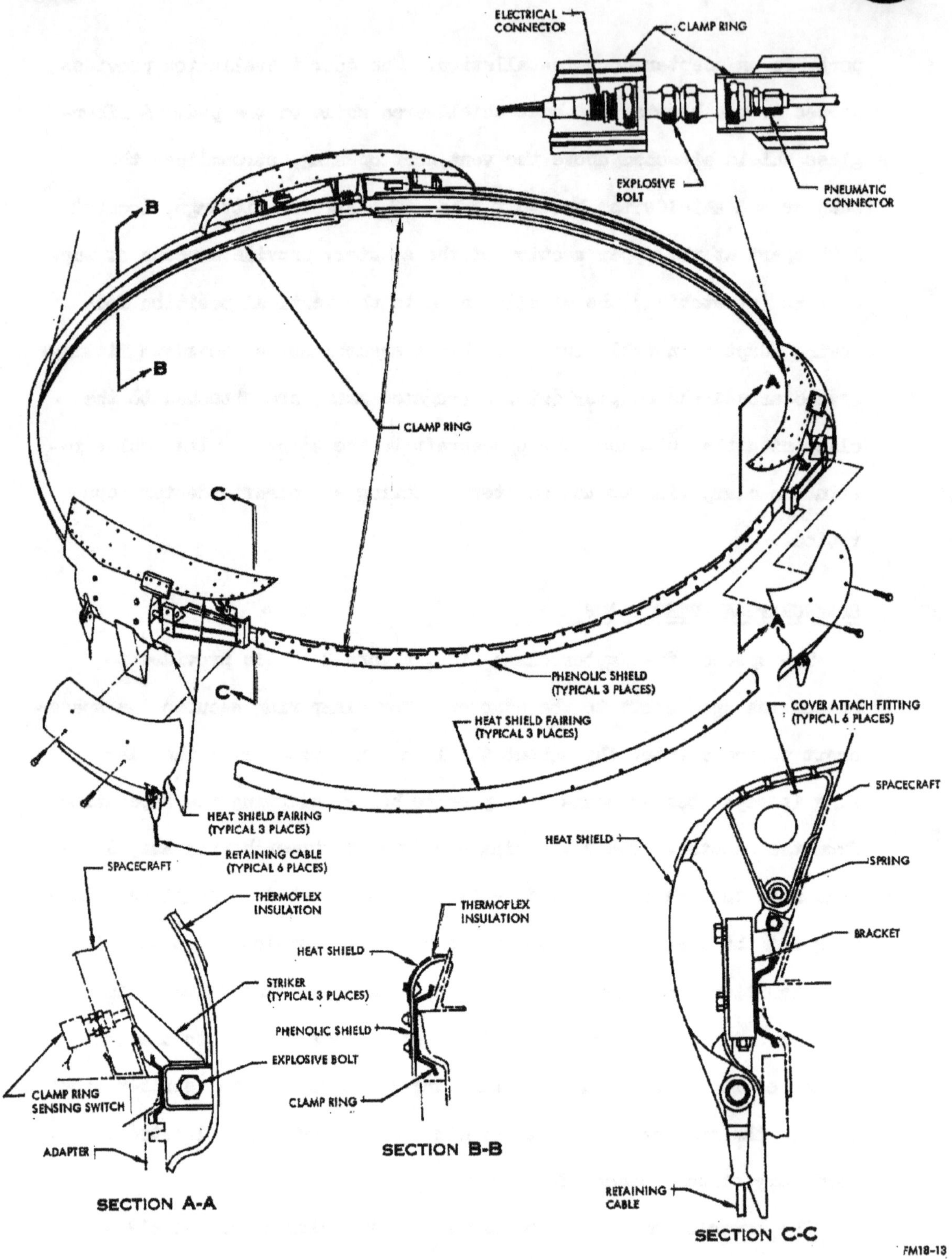

Figure 3-10 Adapter Clamp Ring

consists of three fairing assemblies which are located directly over the explosive bolts and three segmented fairing assemblies which cover the remainder of the spacecraft adapter clamp ring. The fairing assemblies which locate directly over the explosive bolts are a three piece installation. The top piece is fabricated of aluminum and the two bottom pieces are made of titanium. These fairing assemblies are fastened to the clamp ring support fittings. The interior of the fairing assemblies is insulated with thermoflex. The three segmented fairing assemblies are of a titanium construction whose interior is insulated with thermoflex. The segmented fairing assemblies are bolted to the adapter clamp ring. Six cable straps are bolted to the spacecraft adapter cable fitting. These straps aid in spacecraft-adapter separation, by retaining the clamp ring to the adapter when the explosive bolts are fired. An electrical cable, clamped around the interior of the adapter, is connected to each of the clamp ring explosive bolts, to two receptacles in the forebody area and to two receptacles on the launch vehicle. A pneumatic line is also connected to one end of the explosive bolt and to a quick disconnect in the forebody.

3-16. RETRO-PACKAGE

The retro package, Figure 3-11, is a jettisonable dome shaped container mounted to the ablation shield by three retention straps. The package, constructed of aluminum alloy contains mounting provisions for the posigrade and retrograde rockets, HF dipole antenna and associated wiring.

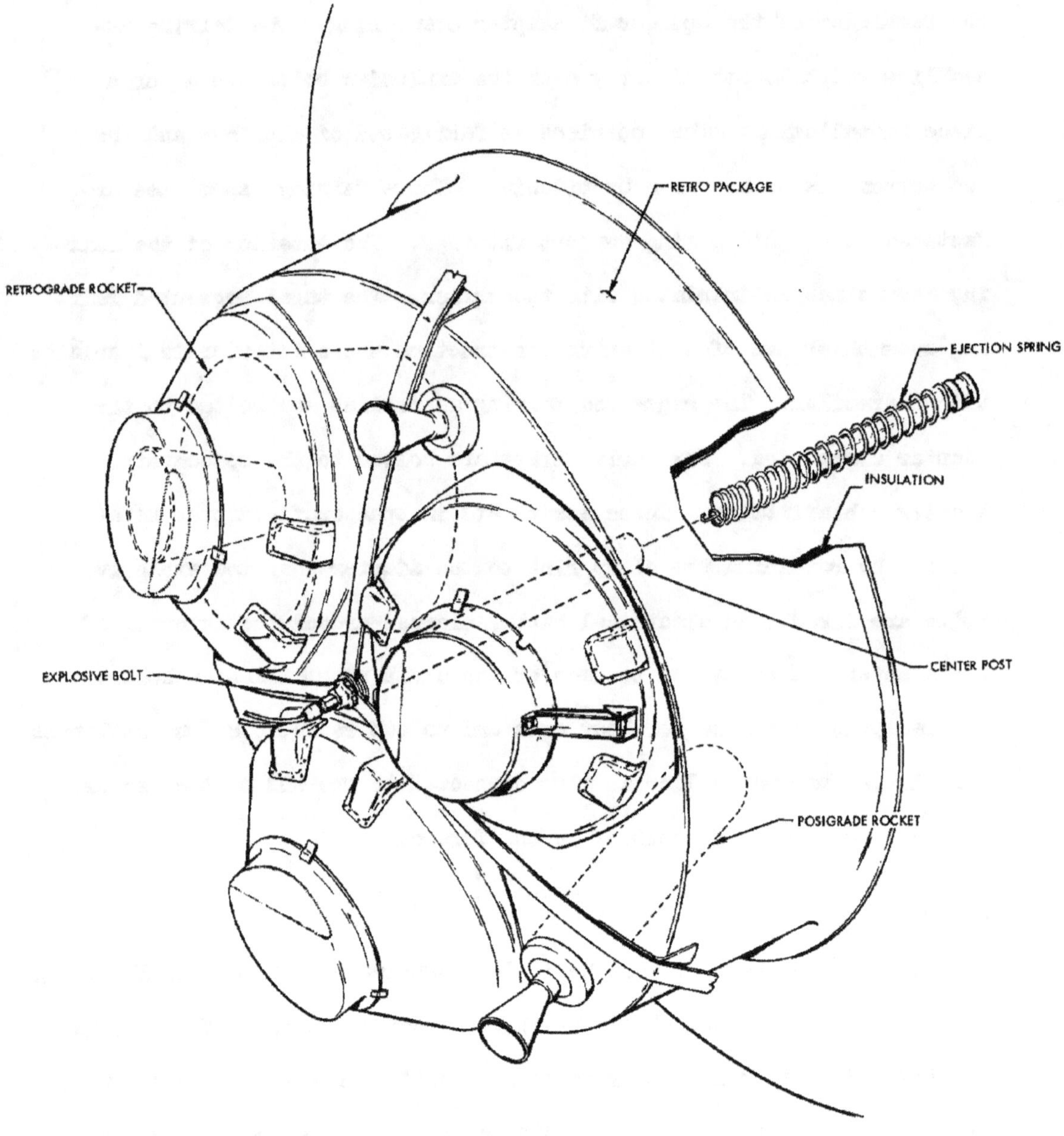

Figure 3-11 Retro Package

Thermoflex insulation is provided on the inner sides of the retro package to protect the rockets from excessive heat. Jettisoning provisions are also provided by means of a post mounted in the center of the retro package containing a spring assembly and explosive bolt. Detonation of the explosive bolt releases the retention straps and the spring assembly ejects the retro package from the spacecraft.

SECTION IV

ENVIRONMENTAL CONTROL SYSTEM

TABLE OF CONTENTS

TITLE	PAGE
Description	4-3
Cabin Environmental Control	4-5
Suit Environmental Control	4-6
Suit Emergency Control	4-7
Oxygen Supply	4-7
Cooling Circuit	4-8
Blood Pressure Measuring System	4-9
Operation	4-12
System Units	4-31

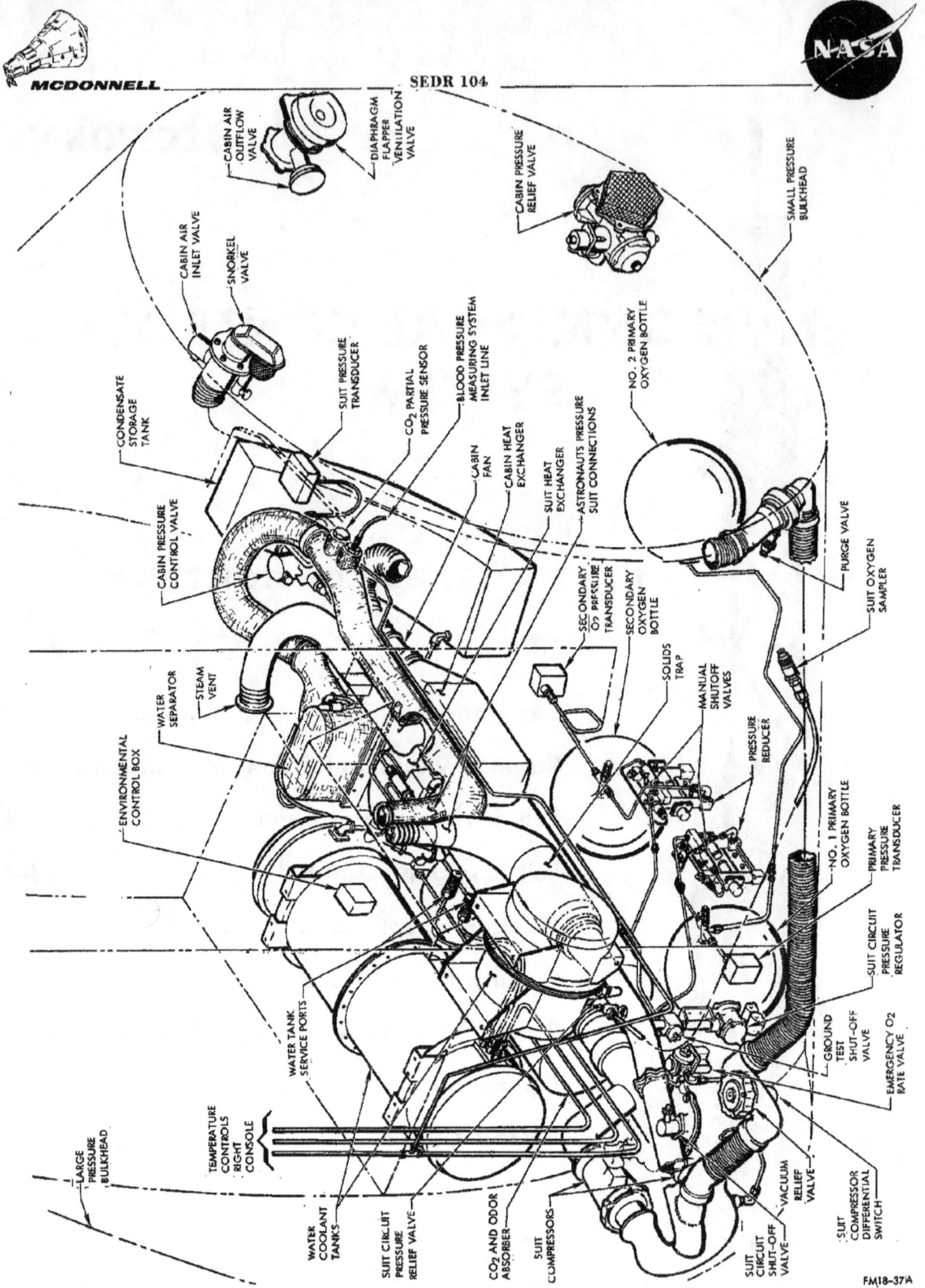

Figure 4-1 Environmental Control System

SEDR 104

IV. ENVIRONMENTAL CONTROL SYSTEM

4-1. **DESCRIPTION**

The environmental control system Figure 4-1, provides the cabin and the astronaut with a 100 per cent oxygen environment to furnish the breathing, ventilation and pressurization gas required during the spacecraft orbital flight and ventilation for a 12-hour post-landing phase. The environmental control system also functions in the following manner: Removes odors, CO_2 and moisture from the astronaut's suit circuit; provides an "emergency rate" usage of oxygen in the event of pressure loss in the suit circuit; maintains cabin and suit temperature at approximately 80°F; provides an emergency fire extinguishing capability and operates in "weightless" or "high g" conditions. System functioning will be automatically controlled during all phases of flight. In the event the system automatic controls malfunction, manual controls are provided to insure system operation.

The environmental control system is designed to be operated in either the suit mode, cabin mode or emergency mode. The system suit mode is normally utilized and enables the astronaut to function in the closed suit circuit during cabin pressurized and emergency (depressurized) conditions. In the event one control mode malfunctions, the remaining control mode will continue to operate. The emergency mode insures astronaut survival in the event both the suit and cabin modes malfunction. (See Figure 4-2).

The environmental control system provides a primary and secondary oxygen supply for both the cabin and suit circuits. Primary and sec-

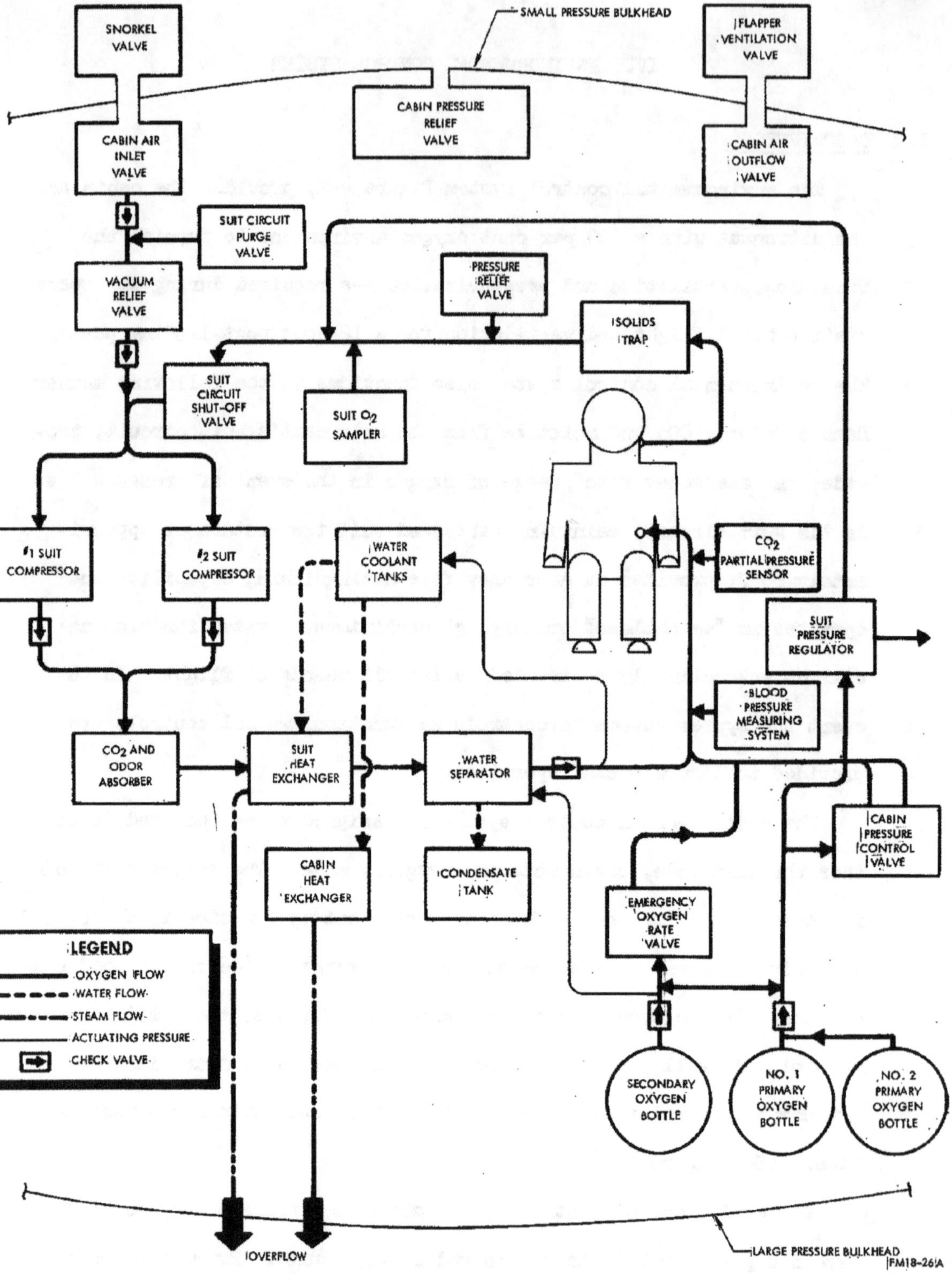

Figure 4-2 Environmental Control System Block Diagram

ondary oxygen systems are basically the same, however, the secondary oxygen regulated pressure is lower than the primary oxygen regulated pressure. A manually controlled cooling circuit, for suit and cabin systems, is provided to control suit and cabin temperature during flight. The environmental control system components are located below the astronaut's support couch adjacent to the large pressure bulkhead and also on the interior of the small pressure bulkhead adjacent to the escape hatch. System manual controls are located on the left and right consoles; system instruments and warning lights are located on the main instrument panel.

4-2.   CABIN ENVIRONMENTAL CONTROL

During normal orbital flight, the environmental control system is operated in both the cabin and suit mode. Operation in the cabin mode permits the astronaut to open his helmet faceplate. The primary and secondary oxygen supply furnishes the cabin with pressurization, breathing, and ventilation gas. The cabin is equipped with automatic and manual controls for cabin ventilation, decompression, pressurization, temperature control, landing and post-landing ventilation.

The cabin is cleared of contaminants and a 100 per cent oxygen environment is made available by purging the cabin prior to launch. During orbital flight, cabin pressure is automatically controlled by a cabin pressure control valve. The cabin pressure relief valve prevents excessive pressure buildup within the cabin and provides a manual means of decompressing the cabin in the event of a fire or buildup of toxic contaminants. Two coolant supply tanks containing 40 pounds and 9 pounds of

water respectively are common to both cabin and suit circuit heat exchangers to provide cabin and suit cooling. The coolant tank containing 40 pounds of water is also a source of drinking water for the astronaut. Cabin temperature is controlled by a manually controlled selector valve, which regulates the amount of water entering the cabin heat exchanger, and in turn provides cabin cooling by means of water evaporation. The cabin fan, located on the inlet side of the heat exchanger, forces cabin air through the exchanger to provide cabin cooling and ventilation. Cabin air inlet and outflow valves, located on the small pressure bulkhead, provide ventilation during the landing and post-landing phase.

4-3. SUIT ENVIRONMENTAL CONTROL

During normal orbital flight, the common oxygen supply furnishes oxygen simultaneously to the suit and cabin environmental control circuits. If a cabin circuit malfunction, such as cabin decompression, should occur at a time when the astronaut has his faceplate removed, the astronaut should immediately close his faceplate. Closing the faceplate initiates the suit mode and confines the astronaut to the closed suit control circuit.

While operating in the suit mode, the suit pressure regulator controls the suit pressure to approximately 5 psia and replenishes oxygen consumed by the astronaut, during normal suit control circuit operation. Suit circuit pressure is utilized as a means of pressurizing the water coolant tanks. The suit environmental control circuit incorporates compressors, filters, absorbers and a temperature control to insure astronaut's maximum comfort. Suit circuit temperature is controlled by means of water evaporation. A water seperator utilizes the common oxygen supply pressure,

MCDONNELL — SEDR 104

to remove moisture from the suit circuit oxygen supply. A compressor, located on the upstream side of the suit circuit heat exchanger, forces the suit circuit oxygen supply throughout the circuit, providing suit circuit ventilation. During the landing and post-landing phase, atmospheric air is drawn in through the cabin air inlet valve to provide suit circuit ventilation.

4-4. SUIT EMERGENCY CONTROL

While operating in the suit mode during orbital flight, the suit emergency control mode automatically activates when suit circuit pressure decreases below $4.0 {}^{+.1}_{-.3}$ psia pressure. A control handle is provided to enable manual selection of the emergency mode. During the landing phase, the emergency mode is automatically activated. The environmental system oxygen rate valve and the suit circuit shutoff valve actuate simultaneously to switch the environmental system from the suit mode to the emergency mode. Actuation of these valves may be either automatic or manual.

An O_2 EMERG light, located on the main instrument panel, and a tone generator indicates when the environmental system is operating in the suit emergency mode. The O_2 Emergency Rate Handle used to activate the suit emergency control, is located on the right console.

4-5. OXYGEN SUPPLY

The environmental system is supplied with oxygen, from primary and secondary bottles. The primary and secondary oxygen bottles are directly interconnected by a supply line, that forms a common oxygen supply to the cabin pressure control valve, suit pressure regulator, emergency oxygen rate valve, and the suit circuit water separator. The primary and secondary oxygen supply lines incorporate shutoff valves, pressure transducers, pressure reducers, and check valves. The pressure transducers transmit oxygen pressure, present in the primary and secondary oxygen bottles, to a dual quantity indicator, tape recorder, and to a telemetry unit. The primary oxygen bottle pressure is reduced to 100 $\pm$ 10 psig, by a primary oxygen pressure reducer. Two primary oxygen pressure reducers are provided for redundancy in the event one pressure reducer fails. The secondary oxygen bottle pressure is reduced to 80 psig by a secondary pressure reducer. The primary oxygen supply pressure, being greater than the secondary oxygen supply reduced pressure, permits the primary oxygen supply to be utilized during normal conditions with the secondary oxygen supply in reserve. The oxygen supply line check valves prevent the total loss of oxygen, in the event either the primary or secondary oxygen pressure supply systems fail.

4-6.   COOLING CIRCUIT

During normal orbital flight, the environmental system cooling circuit furnishes the cabin and the suit circuit with provisions for independently controlling the cabin and suit circuit temperatures. Water is supplied, under oxygen pressure, from water coolant tanks to the cabin and suit circuit heat exchangers. The water absorbs heat from the cabin and suit circuit oxygen which causes the water to evaporate.

MCDONNELL SEDR 104

The cooling circuit basically consists of two water tanks, cabin and suit temperature control valves and heat exchangers. Temperature control valves are located on the right console.

4-7. BLOOD PRESSURE MEASURING SYSTEM

A semi-automatic, independently controlled system is provided to measure the astronaut's blood pressure during orbital flight. The Blood Pressure Measuring System (Figure 4-3) activated by the astronaut, records the blood pressure through transducers and amplifiers which is then telemetered to tracking stations. The system essentially consists of an oxygen bottle, controller, pressure regulator, solenoid operated control valves, occluding cuff and microphone.

Prior to installation of the entrance hatch, the oxygen supply (Figure 4-4) manual shut-off valve is opened allowing an oxygen flow through the regulator to the Fill Valve. The regulator is provided to maintain a Blood Pressure Measuring System differential pressure of $4\frac{1}{4}$ psi in reference to Environmental Control System suit circuit pressure. Depressing the START switch located on the main instrument panel directs electrical power to activate the Blood Pressure Measuring System for 110 seconds. Activation of the system simultaneously opens the Fill valve and closes the Dump valve allowing the astronaut's suit cuff to inflate. A time delay relay con-

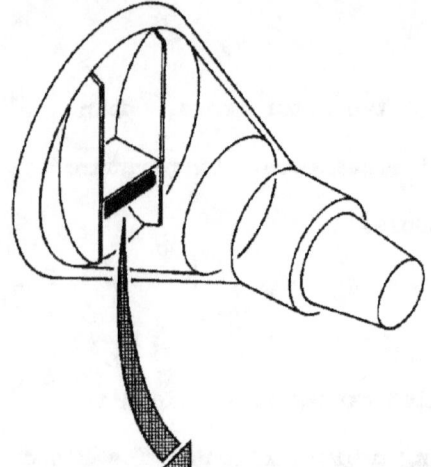
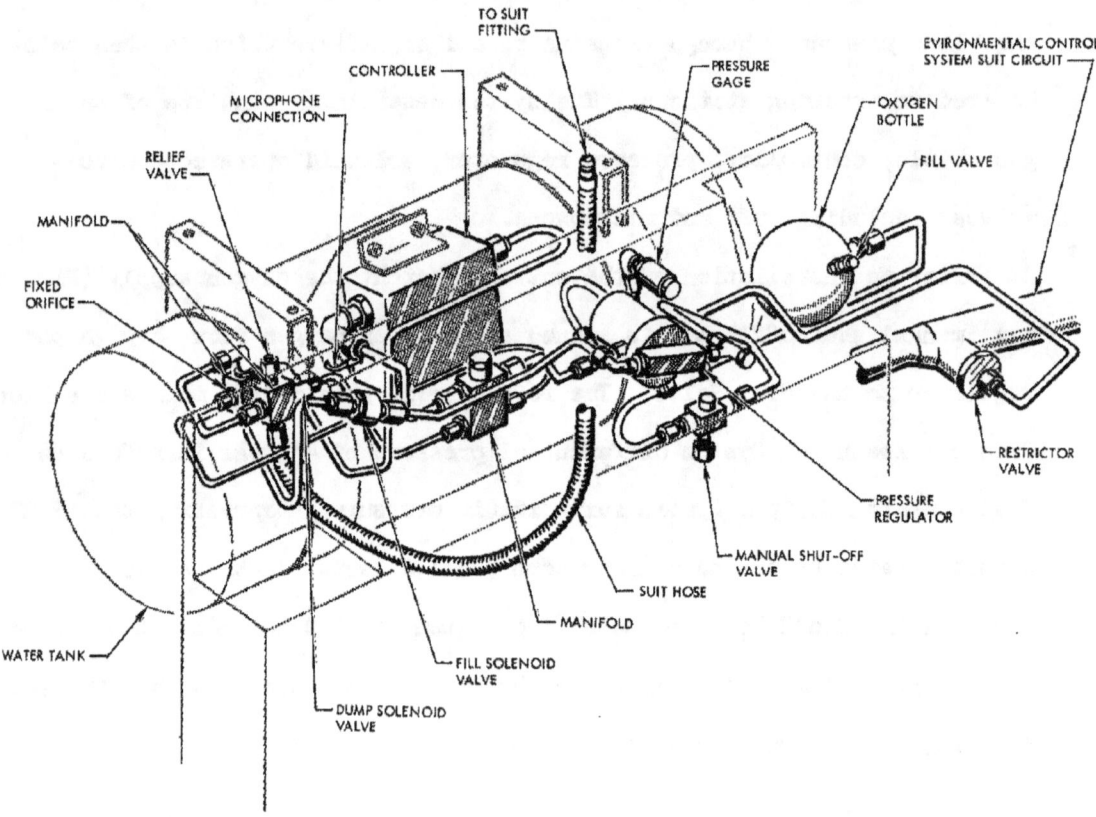

Figure 4-3 Blood Pressure Measuring System

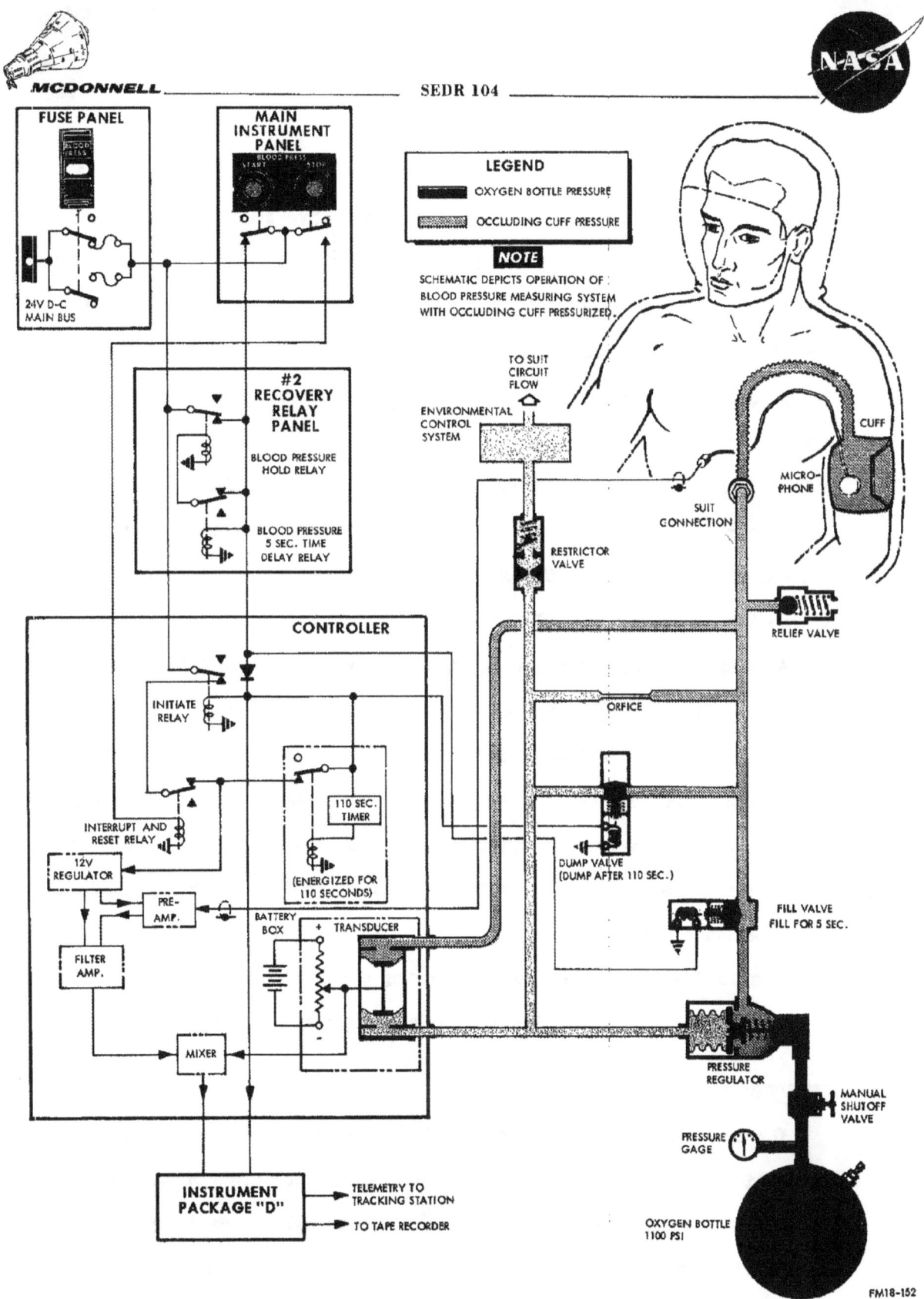

Figure 4-4 Blood Pressure Measuring System Schematic

tained within #2 Recovery Relay Panel actuates five seconds after the START switch is depressed removing power to the Fill Valve solenoid. Pressure in the cuff is dissipated through the orifice on into the ECS suit circuit in approximately thirty seconds. The astronaut's blood pressure is measured and telemetered by means of a transducer in the controller recording inflated cuff pressure, and a microphone placed under the cuff sensing pulse sounds. A timer also contained within the controller de-energizes an initiate relay at the end of 110 seconds, removing electrical power to the system, therefore opening the dump valve allowing system pressure to vent to the Environmental Control System suit circuit.

A restrictor valve provided at the entrance of the suit circuit permits a pressure flow into the suit circuit and a limited flow in the opposite direction. In the event a Blood Pressure Measuring System line should fail, suit circuit pressure would be maintained. A relief valve is provided to dissipate excessive pressure in the event the regulator should fail. If the astronaut desires to discontinue the Blood Pressure Measuring cycle, the STOP switch is depressed. Activating the STOP switch energizes an interrupt relay in the Controller removing electrical power from the system in turn opening the dump valve to discharge system pressure.

4-8. OPERATION

The environmental control system is designed to sequentially operate automatically during the launch, orbit, re-entry and post-landing phases of the flight. The mode in which the environmental system is operated is dependent upon the existing conditions within the cabin and suit circuit.

SEDR 104

During the pre-launch phase of operations, the oxygen and water supply are fully serviced. Refrigerated air is ducted through the spacecraft hatch to precool the cabin and structure during spacecraft pre-flight. The refrigerated air supply is removed when an external supply of freon coolant is directed to the cabin and suit circuit heat exchangers, through the umbilical, to continue precooling the spacecraft structure and cabin equipment after the flight hatch is installed. The oxygen supply manual shut-off valves are opened and the astronaut is connected to the spacecraft suit circuit by attaching the suit circuit personal leads (flex hoses) to the astronaut's pressure suit. The suit compressor and cabin fan are activated at this time. The suit circuit is purged with an external source of low pressure oxygen applied through the suit circuit purge valve. Following the purging operation, a suit circuit leakage check is performed. The spacecraft entrance hatch is bolted into position and the cabin is then checked for leakage and purged with oxygen. The suit circuit incorporates provisions for obtaining launch purge oxygen samples.

Forty-five seconds prior to launch, the ground umbilical plug is disconnected and freon coolant supply to the spacecraft ceases. During launch and orbit, the cabin pressure relief valve maintains cabin pressure at approximately 5.5 differential (cabin/ambient) psi. During spacecraft launch, the suit circuit pressure regulator maintains the suit circuit pressure approximately equivalent to cabin pressure. The suit circuit oxygen is kept free of contaminants by a solids trap and a CO_2 and odor absorber. The solids trap removes foreign particles such as food particles, nasal excretions, hair, etc. The CO_2 and odor absorber filters

odors and CO_2 from the circulating oxygen. Moisture from the suit circuit oxygen is removed from the system by a water separator. The pneumatically activated water separator deposits the moisture into a condensate tank. Cabin and suit circuit temperatures are controlled by manually operated metering valves, that regulate the water flow rate from the water coolant tanks to the cabin and suit circuit heat exchangers. Upon reaching altitudes where the saturation temperature of water is lower than the cabin and suit circuit gas temperature, the cabin and suit circuit heat exchangers will provide cooling by water evaporation.

When the spacecraft descends to an altitude of approximately 21,000 feet, the snorkel explosive door is ejected. (Door is located on spacecraft exterior.) At an altitude of approximately $17,000 \pm 3000$ feet, the cabin air inlet and outflow valves open barometrically venting the cabin to the atmosphere. Operation of the suit circuit compressor draws outside air into the suit circuit through the ejected snorkel door opening, the snorkel valve and the open cabin air inlet valve. The air, circulating through the suit circuit, is relieved into the cabin and in turn flows out through the cabin air outflow valve. Simultaneously, with the opening the cabin air inlet and outflow valves, the environmental system mode of operation switches to the emergency mode, but the suit compressor continues to operate. Switching to the emergency mode provides a greater cooling capacity for the astronaut. An inlet air snorkel valve and an outflow air diaphragm flapper ventilation valve located on the unpressurized side of the small pressure bulkhead, prevent water from entering into the cabin

in the event the spacecraft submerges after landing in water environment. A vacuum relief valve, located in the flexible ducting between the cabin air inlet valve and suit circuit, prevents a vacuum to occur which may cause the snorkel valve to stay closed. During the post-landing phase, the astronaut may continue to operate his suit circuit compressor to provide suit circuit ventilation. The suit circuit compressor draws atmospheric air into the suit circuit, through the cabin air inlet valve.

4-9. CABIN ENVIRONMENTAL CONTROL

Operation of the environmental control system in the cabin mode, (Figure 4-5), after the spacecraft has entered the orbital flight path permits the astronaut to open his helmet faceplate and be exposed to cabin environment. The cabin pressure relief valve relieves cabin pressure in excess of 5.5 psia. In the event cabin pressure tends to exceed the 5.5 psia pressure (cabin over ambient), the relief valve will open to relieve the excessive pressure. When the cabin pressure decays to 5 psia, the cabin pressure control valve will open to maintain approximately 5 psia cabin pressure. Cabin make-up oxygen flow into the suit circuit. The suit pressure regulator will sense the increase in suit pressure, and relieve excess gas into the cabin. Routing the cabin pressure control valve oxygen supply through the suit circuit, provides a constant purging of the suit circuit. Cabin pressure control valve maintains cabin pressures to $5.1 ^{+.2}_{-.3}$ psia.

During orbital flight, cabin gas is circulated throughout the cabin by the cabin fan, located at the inlet to the cabin heat exchanger. The

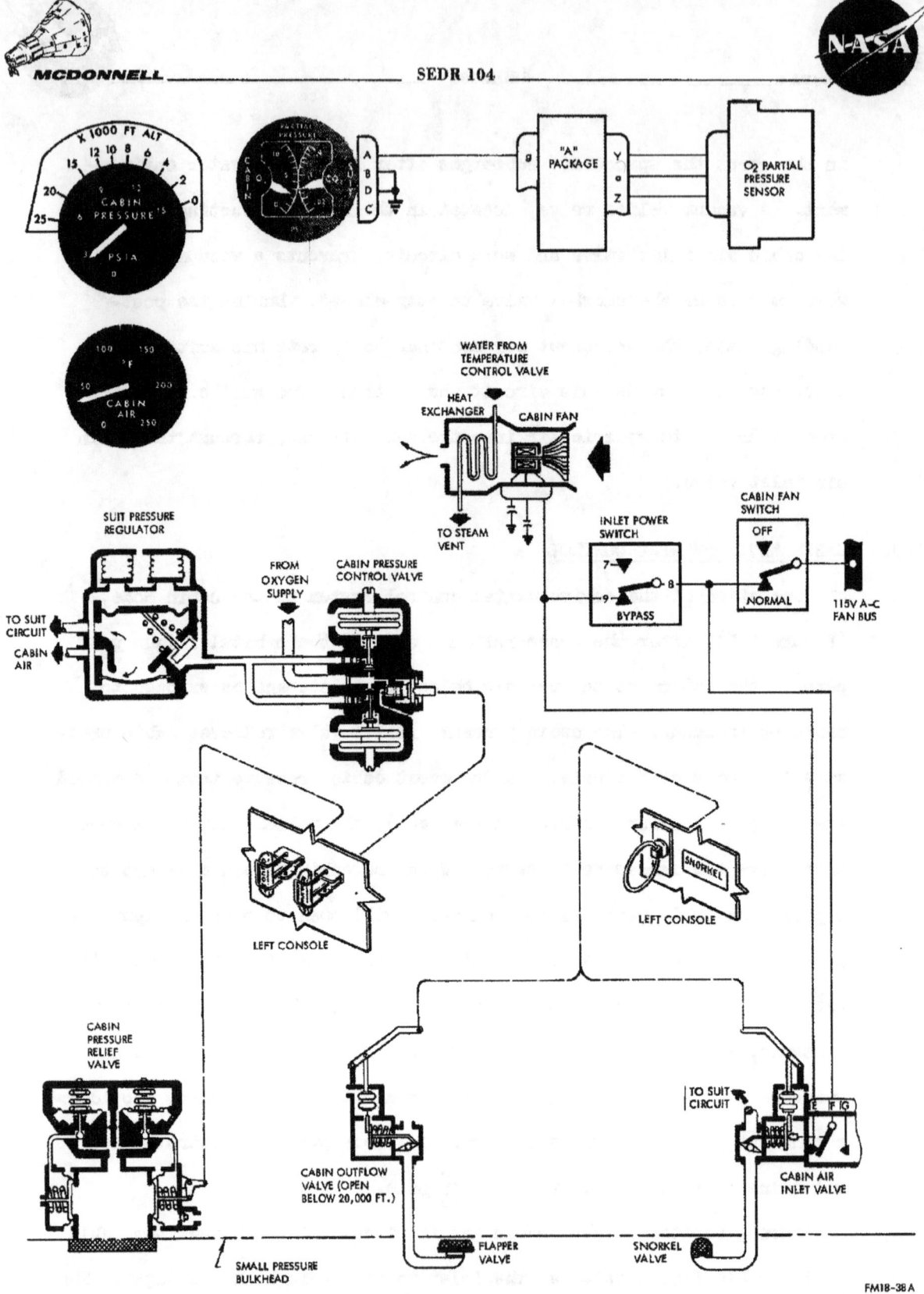

Figure 4-5 Cabin Invironmental Control System

cabin gas absorbs the heat generated by the cabin electronic equipment and in turn is cooled when the gas passes through the cabin heat exchanger. Water from the water coolant tanks evaporates in the heat exchanger which absorbs the heat from the cabin gas. The steam produced then passes overboard through the large pressure bulkhead steam vent. A cabin temperature control valve, located on the right console, is manually operated by the astronaut to control cabin temperature by controlling coolant water flow rates.

In the event of a fire or a buildup of toxic contaminants, within the cabin, the astronaut may manually decompress the cabin by actuating the DECOMPRESS "T" handle, located on the left console. The decompression handle is connected to the cabin pressure relief valve with a cable. During decompression of the cabin, the cabin pressure control valve closes when the cabin pressure decreases to 4.1 psia. Following the extinguishing of the fire, or the removal of toxic contaminants, the astronaut may repressurize the cabin by closing the DECOMPRESS "T" handle and actuating the REPRESS "T" handle. The REPRESS "T" handle is connected to the cabin pressure control valve with a cable. When the cabin has been repressurized to 5.0 psia, the REPRESS "T" handle must be manually closed. In the event of a cabin decompression, due to a meteorite penetration or excessive cabin leakage, the cabin pressure control valve will close automatically at approximately 4.1 psia and prevent oxygen flow to the cabin. Closing of the cabin pressure control valve reserves the remaining oxygen supply for the suit environmental control circuit, enabling the astronaut to continue the mission.

Prior to re-entry, the astronaut should assure that his helmet faceplate is closed. During spacecraft descent, cabin pressure is maintained at approximately 5 psia pressure. At 27,000 feet altitude the cabin pressure relief valve opens allowing atmospheric air to enter the cabin and equalize capsule internal and external pressures within 10-15 "H_2O. When the spacecraft reaches 17,000 ± 3,000 feet altitude, the cabin air inlet and outflow valves open providing outside air ventilation for the suit circuit. Suit circuit air is then vented to the cabin and out through the cabin outflow valve. If the cabin air inlet and outflow valves fail to open at 17,000 ± 3,000 feet altitude, the astronaut should actuate the SNORKEL pull ring to open the valves. A snorkel valve, provided on the inlet side of the cabin air inlet valve and a diaphragm flapper ventilation valve provided on the outlet side of the cabin air outflow valve, prevent water from entering the cabin when the spacecraft lands in the water.

A cabin pressure indicator and cabin temperature indicator are provided on the main instrument panel. An alternate method of determining altitude is incorporated in conjunction with the cabin pressure indicator. A placard with altitude markings mounted over the cabin pressure indicator provides a direct reference to cabin pressure in the event the altimeter should fail. A partial pressure sensor located below the correlation clock is utilized to measure the oxygen content circulating throughout the cabin. Pressure measured by the sensor is transmitted to an indicator marked "PARTIAL PRESSURE" located on the main instrument panel.

4-10. SUIT ENVIRONMENTAL CONTROL

The suit environmental control circuit, Figure 4-6, is supplied oxygen from the environmental system oxygen supply, through the suit pressure

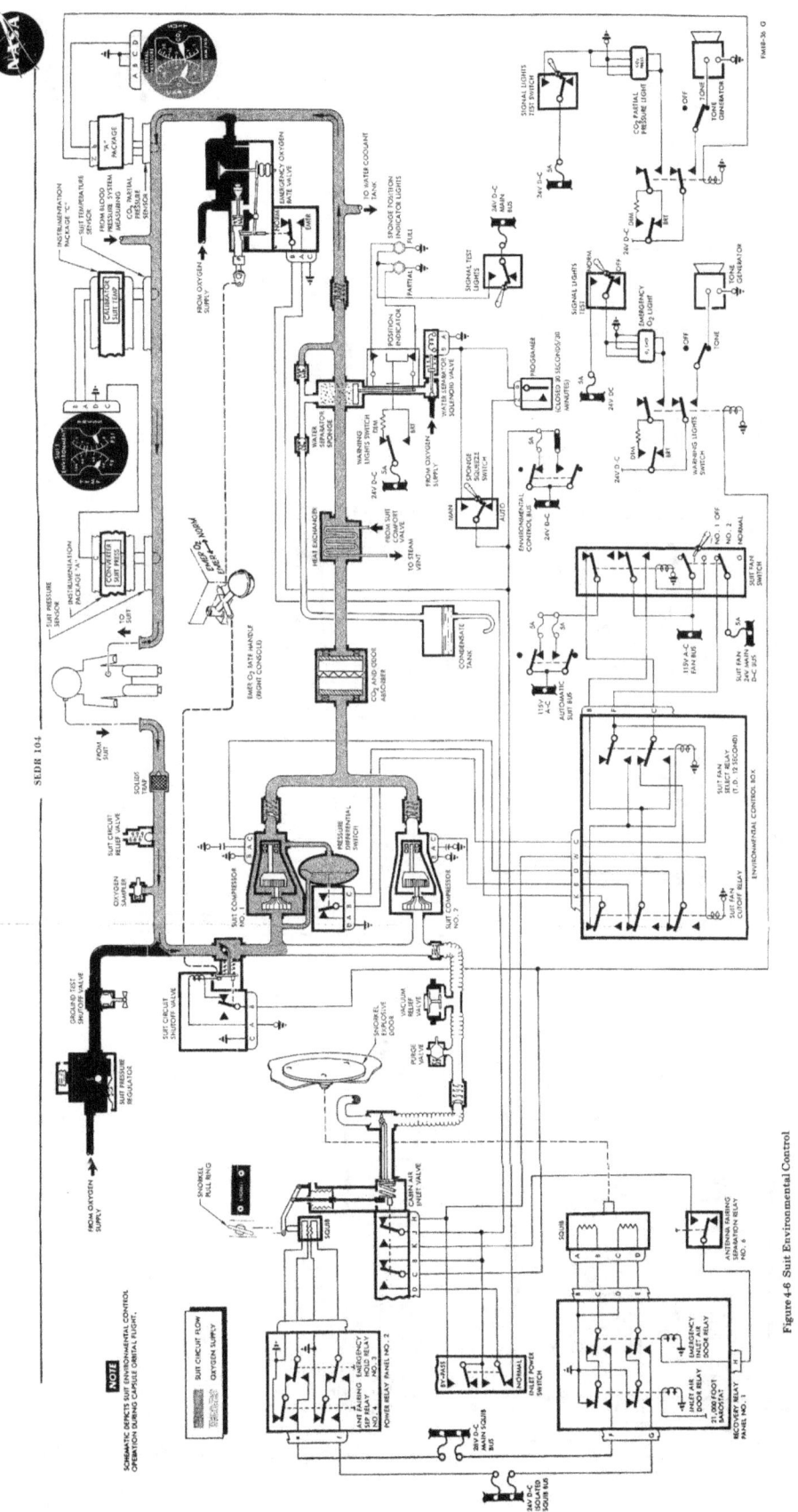

Figure 4-6 Suit Environmental Control

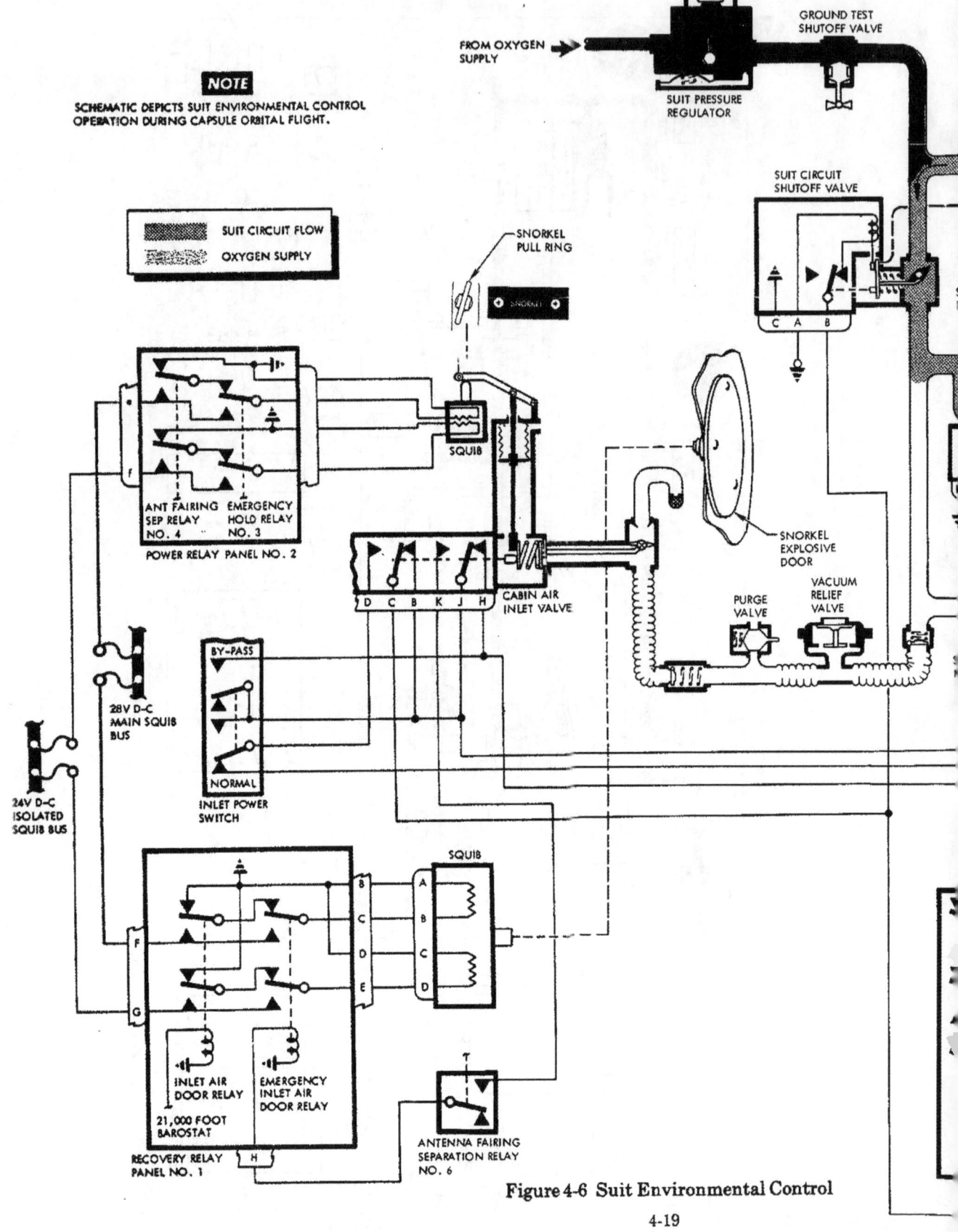

Figure 4-6 Suit Environmental Control

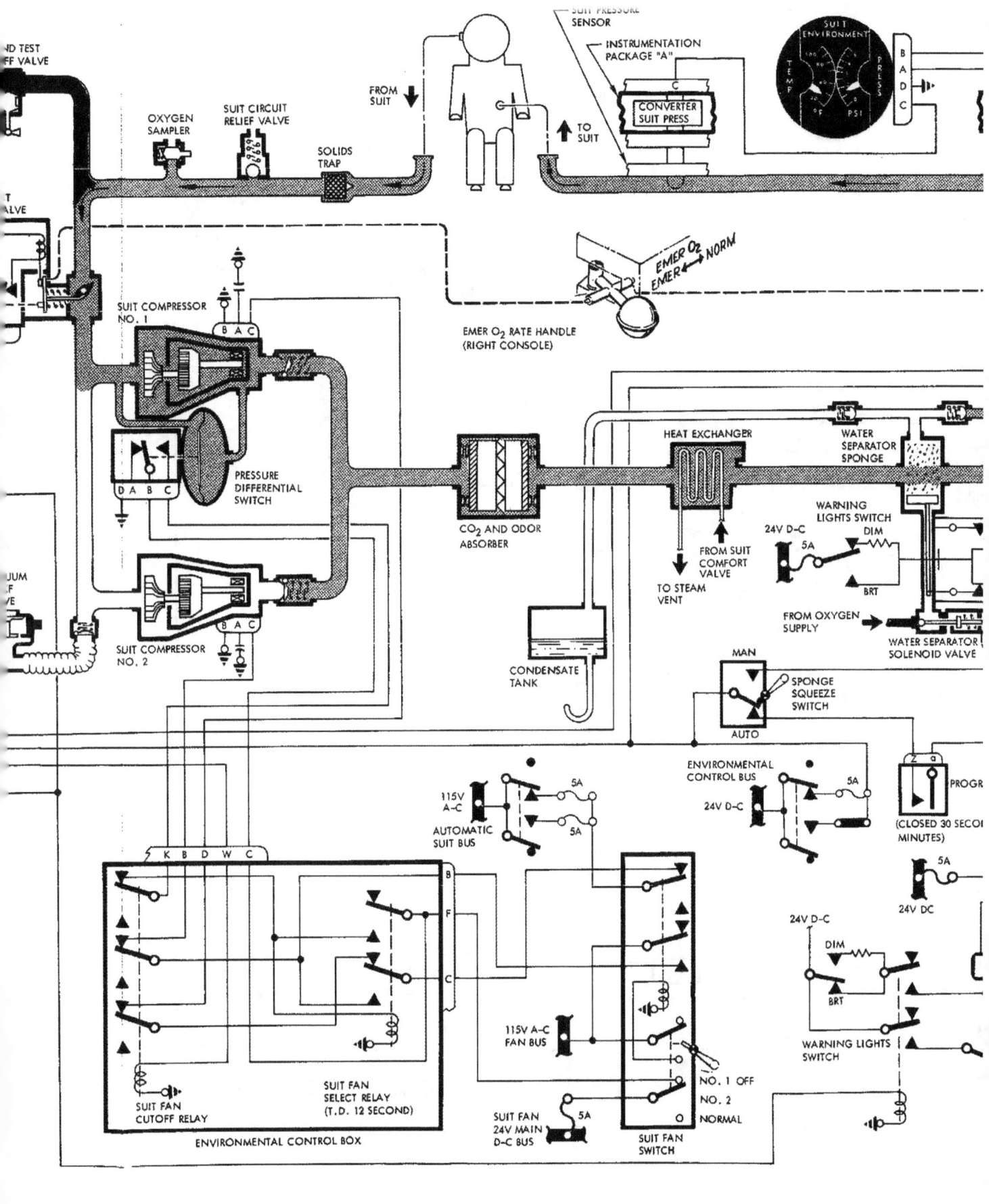

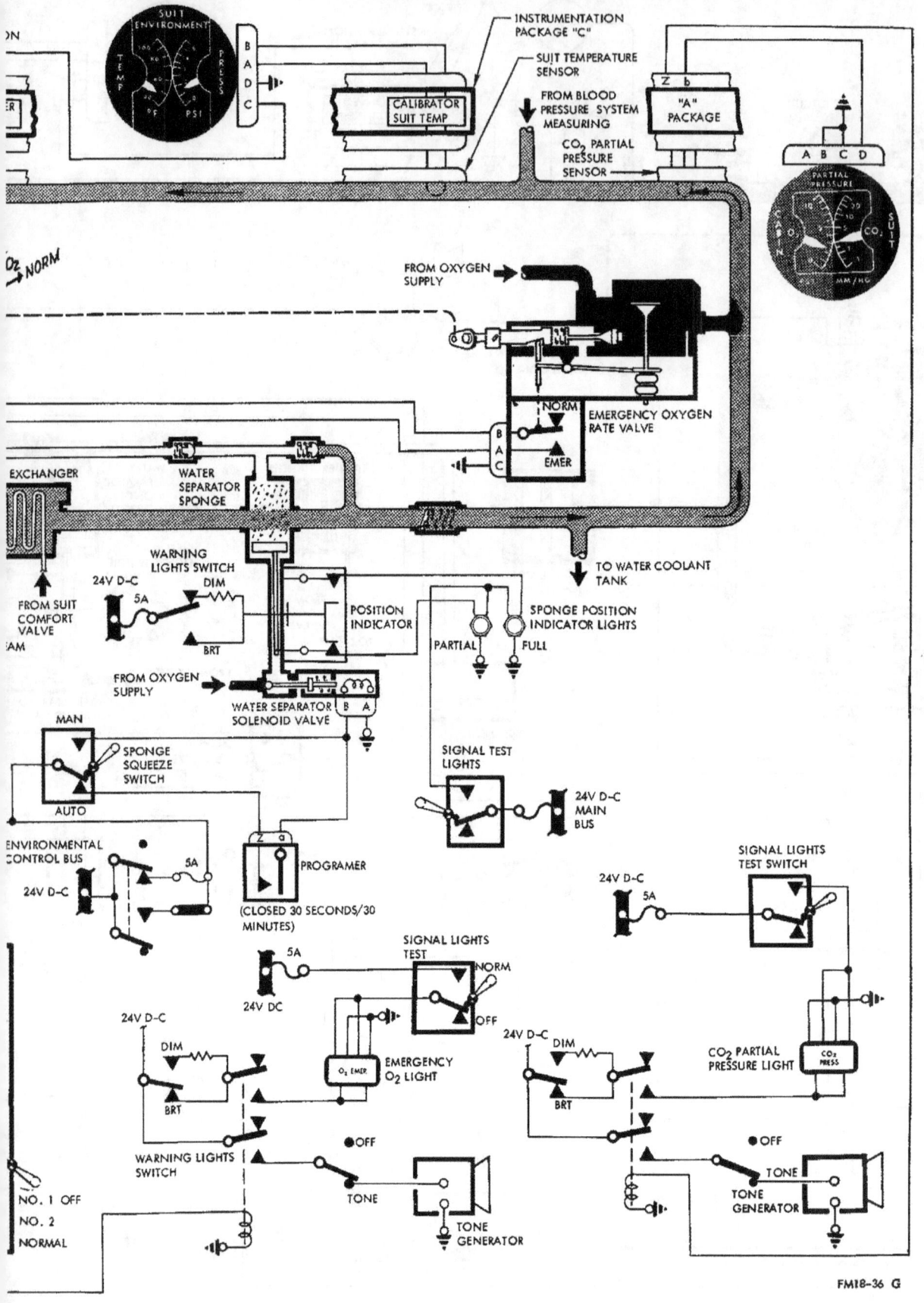

regulator and the cabin pressure control valve. During the launch and re-entry phases, when the astronaut's helmet faceplate is closed, the suit pressure regulator utilizes cabin pressure as a reference to control the suit circuit pressure. While operating in the suit environmental control mode, (helmet faceplate closed), oxygen from the suit pressure regulator enters the suit circuit and is recirculated through the suit compressor, CO_2 and odor absorber, suit heat exchanger, water separator, astronaut's pressure suit, and the suit circuit solids trap. The suit pressure regulator will maintain the suit circuit pressure within 2.5 to 3.5 inches - water of cabin pressure.

The suit circuit incorporates two compressors installed parallel to each other. During normal suit circuit operation the #1 suit compressor circulates throughout the suit circuit. A differential pressure switch is vented to the inlet and outlet ducting of the #1 suit compressor. In the event the #1 suit compressor malfunctions, the differential switch senses the loss of pressure across the #1 suit compressor, and in turn directs power to operate the #2 suit compressor. A SUIT FAN switch is provided on the main instrument panel, to enable selection of either compressor.

Oxygen flowing from the compressors passes through the CO_2 and odor absorber. The absorber is divided into individual sections that contain activated charcoal and lithium hydroxide removing odors and carbon dioxide from the oxygen. Filters, incorporated in the absorber retains charcoal and lithium hydroxide dust.

SEDR 104

Suit circuit temperature, controlled by a suit heat exchanger, removes heat from the suit circuit oxygen flowing through the heat exchanger. Waterflow to the heat exchanger is controlled by a suit temperature control valve, located on the right console.

A water separator containing a sponge, collects moisture from the suit circuit oxygen flowing through the separator. At timed intervals, the sponge is pneumatically compressed and water contained within the sponge is transferred into a condensate storage tank. A sponge position indicator is attached to the piston housing of the water separator. As the water separator piston is in motion, the "PARTIAL" sponge position indicator light illuminates. Both "PARTIAL" and "FULL" indicator lights illuminate upon completion of piston upward travel. As the piston returns to its original position, both indicator lights extinguish. A sponge squeeze switch located on the main instrument panel is provided to actuate the water separator prior to the normal programmed sequence of operations. Suit pressure and temperature sensors, located in the suit circuit, transmit suit circuit pressure and temperature to the SUIT ENVIRONMENT indicator. The dual faced SUIT ENVIRONMENT indicator is located on the main panel. The content of CO_2 in the suit circuit is measured by means of a calibrated partial pressure sensor. Pressure measured by the sensor is transmitted to an indicator marked "PARTIAL PRESSURE" located on the main instrument panel. An indication of excess CO_2 in the suit circuit, detected by the CO_2 partial pressure sensor, directs electrical power to illuminate the "CO_2 PRESSURE" warning light and operate a tone generator.

During the pre-launch phase, the suit circuit is purged with oxygen from an external low pressure source. Suit heat exchanger is also supplied with a freon coolant, from an external ground supply, to provide suit circuit cooling. The suit circuit oxygen circulates through the suit circuit, during the suit mode operation. During spacecraft flight, the pressure within the suit circuit is automatically maintained at approximately 5 psia by the pressure regulator. During the descent phase, the 21,000 feet barostats actuate to cause 24V d-c electrical power to energize the inlet air door relay. Energizing the inlet air door relay directs power to ignite an explosive squib, which in turn ejects the snorkel explosive door. (Door is located on spacecraft exterior.) On descending to an altitude of 17,000 ± 3,000 feet, the cabin air inlet and outflow valves open barometrically. The suit compressor draws atmospheric air into the suit circuit through the ejected snorkel door opening, the snorkel valve and the open cabin air inlet valve. In the event the cabin air inlet and outflow valves fail to open, the astronaut may manually open the valves by actuating the SNORKEL pull ring, located on the left console. Opening of the cabin air inlet valve automatically switches the environmental system mode of operation to the emergency mode, but the suit circuit compressor continues to operate to provide suit circuit ventilation. Also, opening of the cabin air inlet valve directs electrical power to close the suit circuit shutoff valve, which in turn mechanically opens the emergency oxygen rate valve. The emergency air inlet door relay is energized which in turn directs electrical power to ignite the snorkel explosive door squib and eject the snorkel explosive door. (This provision insures the ejection of the snorkel explosive door, in the event the door fails to eject at 21,000 feet.) Air circulating

through the suit circuit is vented through the suit pressure regulator to the cabin, and in turn is vented out of the spacecraft through the cabin outflow valve. In the event the spacecraft submerges momentarily following a water landing, the ball float in the cabin air inlet valve and the diaphragm flapper valve in the cabin air outflow valve will seat. Seating of the valves (snorkel and flapper), prevents water from entering into the suit circuit and cabin, through the open cabin air inlet and outflow valves. Operation of the suit circuit compressor with the snorkel valve closed will create a vacuum in the flexible ducting, located between the cabin air inlet valve and suit circuit. The vacuum relief valve, located in the flexible ducting, will open when the pressure differential between the cabin and flexible ducting is 10-15 inches of water. Opening of the vacuum relief valve reduces the vacuum to permit the air inlet snorkel valve ball float to unseat if the snorkel valve is above the water. This action in turn allows outside air to enter into the suit circuit for ventilation. During spacecraft submersion, cabin air entering the open vacuum relief valve provides suit circuit ventilation.

4-11. SUIT EMERGENCY CONTROL

The suit emergency control, Figure 4-7, is provided to insure the astronaut's survival in the event the cabin and suit environmental control circuits malfunction. Operation in the emergency mode basically consists of opening the emergency oxygen rate valve, to supply oxygen at approximately .05 lbs/min and closing of the suit circuit shutoff valve. Illumination of the O_2 EMERG light and the movement of the EMERG O_2 rate handle to EMERG position, indicates emergency mode of operation.

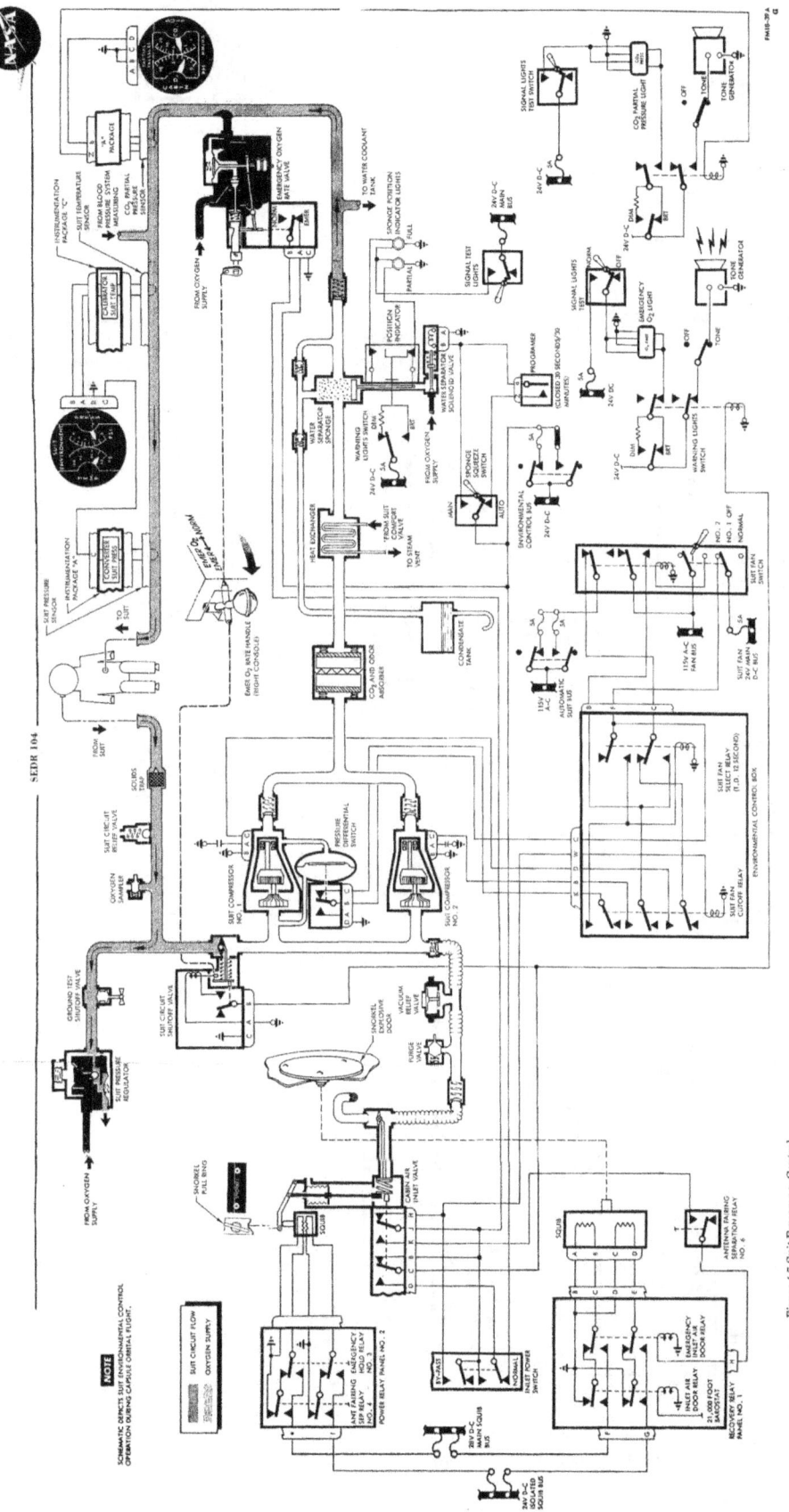

Figure 4-7 Suit Emergency Control

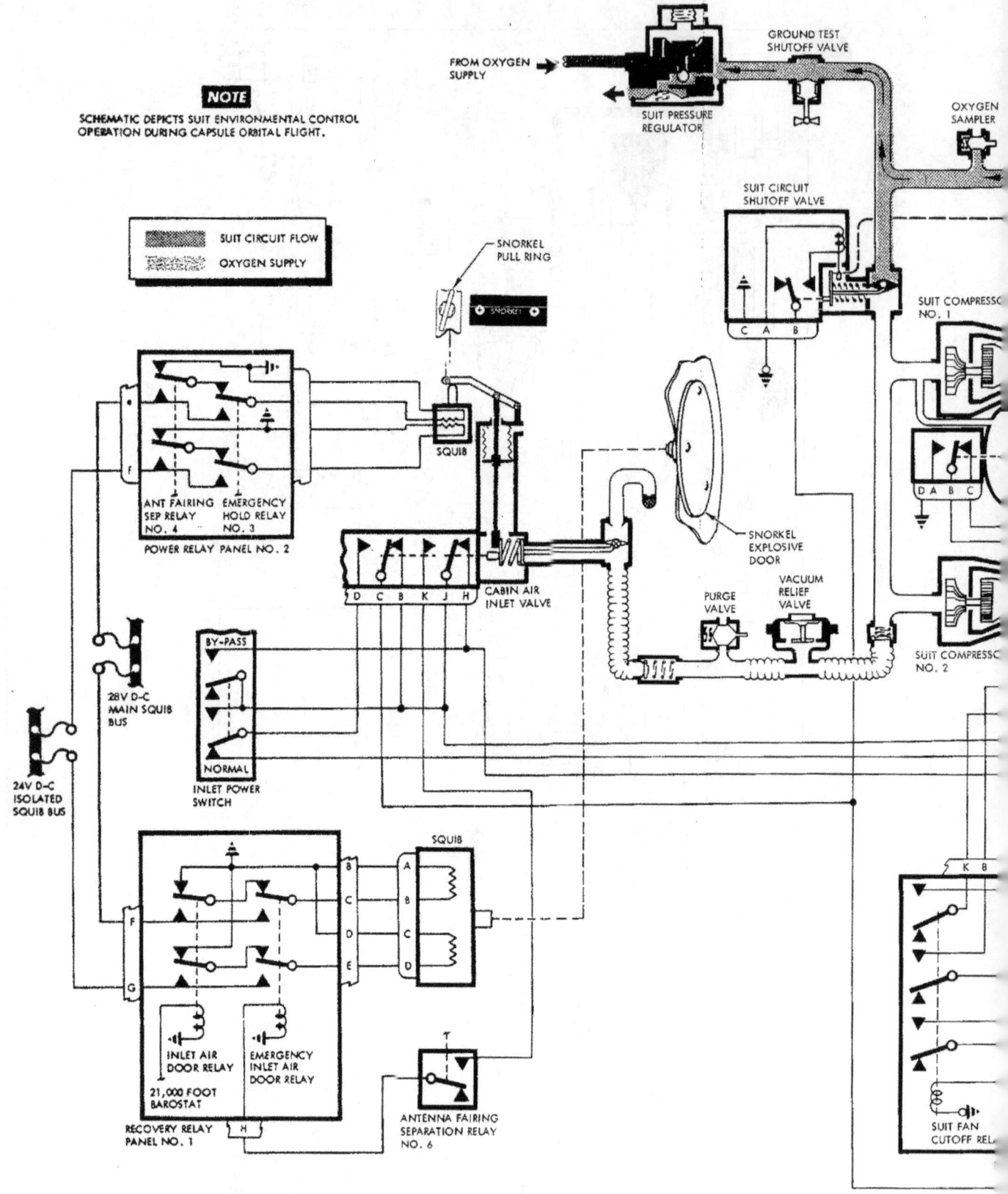

Figure 4-7 Suit Emergency Control

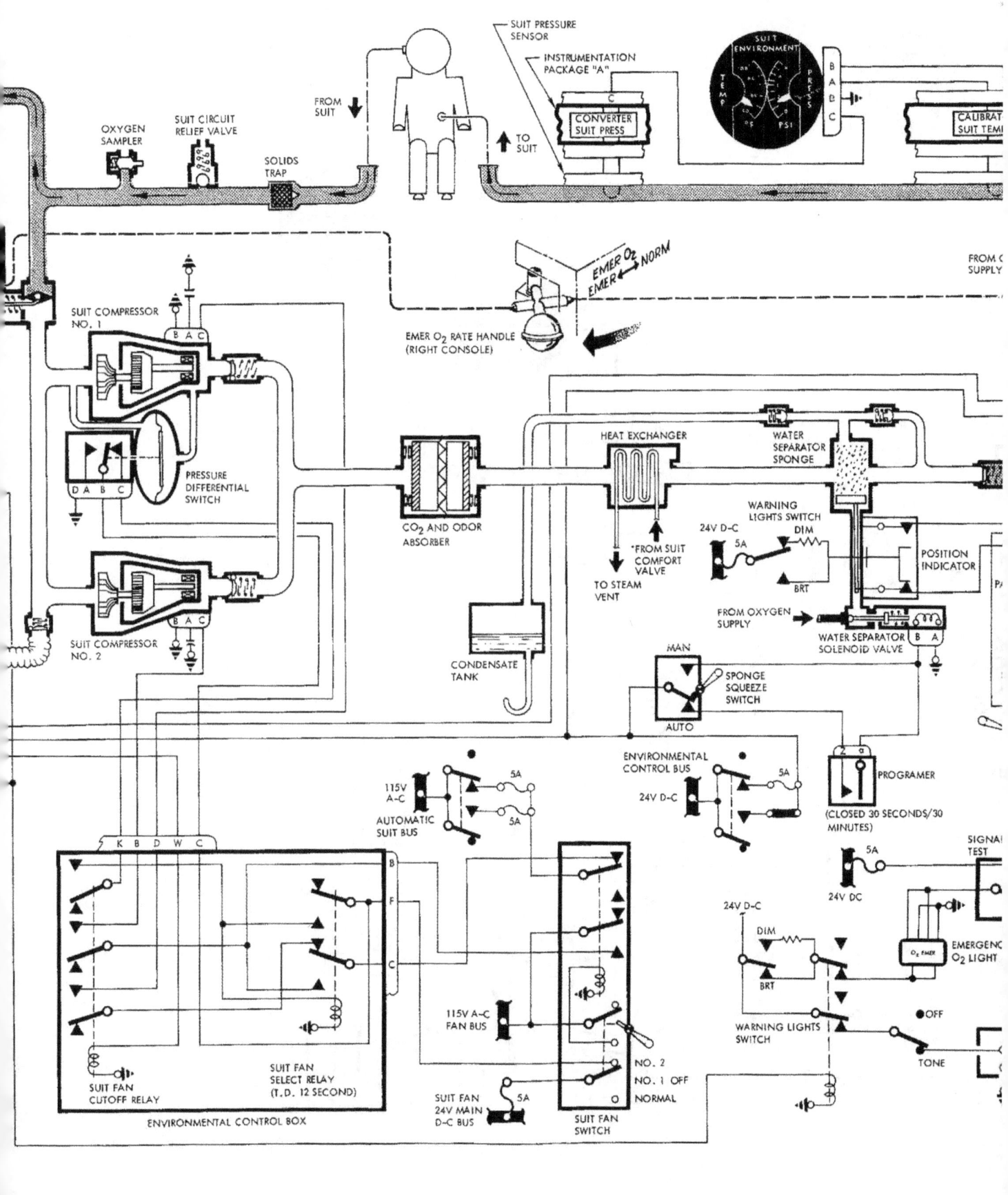

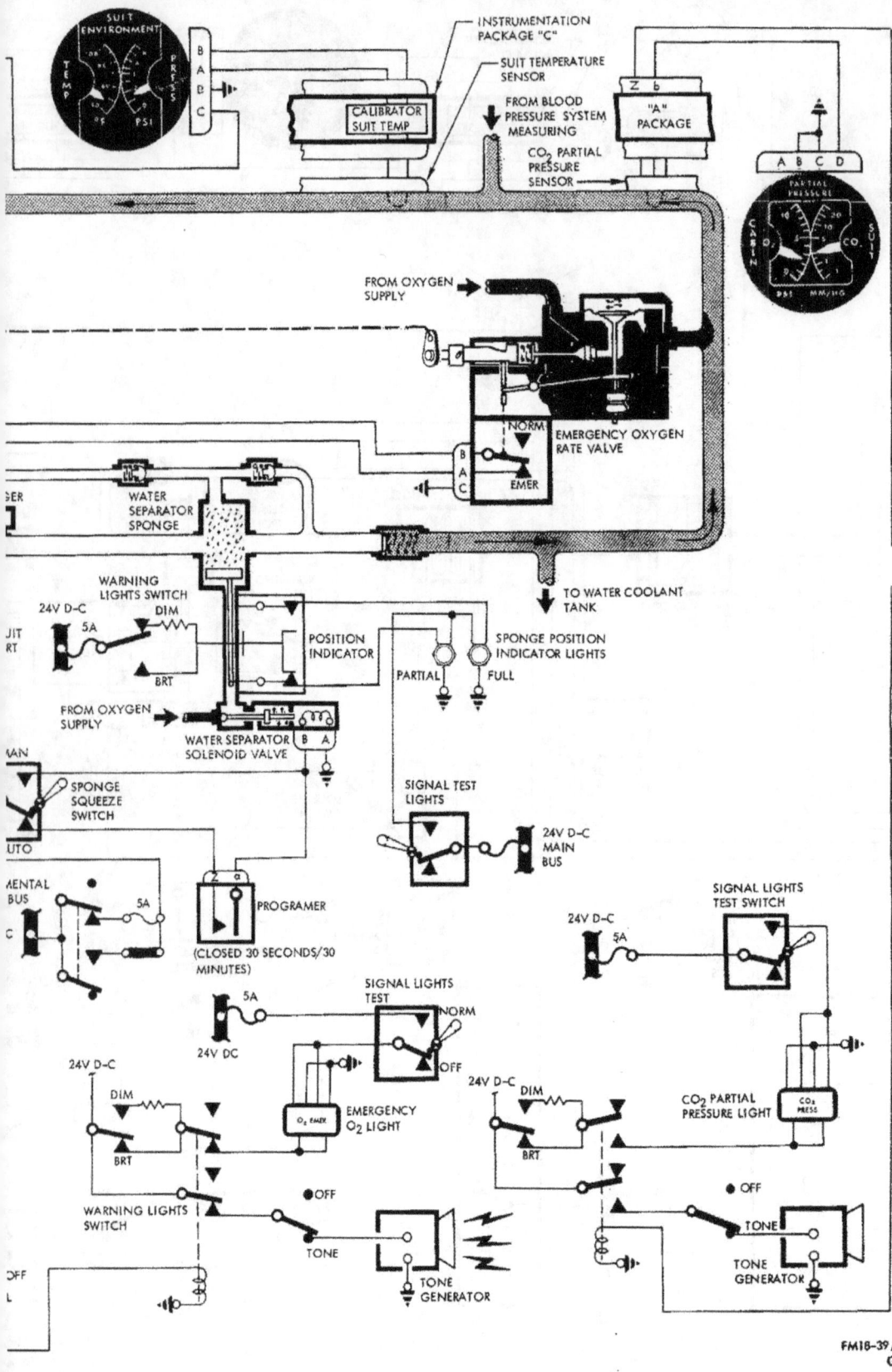

When operating in the normal mode, (See Figure 4-6), during normal orbital flight, the emergency oxygen rate valve is closed, the suit circuit shutoff valve is open, suit compressors are operative, and the suit circuit pressure regulator is controlling oxygen flow to the suit circuit. The emergency oxygen rate valve remains closed as long as suit circuit pressure remains at approximately 5 psia, pressure. In the event the suit circuit pressure drops to $4.0 ^{+.1}_{-.3}$ psia, the rate valve internal aneroid extends, to offseat a poppet, and allows oxygen from the oxygen supply to flow through the rate valve and into the suit circuit. The extension of the rate valve aneroid, due to low pressure, actuates a limit switch that provides electrical power to energize the suit circuit shutoff valve solenoid and the suit fan cut-off relay, illuminate the O_2 EMERG light, and operate a tone generator. Energizing the suit fan cut-off relay removes the 115V a-c electrical power to operate the suit circuit compressor. (At an altitude of 17,000 $\pm$ 3,000 feet, the cabin air inlet relay will open. Opening of the cabin air inlet relay de-energizes the suit fan cutoff relay. The de-energized suit fan cutoff relay routes power to the #1 suit circuit compressor. If the #1 suit circuit compressor fails to operate within 12 seconds, the suit fan selector relay will energize and allow power to be directed to the suit fan cutoff relay and then on to the #2 suit circuit compressor.) Energizing the shutoff valve solenoid releases the shutoff valve shaft arm, and mechanically moves the EMERG O_2 handle, right console, to the EMERG position. Movement of the EMERG O_2 handle mechanically actuates the emergency oxygen rate valve to the open position. With the emergency oxygen rate valve open and the suit circuit shutoff valve closed, oxygen from the oxygen supply flows into the pressure suit and is discharged through the suit pressure regulator relief valve.

Actuating the EMERG O_2 handle to the NORM position resets the shutoff valve to the open position, the emergency oxygen rate valve to the close position, starts suit compressor operation, extinguishes the O_2 EMERG light, and in turn switches the suit circuit operation to the suit normal control mode.

The emergency mode is also automatically selected during spacecraft landing phase, when the spacecraft has descended to an altitude of 17,000 feet. At 17,000 feet the cabin air inlet valve opens. Opening of the cabin air inlet valve actuates a limit switch that provides electrical power to operate the suit circuit compressor and close the shutoff valve, which in turn mechanically opens the emergency oxygen rate valve. An inlet power switch, located on the main instrument panel allows operation in the suit environmental control mode in the event the cabin air inlet valve prematurely opens (See Figure 4-6). Premature opening of the cabin air inlet valve deactivates the cabin fan and closes the suit circuit shutoff valve which in turn opens the emergency oxygen rate valve. The suit circuit is now operating in the emergency mode. To initiate transition back to the suit mode, the inlet power switch is placed in the BY-PASS position. With the inlet power switch in the BY-PASS position, the cabin fan is activated (See Figure 4-5) and the suit circuit shutoff valve is deactivated. The EMERG O_2 handle right hand console, is now

placed in the NORM position; placing of the EMERG O_2 handle to the NORM position opens the suit circuit shutoff valve and closes the emergency oxygen rate valve. The environmental control system is now operating in the suit environmental control mode. To prevent snorkel door separation upon premature opening of the cabin air inlet valve, the emergency inlet air door relay is interconnected to the antenna fairing separation relay during descent. After opening of the cabin air inlet and outflow valves, the inlet power switch is placed in the NORMAL position.

4-12. OXYGEN SUPPLY

During the pre-launch phase and prior to installation of the entrance hatch, primary and secondary oxygen system shutoff valves are manually opened, by ground crewmen, to activate the oxygen supply. Opening of the shutoff valves, Figure 4-8, provides oxygen to the cabin pressure control valve, suit pressure regulator, suit emergency oxygen rate valve and the suit circuit water separator solenoid valve.

During operation, when the primary oxygen bottle pressure drops below approximately 200 psig, due to near depletion of the primary oxygen supply; the secondary oxygen supply line pressure will override the primary oxygen pressure and continue to supply the environmental system with oxygen. Audible and visual means are provided on the Main Instrument Panel to warn the astronaut of a diminishing oxygen supply. As the secondary O_2 supply decreases to 6500 psi, the secondary oxygen supply pressure transducer activates the "O_2 QUANTITY" telelight and the tone generator. A quantity indicator gage, located on the main instrument panel, is provided to indi-

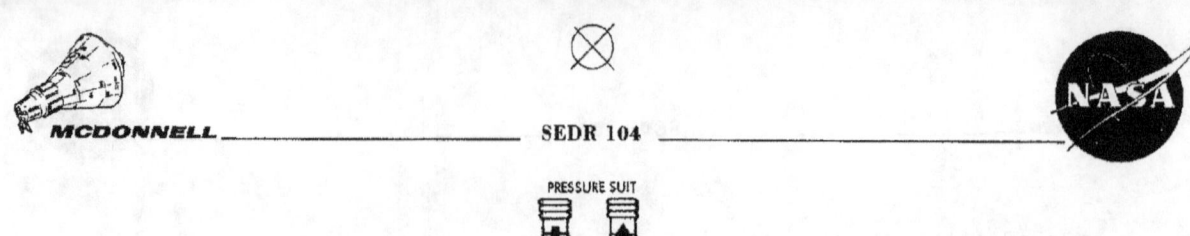

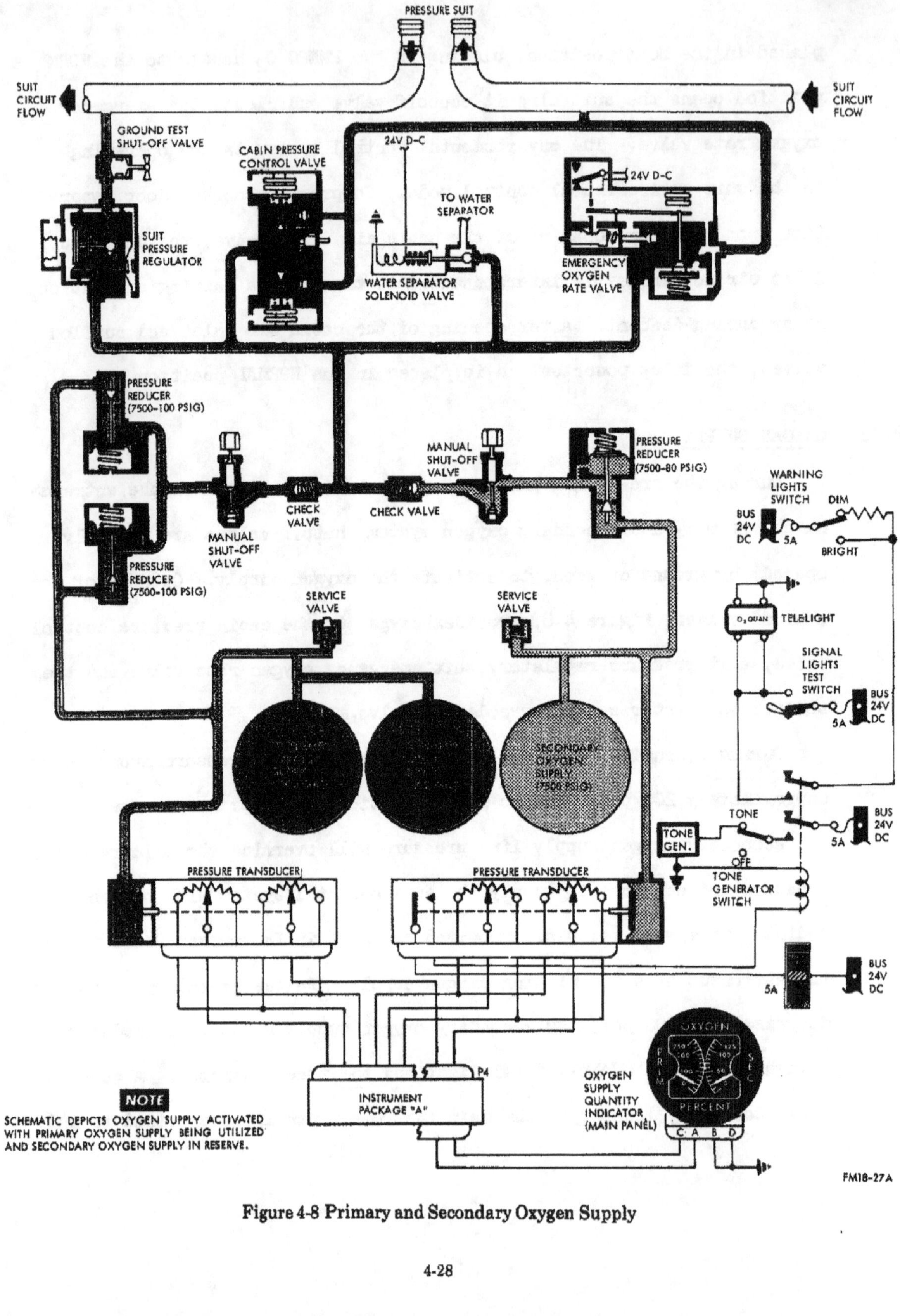

Figure 4-8 Primary and Secondary Oxygen Supply

cate remaining oxygen supply. Two transducers, primary and secondary supply, are provided to enable telemetering of oxygen quantity remaining.

4-13. COOLING CIRCUIT

During pre-launch, cabin and suit circuit cooling, (Figure 4-9) is achieved by supplying freon (F-114) through the umbilical connector and into the cabin and suit heat exchangers. The freon coolant evaporates in the cabin and suit heat exchangers and is discharged overboard through the environmental system steam vents, located in the large pressure bulkhead. Prior to launching, the freon coolant supply is discontinued. When the spacecraft reaches approximately 115,000 feet altitude, cabin and suit circuit cooling is achieved by water evaporation, that occurs within the suit and cabin heat exchangers.

Water from the water coolant tanks is supplied through the temperature control valves, to the suit and cabin heat exchangers. Oxygen, from the suit circuit is utilized to pressurize both water coolant tanks. Oxygen pressure moving each tank diaphragm, forces the water supply out of the tanks at a rate dependent upon the position of the temperature control valves. The temperature control valves control the amount of water entering the heat exchangers, and in turn controls cabin and suit temperatures. The steam flows out through the steam vents, located in the large pressure bulkhead. A dual faced heat exchanger temperature indicator, and warning light is provided on the main instrument panel to indicate temperature conditions in the cabin and suit heat exchangers. An excessive amount of water flowing through the cooling circuit is detected by sensors attached

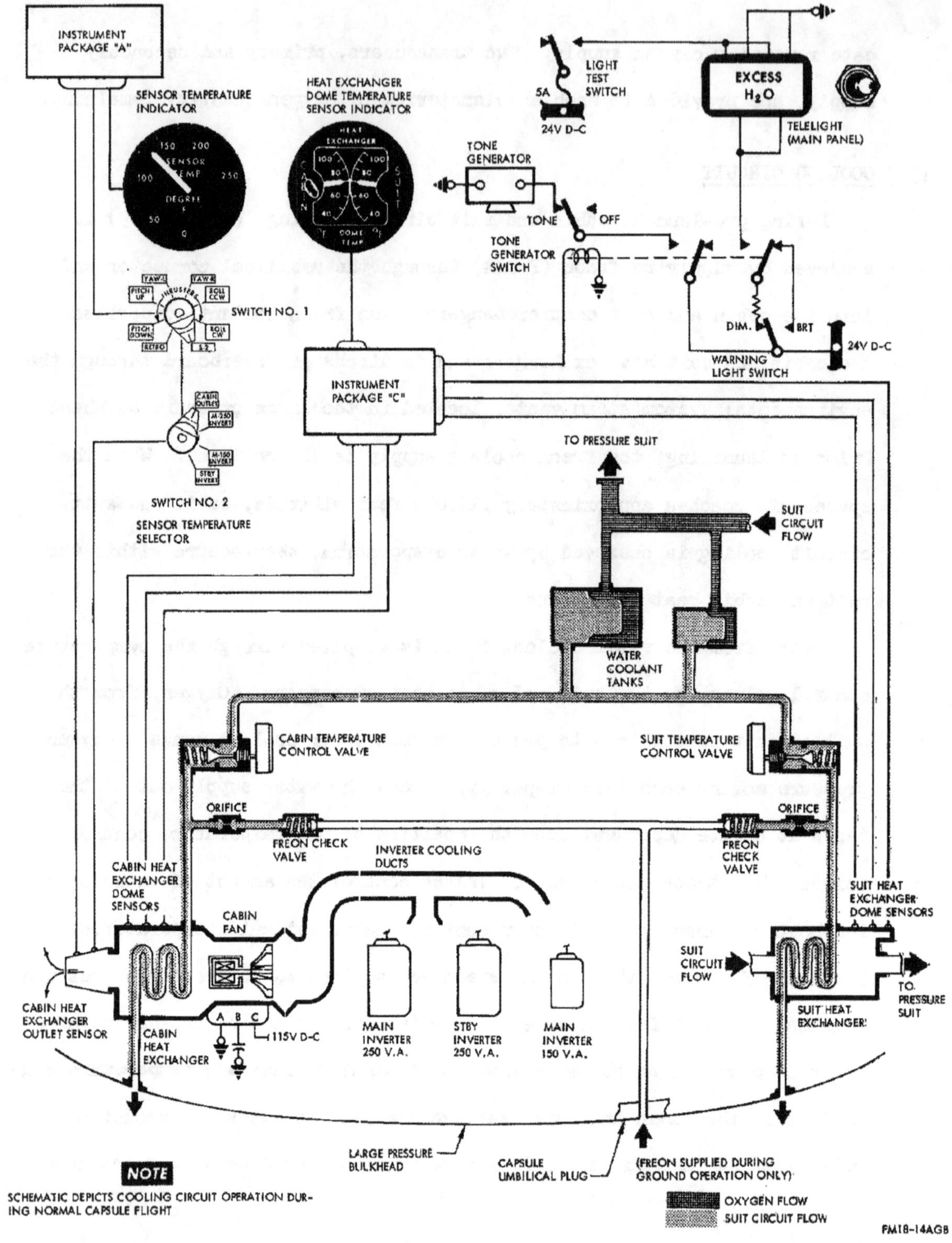

Figure 4-9 Cooling Circuit

to the cabin and suit heat exchangers directing electrical power to illuminate the "EXCESS H$_2$O" warning light and also operate the tone generator. The astronaut must then position the cabin or suit temperature control valve to a warmer setting to prevent wasting coolant and to achieve efficient operation of the heat exchanger.

4-14. **SYSTEM UNITS**

4-15. **PRIMARY AND SECONDARY OXYGEN BOTTLES**

The primary and secondary spherical shaped oxygen bottles are located beneath the astronaut's support couch adjacent to the spacecraft conical section and large pressure bulkhead. Each bottle has a capacity of 4 pounds oxygen, stored under a 7500 psig pressure at 70°F temperature. Servicing of the oxygen bottles is accomplished through a quick disconnect filler coupling.

4-16. **SUIT CIRCUIT PRESSURE REGULATOR**

The suit circuit pressure regulator, Figure 4-10, is provided to regulate oxygen pressure to the suit circuit and to replenish suit circuit oxygen consumed by the astronaut, or lost through leakage. The regulator is a demand type diaphragm operated regulator that controls suit circuit pressure in reference to cabin pressure. Suit circuit pressure is maintained approximately 2.5 - 3.5 inches of water below cabin pressure during normal system operation, under ideal (no cabin leakage) conditions. Cabin pressure is sensed on the upper side of the regulator control diaphragm and

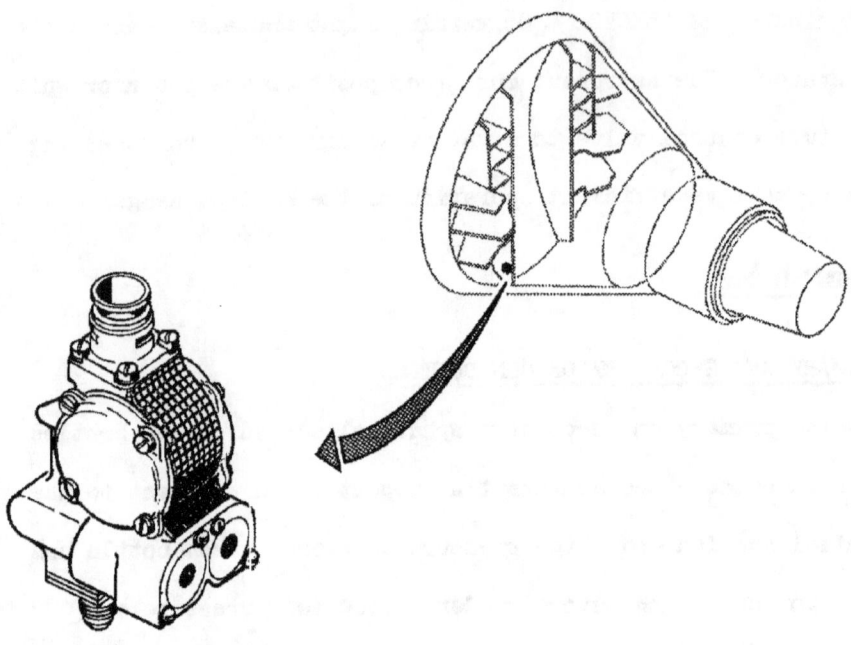

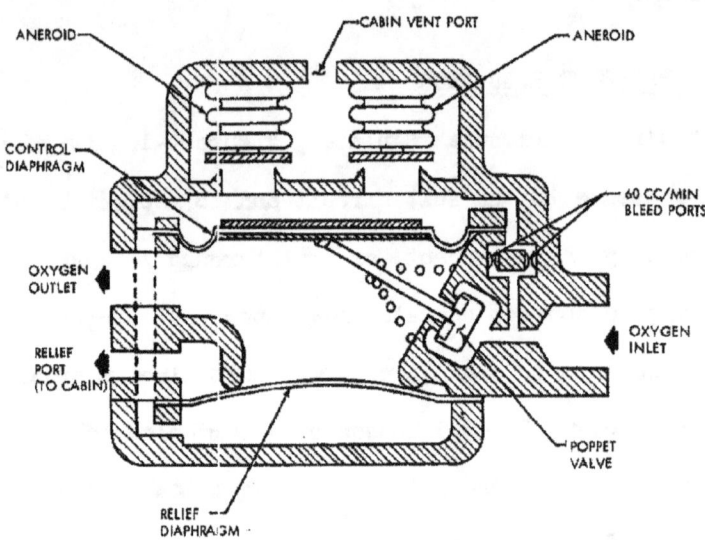

Figure 4-10 Suit Pressure Regulator

suit circuit pressure is sensed on the lower side of the diaphragm. The regulator also contains a resilient type diaphragm that is used to relieve excessive suit circuit pressures. Two aneroids are provided to shut off cabin vent port of regulator in the event cabin pressure decreases below 4.6 + .2 psia. In the event of cabin leakage, not decreasing below $4.0 \pm ^{.2}_{.1}$ psia, the cabin pressure control valve will open to replenish cabin pressure to $5.1 \pm ^{.2}_{.1}$ psia. Make up oxygen from the cabin pressure control valve will flow through the suit circuit and out through the suit pressure regulator relief valve and into the cabin. At this time, suit circuit pressures will exceed cabin pressures due to pressure differential (suit above cabin) required to open the suit circuit pressure regulator relief valve.

During normal ascent, cabin pressure decreases, and the regulator relief diaphragm relieves suit circuit pressure to within 2 - 9 inches H_2O above cabin pressure. During normal orbital flight, the control diaphragm will regulate suit circuit pressure in relationship to cabin pressure. An increase in cabin pressure will act on the diaphragm to unseat a poppet valve and allow suit circuit pressure to increase to within 2.5 - 3.5 inches of H_2O below cabin pressure. In the event cabin pressure decreases below 4.6 ± .2 psia, the aneroids will extend and close off cabin vent port of regulator. Two 60 cc/min bleed ports provide oxygen to pressurize the reference chambers and permit the regulation of suit circuit pressure to 4.6 ± .2 psia. Two aneroids and two bleed ports are provided to insure redundancy in the event either aneroid or either bleed port fails to function. Descent operation of the regulator is the same as an increase in cabin pressure during normal orbital flight.

4-17. SUIT CIRCUIT SHUTOFF VALVE

The suit circuit shutoff valve, Figure 4-11, is designed to shut off oxygen flow to the suit environmental circuit accessory components, whenever the suit circuit is operating in the emergency mode. The shutoff valve, spring loaded to the close position, is latched in the open position during normal suit circuit operation. Valve is maintained in the open position by a solenoid controlled detent pin engaged into the valve spoon arm. A microswitch, depressed by the valve arm, completes the solenoid circuit when the valve is latched open. Opening of either the emergency oxygen rate valve or the cabin air inlet valve directs an electrical signal to energize the shutoff valve solenoid. Energizing the solenoid retracts the detent pin and allows the valve spring to rotate the valve spoon to the close position. Closing of the valve, through an inter-connecting linkage. The shutoff valve is mechanically opened by the EMER O_2 control handle, located on the right-hand console. The shutoff valve is interconnected to the emergency rate valve, so that when the emergency rate valve closes, the shutoff valve opens.

4-18. EMERGENCY OXYGEN RATE VALVE

The emergency oxygen rate valve, Figure 4-12, is provided to supply a regulated amount of oxygen directly into the astronaut's pressure suit, in the event malfunction occurs in the suit circuit operation. The rate valve is designed to operate automatically and contains provisions for manual operation. The valve, closed during normal suit circuit operation, contains an aneroid that senses suit circuit pressure. Whenever suit cir-

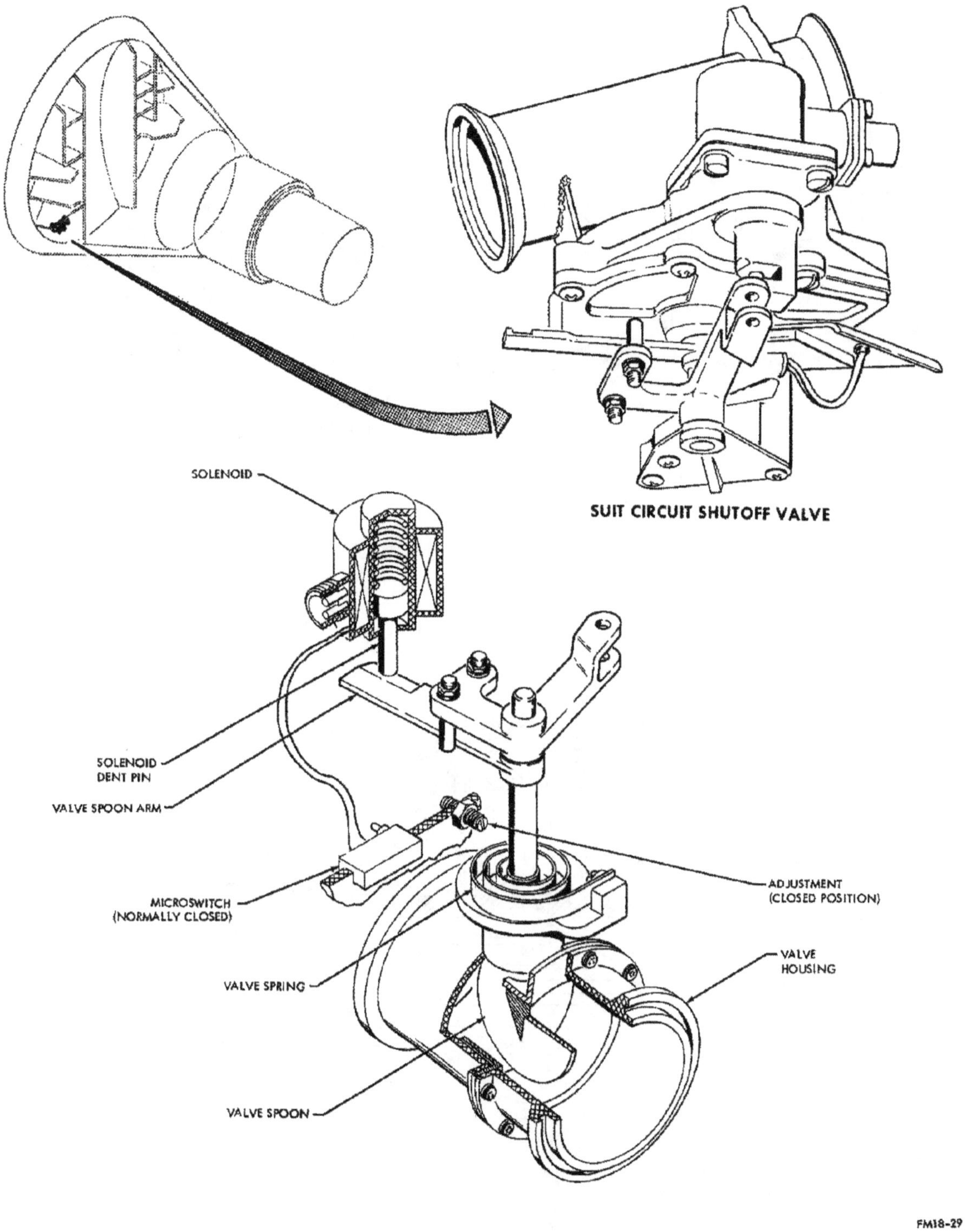

Figure 4-11 Suit Circuit Shutoff Valve

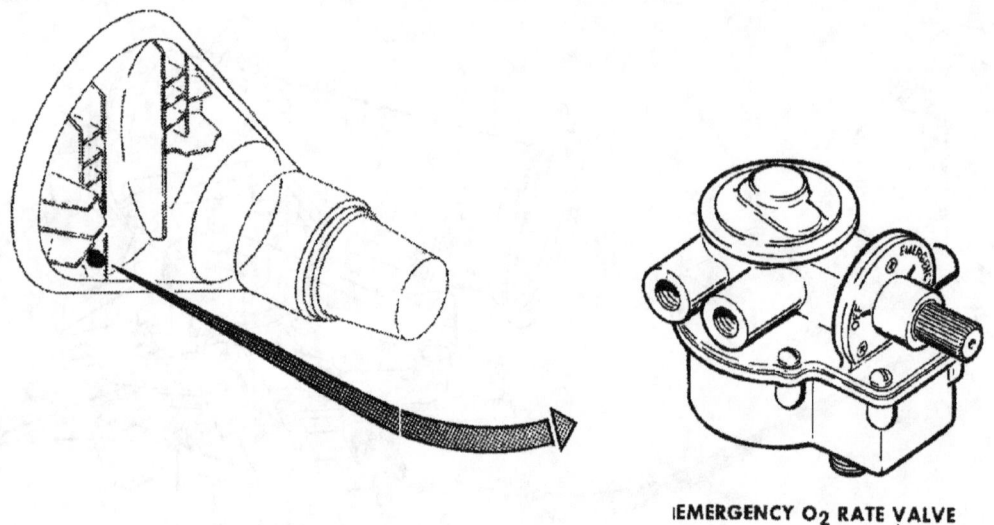

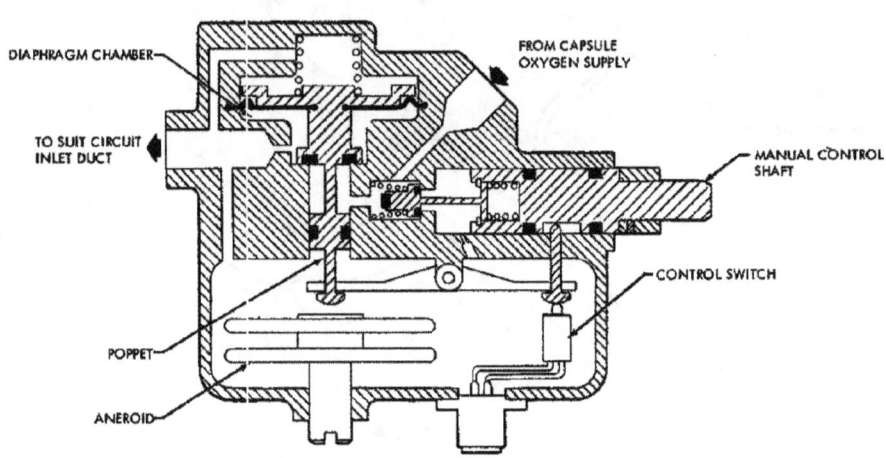

Figure 4-12 Emergency O_2 Rate Valve

cuit pressure drops below $4.0 \genfrac{}{}{0pt}{}{+ .1}{- .3}$ psia, the aneroid extends to offseat a spring loaded poppet and allow oxygen to enter the diaphragm chamber. The pressure in the diaphragm chamber increases and fully strokes the poppet, allowing oxygen to flow into the astronaut's suit at a fixed flow of .049 to .051 #/min. Simultaneously with the offseating of the poppet, a control switch is actuated through a lever mechanism, and directs electrical power to close the suit circuit shutoff valve, illuminate the O_2 EMERG light, and stop suit circuit compressor operation during orbital flight. Suit circuit shutoff valve is interconnected with emergency oxygen rate valve. Therefore, closing of the shutoff valve actuates the emergency oxygen rate valve manual control shaft to close off oxygen flow to valve poppet inlet. Oxygen then flows directly into suit circuit through the valve aneroid chamber.

Emergency oxygen rate valve may be opened manually by selecting EMER position with EMERG O_2 control handle. Whenever the EMER O_2 control handle is moved to NORM, the suit circuit shutoff valve opens and emergency oxygen rate valve closes.

4-19. SUIT CIRCUIT COMPRESSORS

The suit circuit environmental control system utilizes two electric motor driven, single-stage, centrifugal compressors (See Figure 4-6). One compressor is a standby compressor used in the event of normal compressor failure. If the normal compressor fails, the standby compressor is activated by a pressure differential switch which directs power to the standby compressor electrical connections. The only time the suit compressor is inoperative is during orbital flight when the astronaut is

SEDR 104

utilizing oxygen from the emergency oxygen rate valve. When supplementary oxygen from the emergency oxygen rate valve is being used below 20,000 feet, the suit circuit compressor will continue to operate to circulate ambient air to the astronaut.

4-20. CO_2 AND ODOR ABSORBER

The CO_2 and odor absorber, Figure 4-13, is provided to remove astronaut emitted odors and carbon dioxide from the suit circuit. The absorber is basically a metal cannister divided into two sections. The inlet section contains activated charcoal that removes objectional odors from the suit circuit oxygen. Lithium hydroxide, located in the center sections removes carbon dioxide. The outlet section is an exit filter, provided to retain lithium hydroxide in the cannister. The charcoal and lithium hydroxide granules are compressed by a spring force. The useful life of the CO_2 and odor absorber is forty-four hours.

4-21. SUIT CIRCUIT HEAT EXCHANGER

The suit circuit heat exchanger (Figure 4-14) is of a plate fin construction with rectangular offset fins, double sandwich, one pass on the oxygen side and two pass, single sandwich on the water side. The function of the heat exchanger is to cool the gases circulating throughout the suit circuit. Water from the water cooling tanks is routed to the inlet side of the heat exchanger which contains a high density woven felt pad. The function of the felt pad is to evenly distribute the water through the core of the heat exchanger. As water passes through the felt pad, it comes into contact with the heat transfer surfaces on the water

 SEDR 104

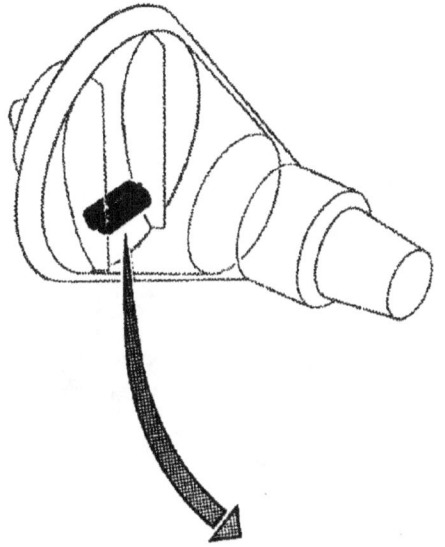

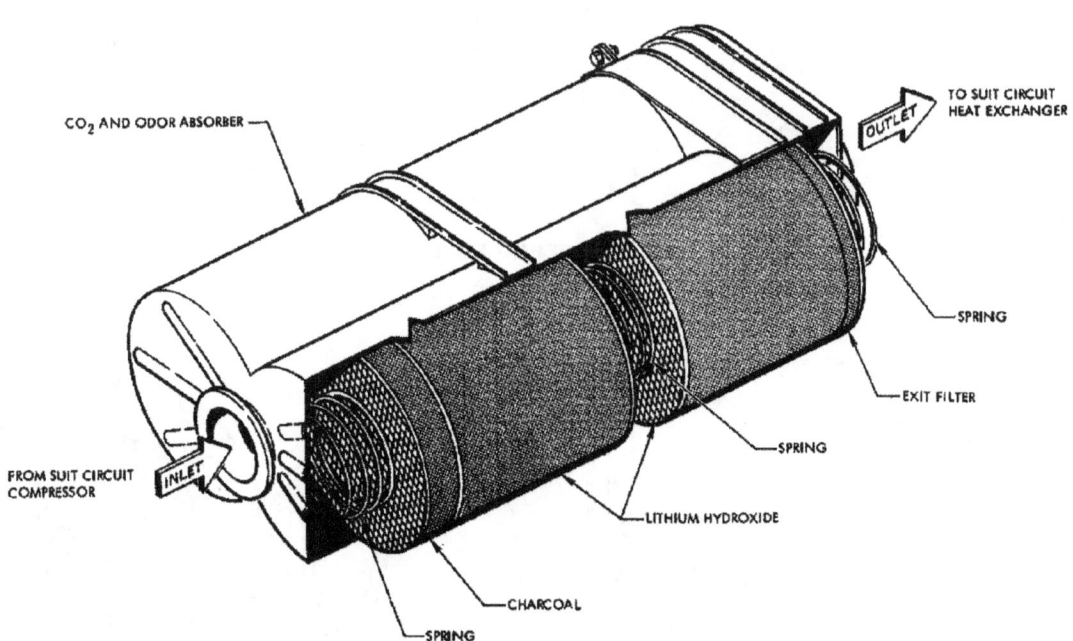

Figure 4-13 CO₂ and Odor Absorber

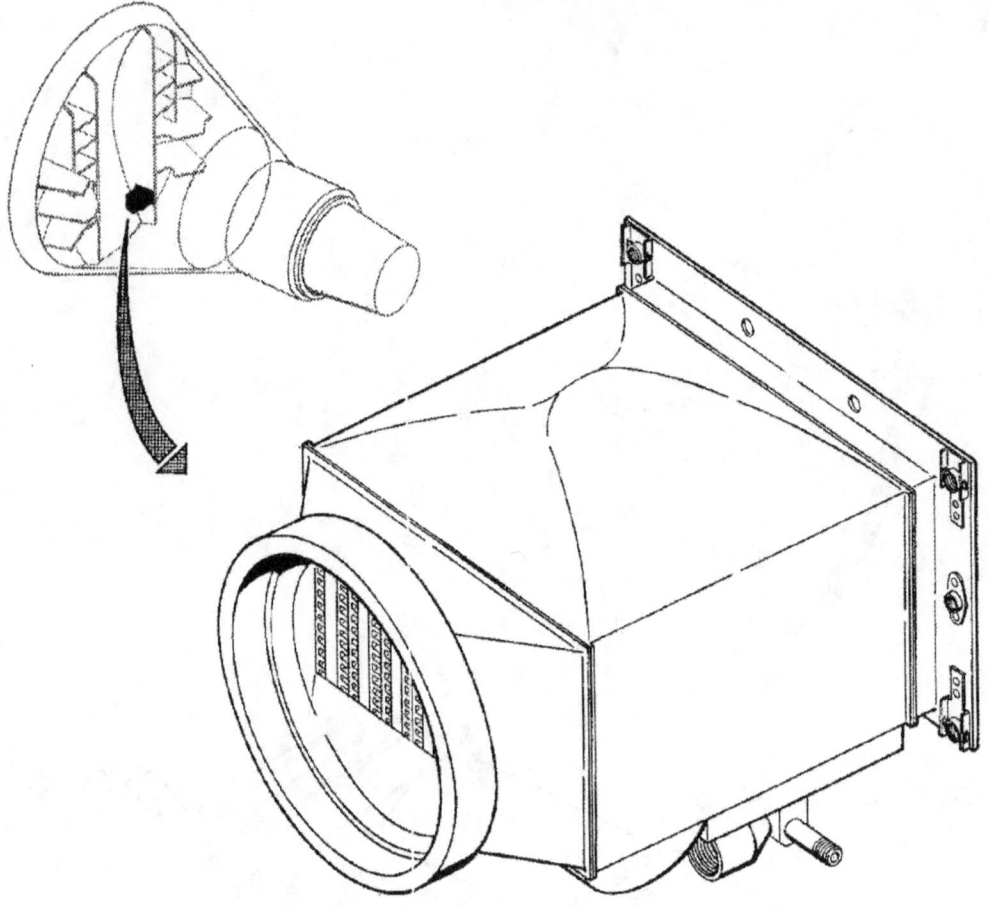
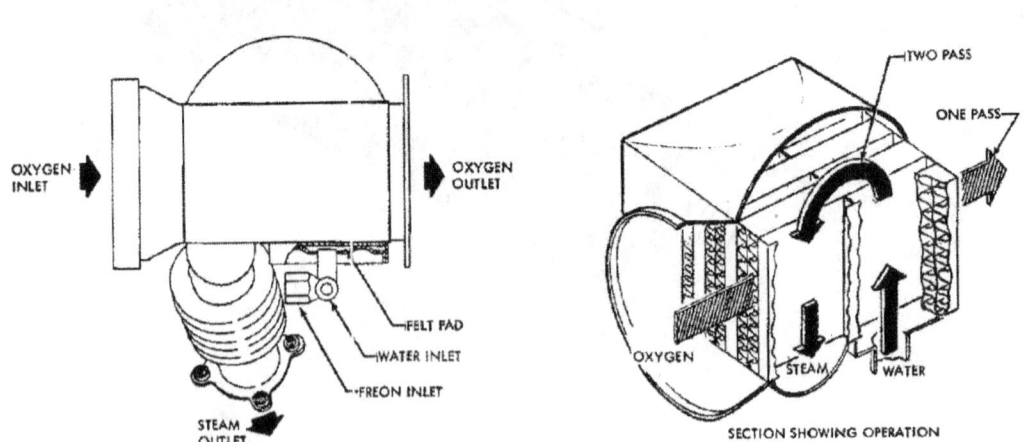

Figure 4-14 Suit Heat Exchanger

side of the heat exchanger. The water evaporates and the steam is vented overboard.

4-22. WATER SEPARATOR

The water separator, Figure 4-15, is provided to remove condensate from the suit circuit oxygen. The separator contains a sponge that collects liquid from the oxygen passing through it. The liquids are squeezed from the sponge and deposited into a storage tank. Once every 30 minutes, for a duration of 30 seconds, the spacecraft programmer supplies electrical power to energize the water separator solenoid valve. Energizing the normally closed solenoid valve opens the valve and directs oxygen from either the primary or secondary supply to the piston stem and the piston plate chambers. A sponge squeeze switch, located on the main instrument panel, is provided to permit actuation at any time.

The piston raises the sponge out of the suit circuit oxygen flow and compresses the sponge against the separator housing plate. Water squeezed out of the sponge is forced into the condensate tank. The water separator solenoid valve is de-energized and the solenoid valve closes. Oxygen below the separator piston is vented to cabin through the separator solenoid valve. Oxygen above the piston, entrapped by a check valve, forces the piston down, thus returning the sponge into suit circuit oxygen flow. During squeezing operation, suit circuit oxygen flow will not be affected, as oxygen will continue to flow through area normally occupied by the sponge. A sponge position indicator is attached to the base of the water separator. "PARTIAL" and "FULL" travel indicator lights used in conjunction with the

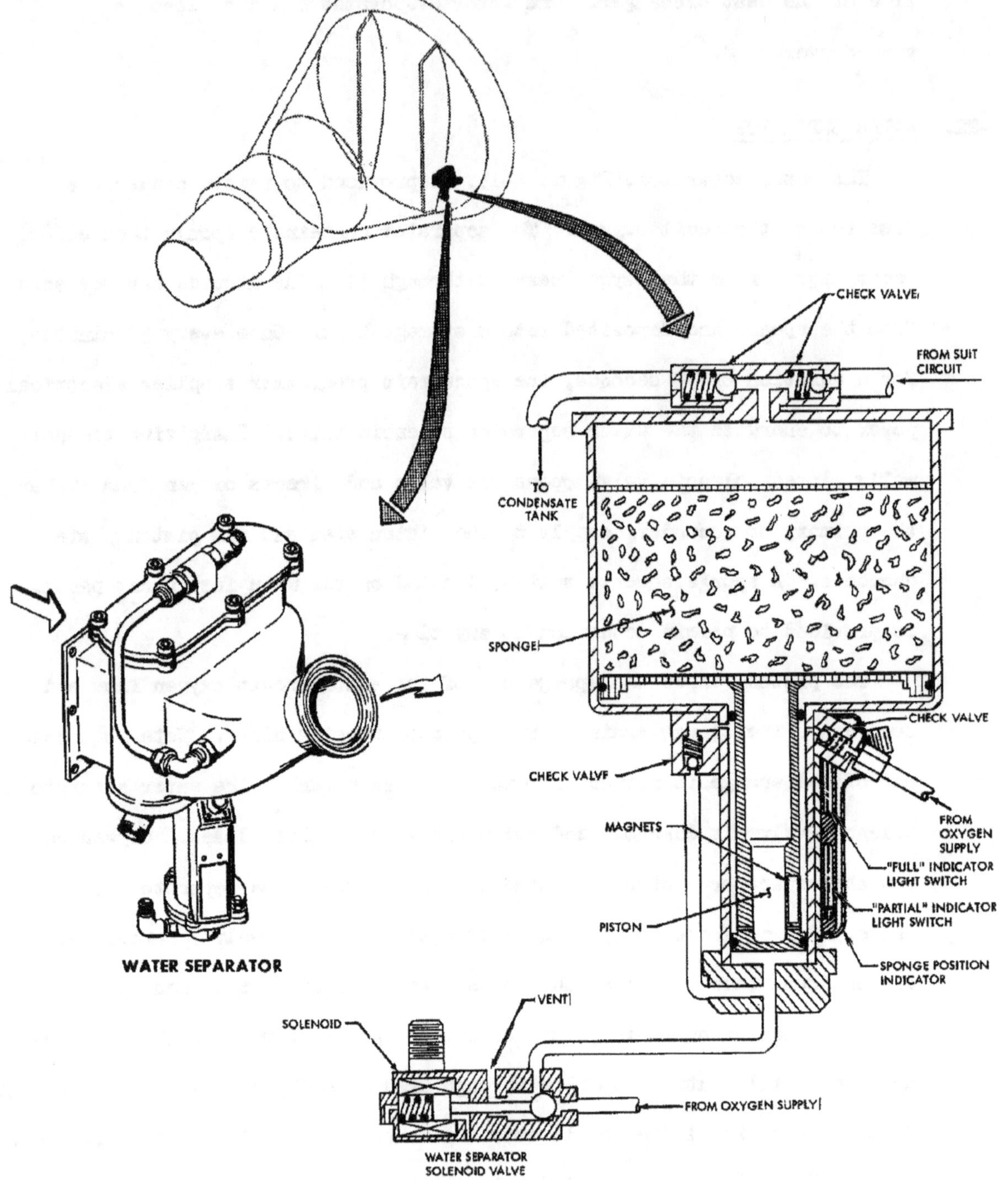

Figure 4-15 Water Separator

position indicator, are located on the Main Instrument Panel. Both "PARTIAL" and "FULL" indicator lights illuminate upon completion of piston upward travel. As the piston returns to its original position both indicator lights extinguish.

4-23. SOLIDS TRAP

The suit circuit solids trap, Figure 4-16, is located in the pilot's suit oxygen outlet duct. The trap consists of a 40 micron mesh screen filter which incorporates an integral bypass to insure operation in the event the trap would become choked with collected solids.

4-24. CABIN HEAT EXCHANGER

The cabin heat exchanger (Figure 4-17), cools the cabin gas in the same manner as the suit circuit heat exchanger. Internal structure is the same as the suit circuit heat exchanger.

4-25. WATER COOLANT TANKS

Two water coolant tanks, (Figure 4-18), are provided in the environmental control system. The larger of the two water tanks, located directly below the astronaut's couch, contains forty pounds of water. The second tank containing nine pounds of water is located to the right of the astronaut's right knee. Water is displaced from each tank by oxygen from the suit circuit acting upon the gas side of a rubber bladder which separates the coolant water from gas. Water is directed to two manual control valves which control the water supply to the suit circuit and cabin heat exchangers. The larger water tank also provides the astronaut with a source of drinking water.

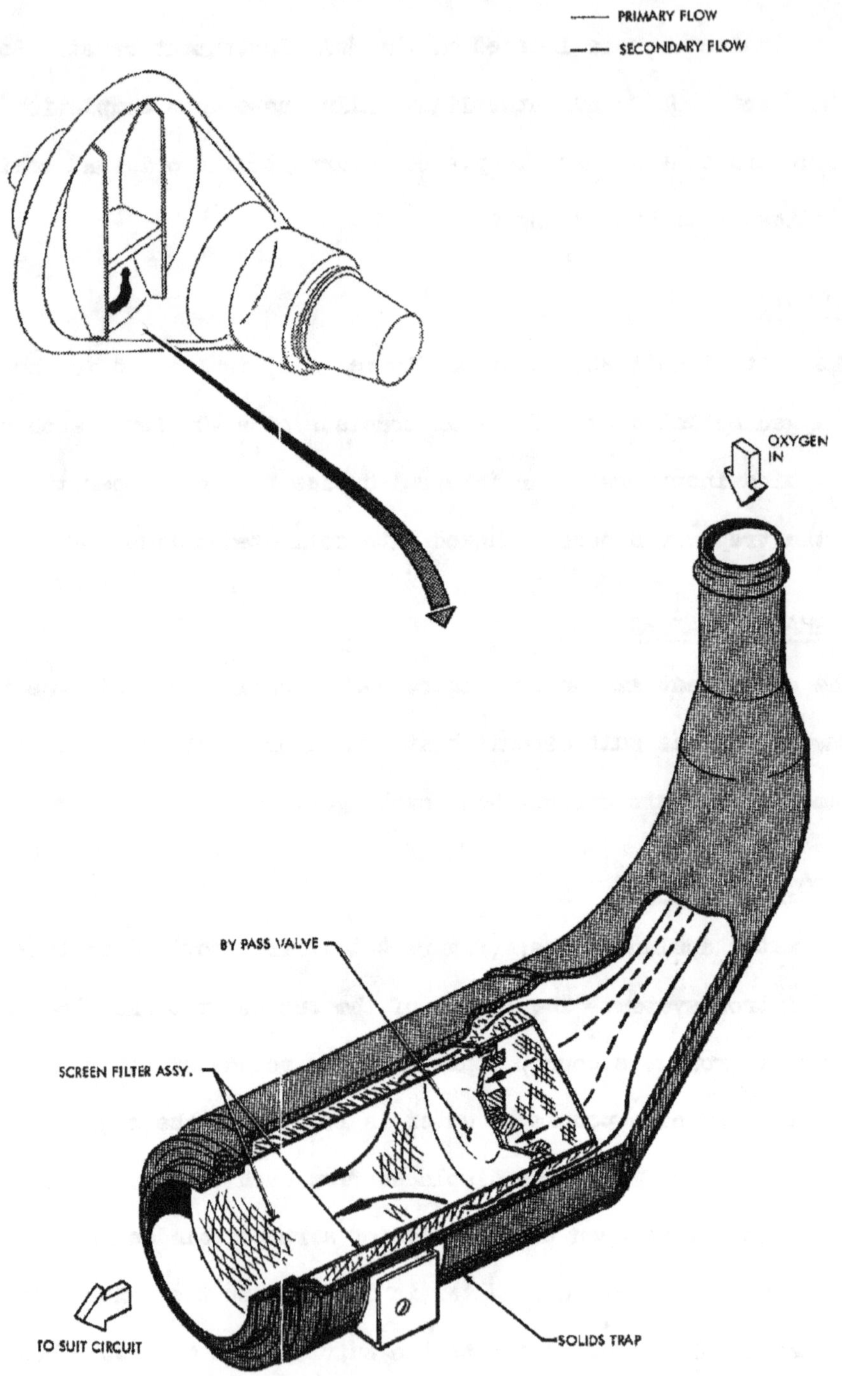

Figure 4-16 Suit Circuit Solids Trap

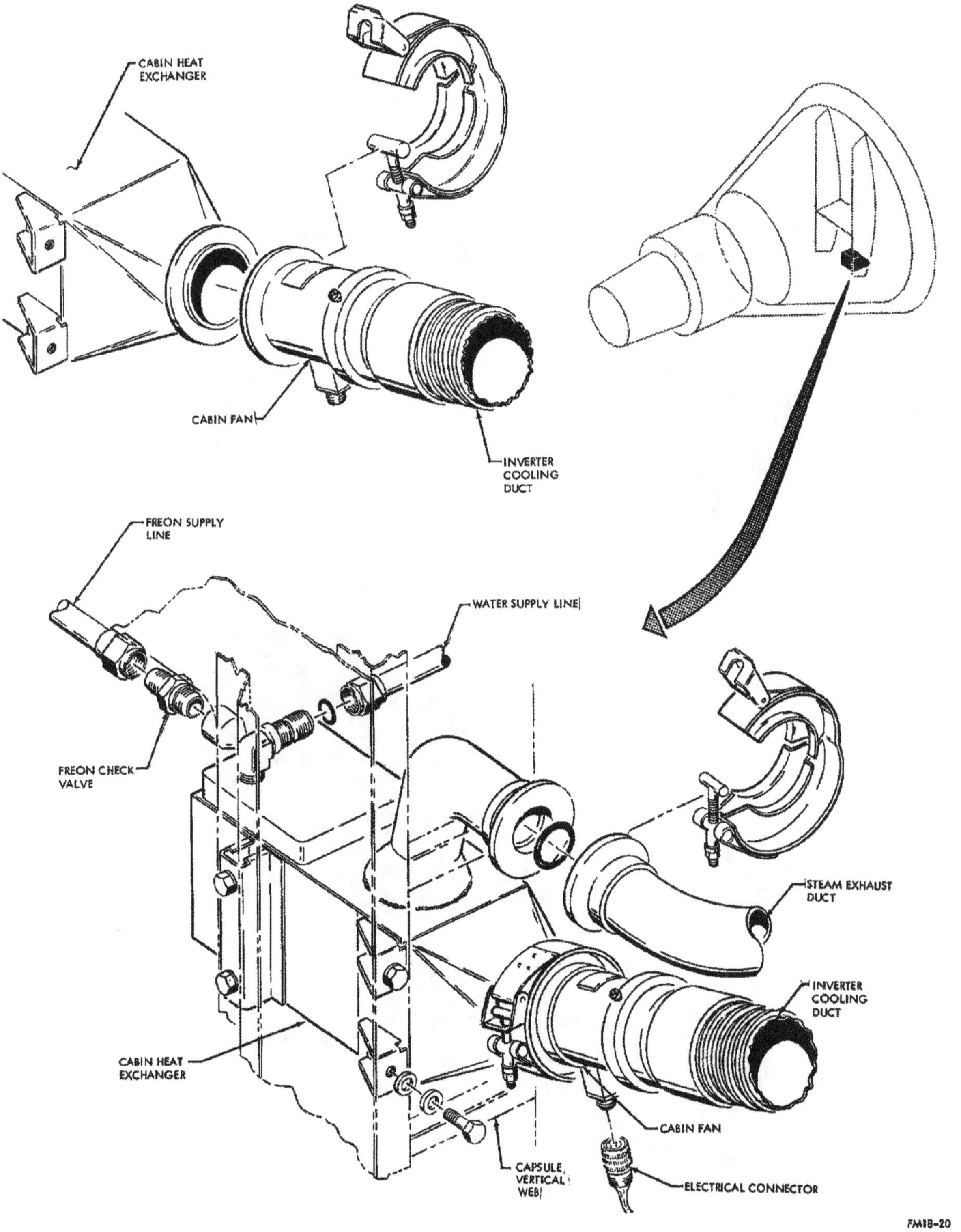

Figure 4-17 Cabin Heat Exchanger and Fan

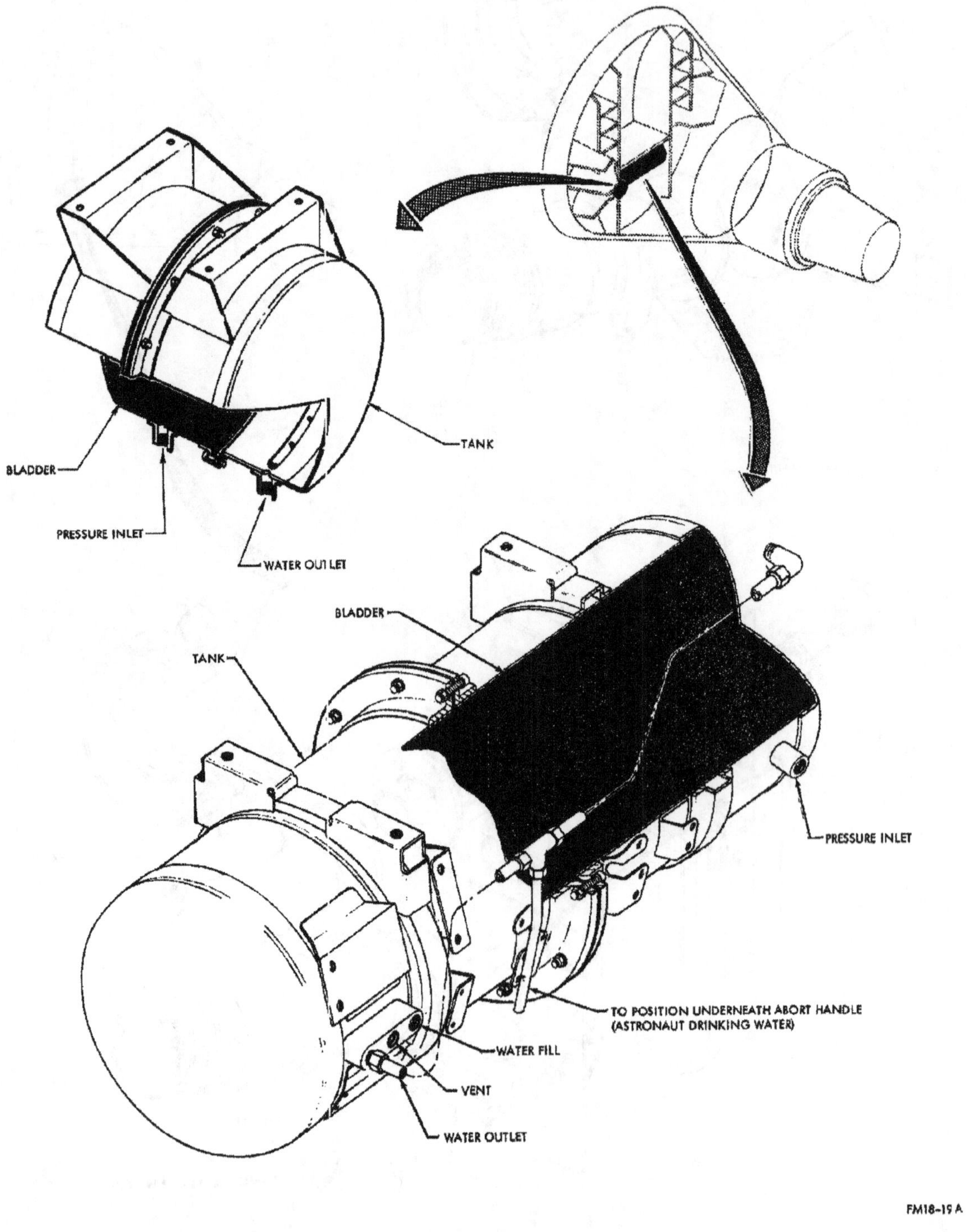

SEDR 104

Figure 4-18 Water Tanks

4-26. **CABIN PRESSURE CONTROL VALVE**

The cabin pressure control valve, Figure 4-19, is provided to maintain cabin pressure to 5.1 ± .3 psia. The control valve contains two aneroids that sense cabin pressure. Whenever cabin pressure drops below 5.1 ± .3 psia, the aneroids partially expand and unseat the spring loaded metering pins, which in turn permit oxygen to flow into the suit circuit. The suit pressure regulator senses the increase in pressure, and relieves suit circuit gas to the cabin. Directing oxygen flow through the suit circuit provides constant purging of suit circuit. When cabin pressure increases to 5.1 ± .3 psia the aneroids contract, allowing the metering pins to seat and shut off the oxygen flow. In the event of cabin decompression, or whenever cabin pressure drops below $4.0 ^{+.2}_{-.1}$ psia, the aneroids fully expand and seat against the inlet port. This stops oxygen flow and reserves the remaining oxygen supply for the suit circuit. Two aneroids are provided in the valve to insure operation in the event that one aneroid fails. A manual control is also provided to enable cabin repressurization. Actuation of the REPRESS "T" handle offseats a spring loaded poppet in the valve and allows oxygen to flow directly into the cabin. REPRESS "T" handle should then be pushed in, following cabin repressurization, to stop repressurization flow.

4-27. **CABIN PRESSURE RELIEF VALVE**

The cabin pressure relief valve, Figure 4-20, automatically relieves cabin pressure relative to ambient pressure during launch, orbit, re-entry and landing phases. In the event of a water landing, the valve

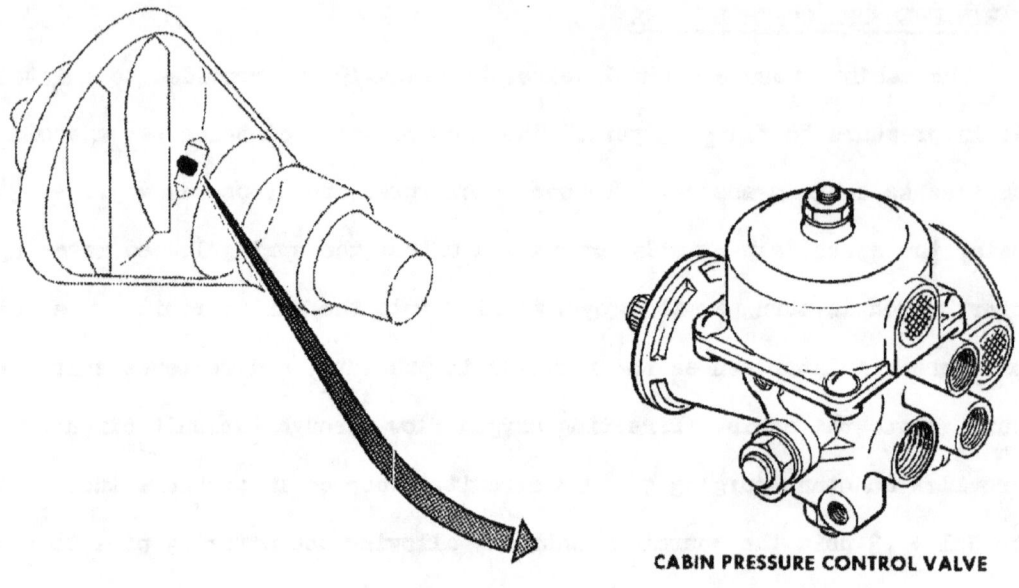

CABIN PRESSURE CONTROL VALVE

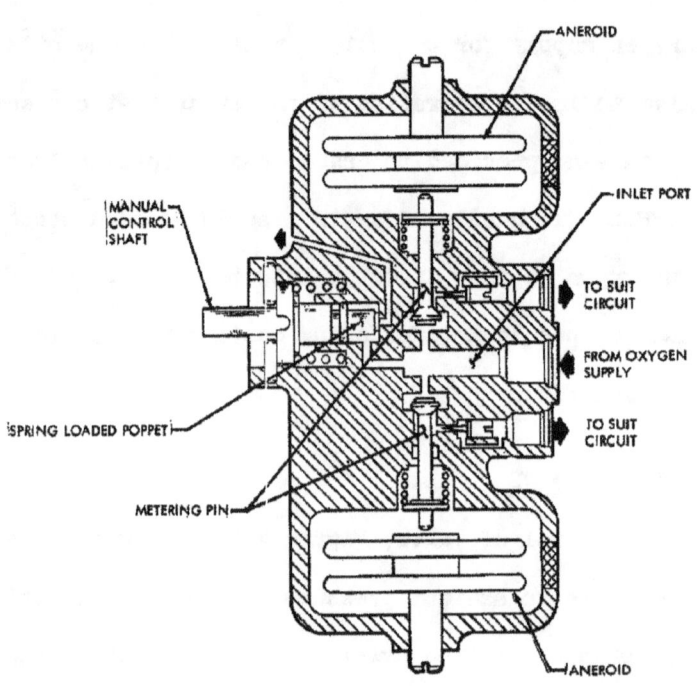

Figure 4-19 Cabin Pressure Control Valve

MCDONNELL ———————————— **SEDR 104** ————————————

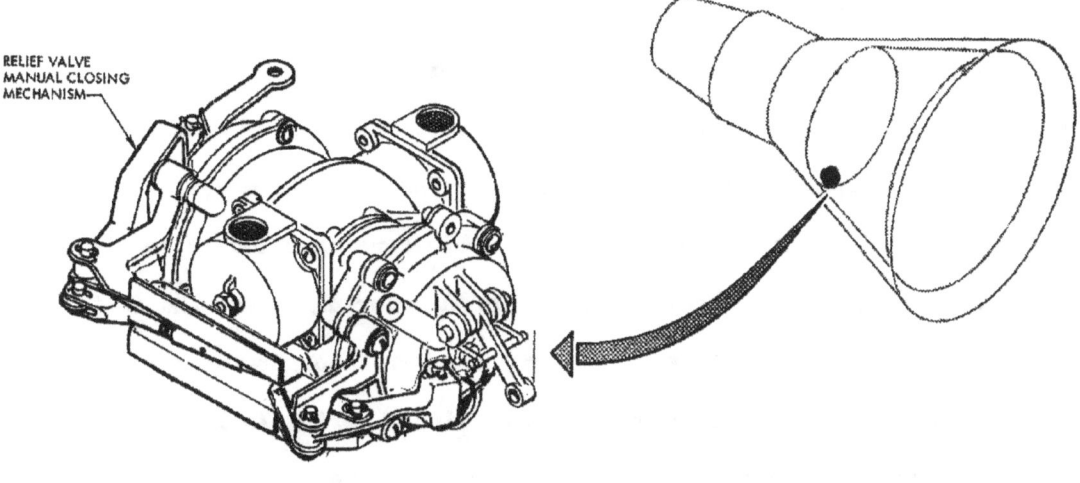

CABIN PRESSURE RELIEF VALVE

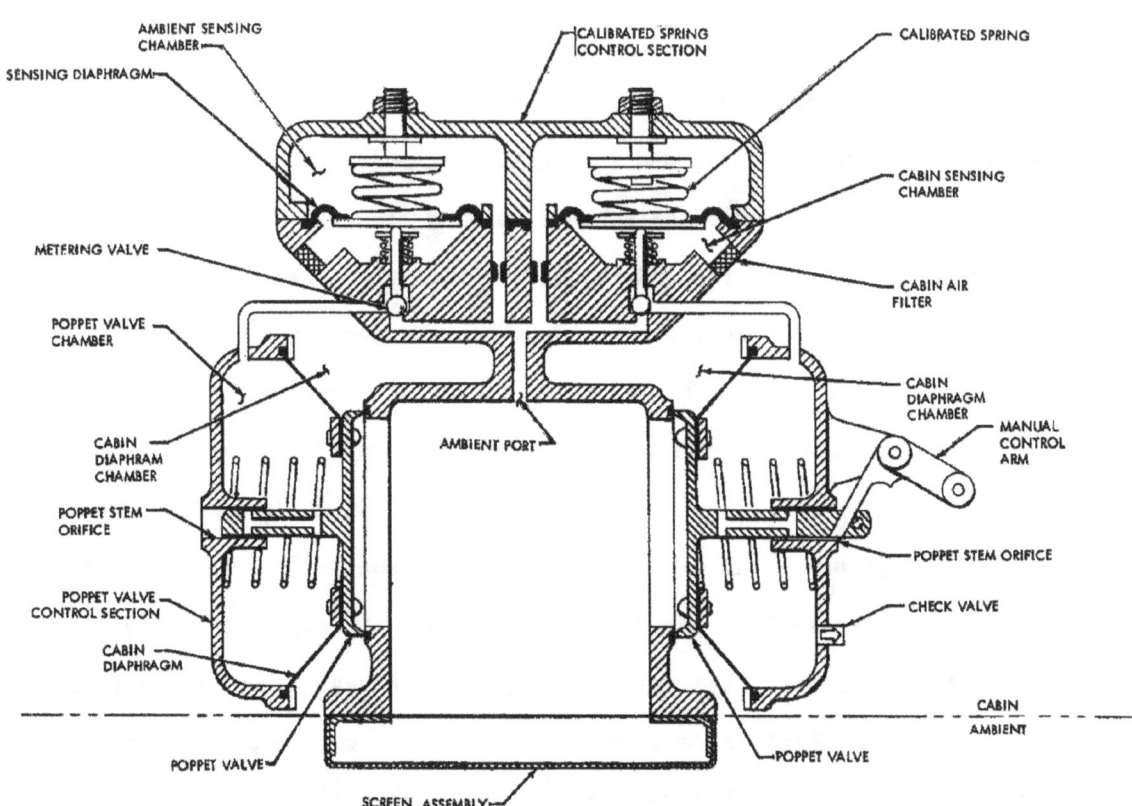

Figure 4-20 Cabin Pressure Relief Valve

incorporates provisions to keep water from entering the cabin. The valve also features means for manually decompressing the cabin. The cabin pressure relief valve consists of a calibrated spring control section and a poppet valve control section. The calibrated spring control section incorporates ambient and cabin sensing chambers separated by a sensing diaphragm, spring loaded metering valves and calibrated springs. The poppet valve control chamber incorporates a manual control arm, a check valve, poppet stem orifices, spring loaded poppet valves, poppet chamber diaphragms and poppet chambers.

During launch, the cabin pressure relief valve will relieve cabin pressure to maintain a pressure differential (cabin/ambient) of 5.5 psia. Cabin gas will be vented, through the poppet stem orifices, into the poppet valve chamber. Cabin gas will also be vented, through the cabin air filters, into the cabin sensing chamber. Ambient gas will be vented, via the ambient port, into the ambient sensing chamber. The calibrated springs are designed to respond to differential pressures in excess of approximately 5.5 psi (cabin/ambient). When the pressure differential between the cabin sensing chamber and ambient sensing chamber exceeds approximately 5.5 psia (cabin/ambient), the calibrated springs will retract. The metering valves will then be lifted from their seats, allowing differential pressures in excess of 5.5 psia to escape through the ambient port. Due to the ambient port being larger than the poppet stem orifices, the dissipation rate of the excessive differential cabin pressure (inside the poppet valve chambers) will exceed the rate of build-up in the poppet valve chambers. This will momentarily cause the cabin pressure to be greater than the poppet valve chamber pressure. The greater cabin pressure will act against the cabin diaphragm, unseating the poppet valves. The poppet valves will then aid in relieving excessive differential pressure. If the astronaut executes a manual decompression of the cabin, the check valve acts as an exhaust for poppet valve chamber pressure.

During orbit, the cabin pressure relief valve will prevent cabin pressure build-up in excess of approximately 5.5 psia. Cabin pressure in excess of approximately 5.5 psia will be exhausted to the outside atmosphere. Upon re-entry, when the ambient pressure becomes 15 inches of water greater than cabin pressure, the poppet valves will commence to open allowing ambient pressure to enter the cabin. Valve relieving operations will then be similar to those during launch. In the event the spacecraft makes a water landing, the poppet valves will not open until water pressure exceeds cabin pressure by 15 inches of water.

4-28. SNORKEL AND DIAPHRAGM FLAPPER VALVES

The cabin inlet air snorkel valve and the cabin outflow diaphragm flapper valve act as water check type valves. During the landing and post-landing phases, (often reaching a pressure altitude of approximately 17,000 feet) ambient air is circulated through the valves. In the event the valve parts were under water, the valves would seat and prevent water from entering the cabin.

4-29. CABIN AIR INLET VALVE

The cabin air inlet valve, Figure 4-21, provides ventilation and cooling for the suit circuit and cabin during landing and post-landing

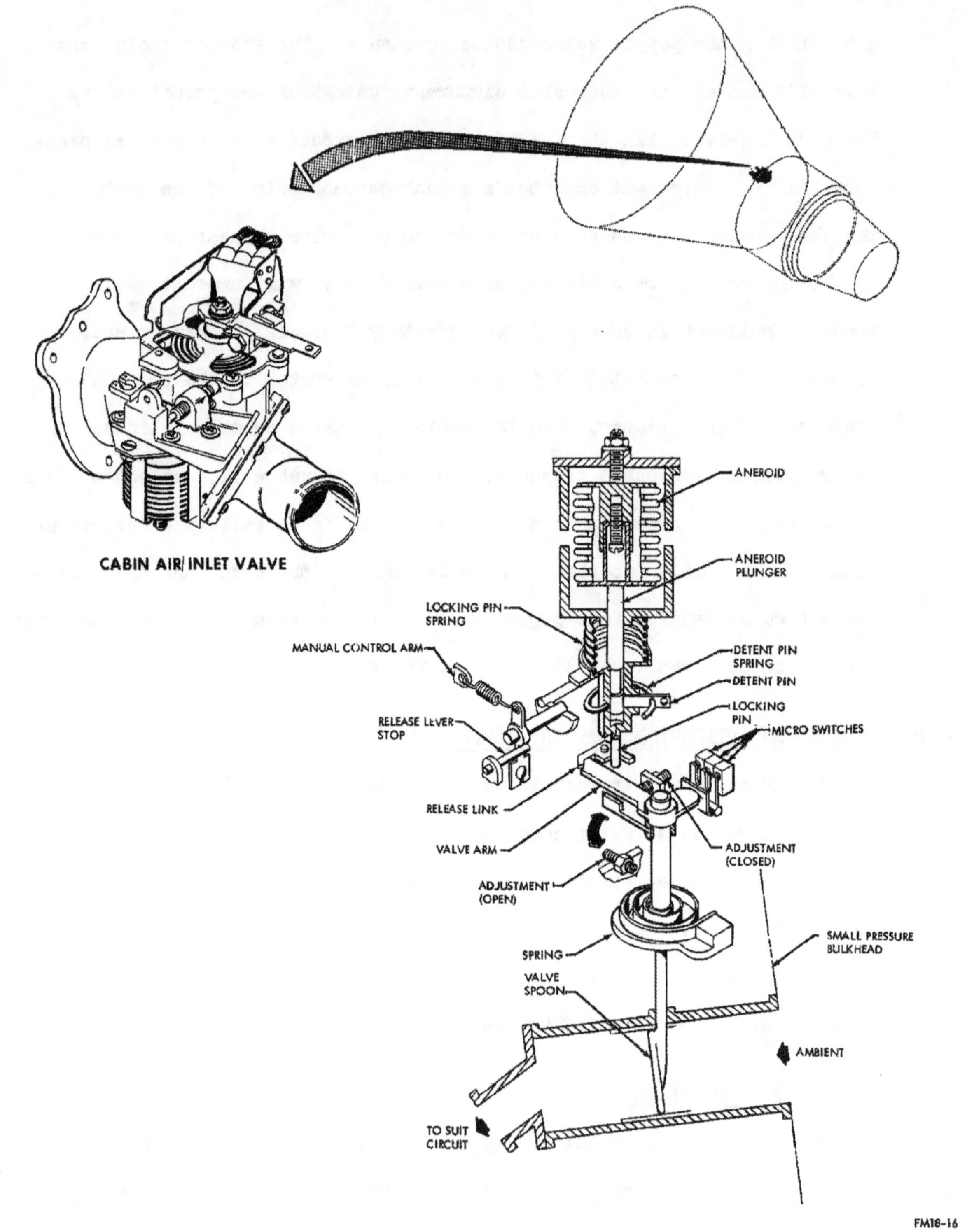

SEDR 104

Figure 4-21 Cabin Air Inlet Valve

phases. It is a spring loaded close, spoon type valve and is barometrically controlled. Prior to launch, the valve is manually latched closed so that one mechanism spring loaded detent pin rides on the large diameter of the aneroid plunger (maximum allowable pull to set detent pin in five (5) pounds); and the valve arm is engaged by the release link, which is engaged by the spring loaded aneroid locking pin. During launch the aneroid expands due to decreasing cabin pressure, and forces the aneroid plunger down. The valve mechanism detent pin then slips off the plunger large diameter onto the plunger small diameter.

During the landing phase, when the spacecraft descends to an altitude of approximately 17,000 $\pm$ 3,000 feet, the aneroid retracts as cabin pressure increases. Retraction of the aneroid moves the aneroid plunger upward, engaging the detent pin against the plunger larger diameter, which in turn compresses the aneroid locking pin spring. This action raises the locking pin from release link and allows spring loaded valve to open. The valve arm is attached to valve shaft and moves with closing thereby disengaging micro-switches. Disengagement of micro-switches directs electrical power to close the suit circuit shutoff valve, which in turn opens the emergency oxygen rate valve. A manual control arm is provided to enable valve opening in the event valve fails to open at specified altitude. Actuation of the manual control arm, mechanically, contacts the locking pin spring and disengages locking pin from release link, allowing valve to open. The valve must be manually reset to the closed position. Opening of the valve enables suit compressor to draw ambient air into suit circuit to provide suit circuit and cabin ventilation.

SEDR 104

The cabin air outlet valve is basically of the same construction and functions in the same manner as the cabin air inlet valve.

4-30. VACUUM RELIEF VALVE

The vacuum relief valve, Figure 4-22, is designed to open at a pressure differential of 10 to 15 inches of water, to provide suit circuit ventilation whenever the inlet snorkel valve closes (ball float seats). The relief valve is located in the flexible ducting, between the cabin air inlet valve and the suit circuit inlet duct. In the event the spacecraft submerges momentarily, following a water landing, the snorkel valves ball floats will seat (close) and prevent water from entering into the suit circuit and cabin. The operation of the suit compressor and the closed air inlet snorkel valve will create a vacuum in the suit circuit air inlet duct (flexible ducting). When cabin pressure exceeds the flex duct pressure, by 10 - 15 inches of water, the vacuum relief valve will open. As the valve opens, cabin pressure acting on the valve poppet surface will be great enough to hold the valve open until the pressure differential (between cabin and duct) is approximately two inches of water or less. Suit circuit ventilation is provided by the cabin air, entering the opened vacuum relief valve, whenever the inlet snorkel valve ball float is seated (closed). Also, the opening of the relief valve removes the vacuum in the flex duct to enable the snorkel valve ball float to unseat (open) whenever the snorkel valve is above water.

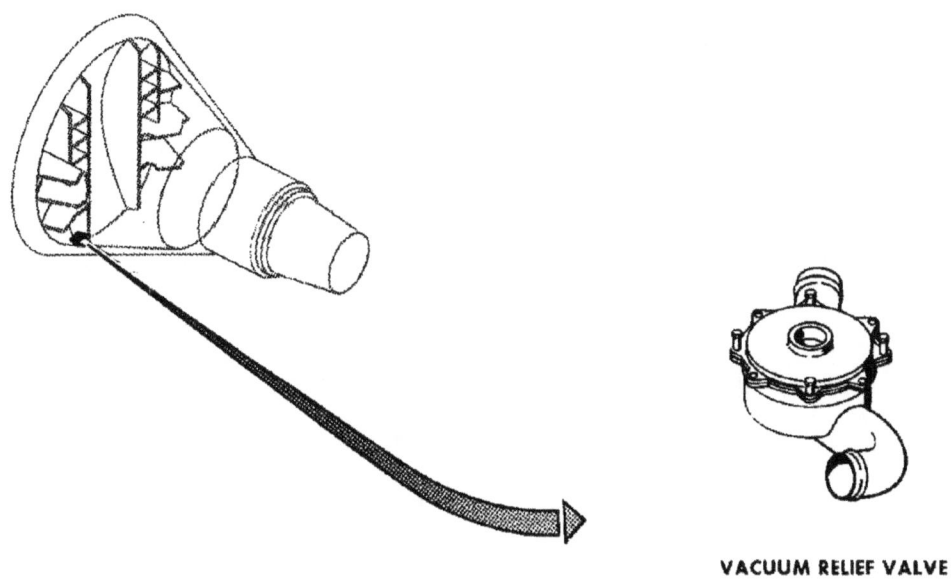
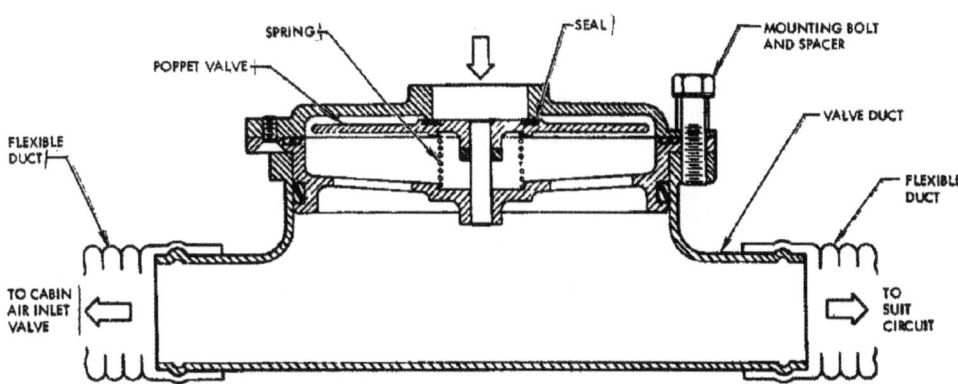

Figure 4-22 Vacuum Relief Valve

SECTION V

STABILIZATION CONTROL SYSTEMS

TABLE OF CONTENTS

TITLE	PAGE
Automatic Stabilization	
Control System	5-3
System Operation	5-12
System Units	5-20
Reaction Control System	
System Description	5-26
System Operation	5-30
System Units	5-38
Horizon Scanner System	
System Description	5-41
System Operation	5-49

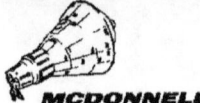

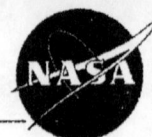

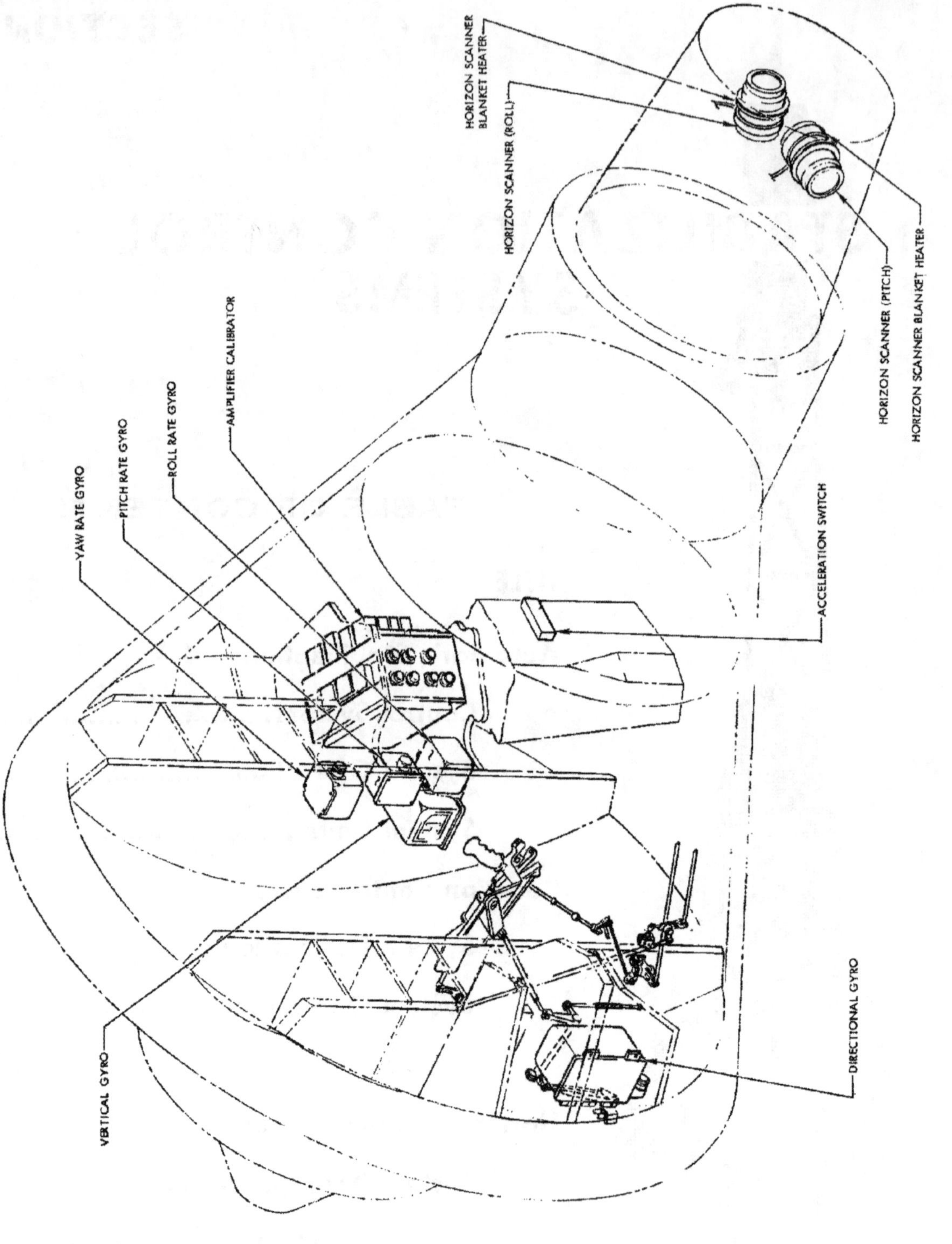

FM18-125A

Figure 5-1 A.S.C.S. Component Location

SEDR 104

V. STABILIZATION CONTROL SYSTEMS

5-1. GENERAL

Stabilization of the spacecraft is accomplished by the Automatic Stabilization Control System in conjunction with two sub-systems, the Horizon Scanners and the Reaction Control System. These systems establish and maintain a stable platform with four basic automatic modes; Damper, Orientation, Attitude Hold and Re-entry. In addition, a visual indication of yaw, roll, and pitch attitude is provided. The following paragraphs describe the individual systems and functions involved.

5-2. AUTOMATIC STABILIZATION CONTROL SYSTEM

5-3. SYSTEM DESCRIPTION

The Automatic Stabilization Control System (ASCS) is composed of a Directional Gyro, Vertical Gyro, .05g accelerometer switch, Rate Gyros (yaw, roll and pitch), and an Amplifier Calibrator Unit. Location of the individual components within the spacecraft is shown in Figure 5-1.

Three switches are provided in conjunction with the ASCS. The SELECT-NORM switch, FLY-BY-WIRE - AUX DAMP switch and the GYRO CAGE, FREE-GYRO SLAVE switch. With the SELECT-NORM switch in the NORM position, stabilization is accomplished in a completely automatic manner, requiring no assistance from the astronaut. With the SELECT-NORM switch in the SELECT position and the FLY-BY-WIRE-AUX DAMP switch in the FLY-BY-WIRE position, the automatic feature is disabled and 24V d-c power is connected to the Fly-By-Wire limit switches on the astronaut's control stick. A FBW THRUST SELECT switch with LOW & HIGH and LOW ONLY provides the astronaut with two thrust selections for fuel conservation purposes. Stabilization is accomplished through an electro mechanical arrangement (See Figure 5-9) by movement of the astronaut's control stick in the desired plane. Low and high thrust actuation occur at approximately 30% and 75% of full travel, for yaw, pitch, or roll. The AUX DAMP position disables both the automatic and fly-by-wire function, permitting rate damping as a singular feature. The GYRO switch is a three position switch incorporating a CAGE, FREE, and NORMAL position. In the CAGE position, the Attitude gyros are mechanically caged and the Horizon Scanner slaving function is disabled. In the FREE position, the Attitude gyros are uncaged; the Horizon Scanner slaving function remains disabled. The NORMAL position uncages the attitude gyros and permits Horizon Scanner slaving.

5-4. <u>ASCS SEQUENCING</u>

The following paragraphs, 5-5 and 5-9, describe the ASCS sequential operation under normal and abort conditions. Figures 5-2, 5-3, and 5-4 are provided for clarity and should be followed closely in conjunction with the text concerning the various modes of operation.

5-5. <u>NORMAL SEQUENCING</u>

In Figure 5-2, the progress of a normal orbital mission is shown divided into eight phases appropriate to the following discussion.

The ASCS is in the "ready" status prior to separation of the escape tower, its gyros are running and all circuits except the final 12 output relays are fully energized. Phase (A), involving gyro slaving to the Ho-

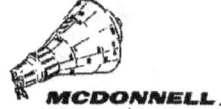

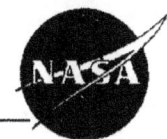

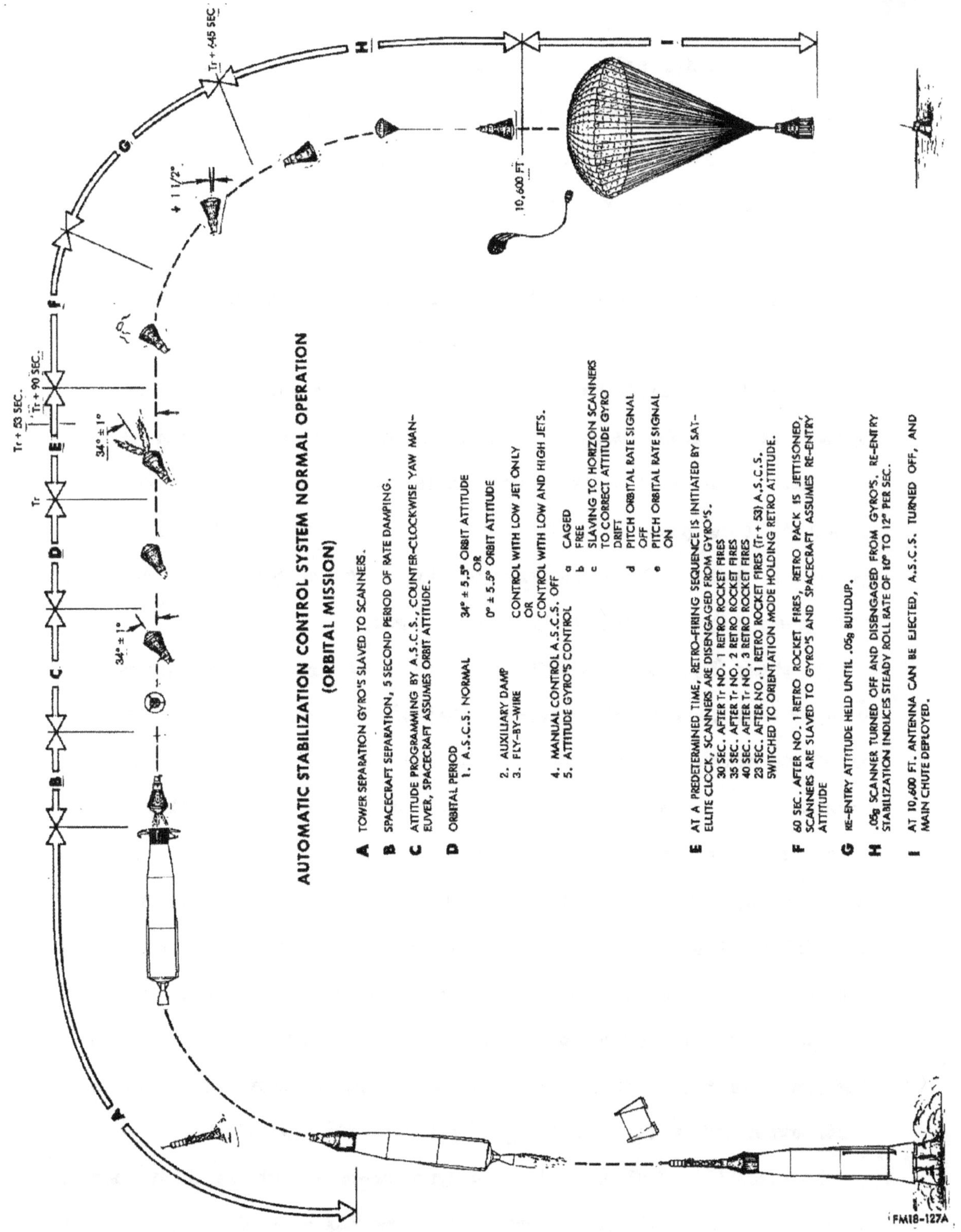
Tr + 53 SEC.
Tr
Tr + 90 SEC.
Tr + 645 SEC
A
B
C
D
E
F
G
H
I
34° ± 1°
34° ± 1°
+ 1 1/2°
10,600 FT
AUTOMATIC STABILIZATION CONTROL SYSTEM NORMAL OPERATION
(ORBITAL MISSION)
A TOWER SEPARATION GYRO'S SLAVED TO SCANNERS.
B SPACECRAFT SEPARATION, 5 SECOND PERIOD OF RATE DAMPING.
C ATTITUDE PROGRAMMING BY A.S.C.S., COUNTER-CLOCKWISE YAW MAN-EUVER, SPACECRAFT ASSUMES ORBIT ATTITUDE.
D ORBITAL PERIOD
1. A.S.C.S. NORMAL 34° ± 5.5° ORBIT ATTITUDE OR 0° ± 5.5° ORBIT ATTITUDE
2. AUXILIARY DAMP
3. FLY-BY-WIRE CONTROL WITH LOW JET ONLY OR CONTROL WITH LOW AND HIGH JETS.
4. MANUAL CONTROL A.S.C.S. OFF
5. ATTITUDE GYRO'S CONTROL
a CAGED
b FREE
c SLAVING TO HORIZON SCANNERS TO CORRECT ATTITUDE GYRO DRIFT
d PITCH ORBITAL RATE SIGNAL OFF
e PITCH ORBITAL RATE SIGNAL ON
E AT A PREDETERMINED TIME, RETRO-FIRING SEQUENCE IS INITIATED BY SAT-ELLITE CLOCK, SCANNERS ARE DISENGAGED FROM GYRO'S.
30 SEC. AFTER Tr NO. 1 RETRO ROCKET FIRES
35 SEC. AFTER Tr NO. 2 RETRO ROCKET FIRES
40 SEC. AFTER Tr NO. 3 RETRO ROCKET FIRES
23 SEC. AFTER NO. 1 RETRO ROCKET FIRES (Tr + 53) A.S.C.S. SWITCHED TO ORIENTATION MODE HOLDING RETRO ATTITUDE.
F 60 SEC. AFTER NO. 1 RETRO ROCKET FIRES, RETRO PACK IS JETTISONED, SCANNERS ARE SLAVED TO GYRO'S AND SPACECRAFT ASSUMES RE-ENTRY ATTITUDE
G RE-ENTRY ATTITUDE HELD UNTIL .05g BUILDUP.
H .05g SCANNER TURNED OFF AND DISENGAGED FROM GYRO'S. RE-ENTRY STABILIZATION INDUCES STEADY ROLL RATE OF 10° TO 12° PER SEC.
I AT 10,600 FT. ANTENNA CAN BE EJECTED, A.S.C.S. TURNED OFF, AND MAIN CHUTE DEPLOYED.
FM18-127A

Figure 5-2. A.S.C.S. Normal Operation

rizon Scanner pitch and roll outputs during ascent, is to minimize gyro errors which may accumulate while the spacecraft is being boosted.

Phase (B) starts after spacecraft separation when a brief, five-second signal commands the ASCS to provide rate damping to stop any tendency to tumble.

Phase (C) is initiated at the completion of five seconds of rate damping. The ASCS is placed in the orientation mode, spacecraft turn around (180° counterclockwise Yaw Rotation) is accomplished, and the spacecraft is pitched-down to the retrograde firing angle within 30 seconds. Pitch, roll and yaw gyro slaving to the Horizon Scanners is provided during phase (C) while the Gyro Select Switch is in the slave position. Both the Free and Caged positions of the Gyro Select Switch will prevent Gyro slaving to Horizon Scanners.

In phase (D) the spacecraft is in orbit. An orbit pitch attitude of -34° (small end down) is held so that the spacecraft is ready for an immediate abort. As in phase (C) attitude gyros will slave the horizon scanners if Gyro Select Switch is in the normal position. During the orbit phase manual control and fly-by-wire control may be utilized as desired. Rate damping may be obtained by placing ASCS mode select switch into AUX DAMP position. Rate gyro run-up is continued throughout phase (D). Another feature utilized in Phase (D) is an automatic return to the orientation mode. If the spacecraft drifts (from orbit attitude) beyond the limits of the retro-interlock sector switches, automatic return to orientation mode will occur at $\pm 15°$ pitch, $\pm 30°$ yaw and roll.

In phase (E) of Figure 5-2, rate gyro run-up is automatically assured by relay switching 10 minutes prior to retrograde attitude. The astronaut

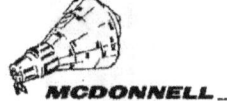

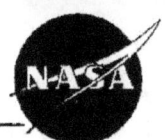

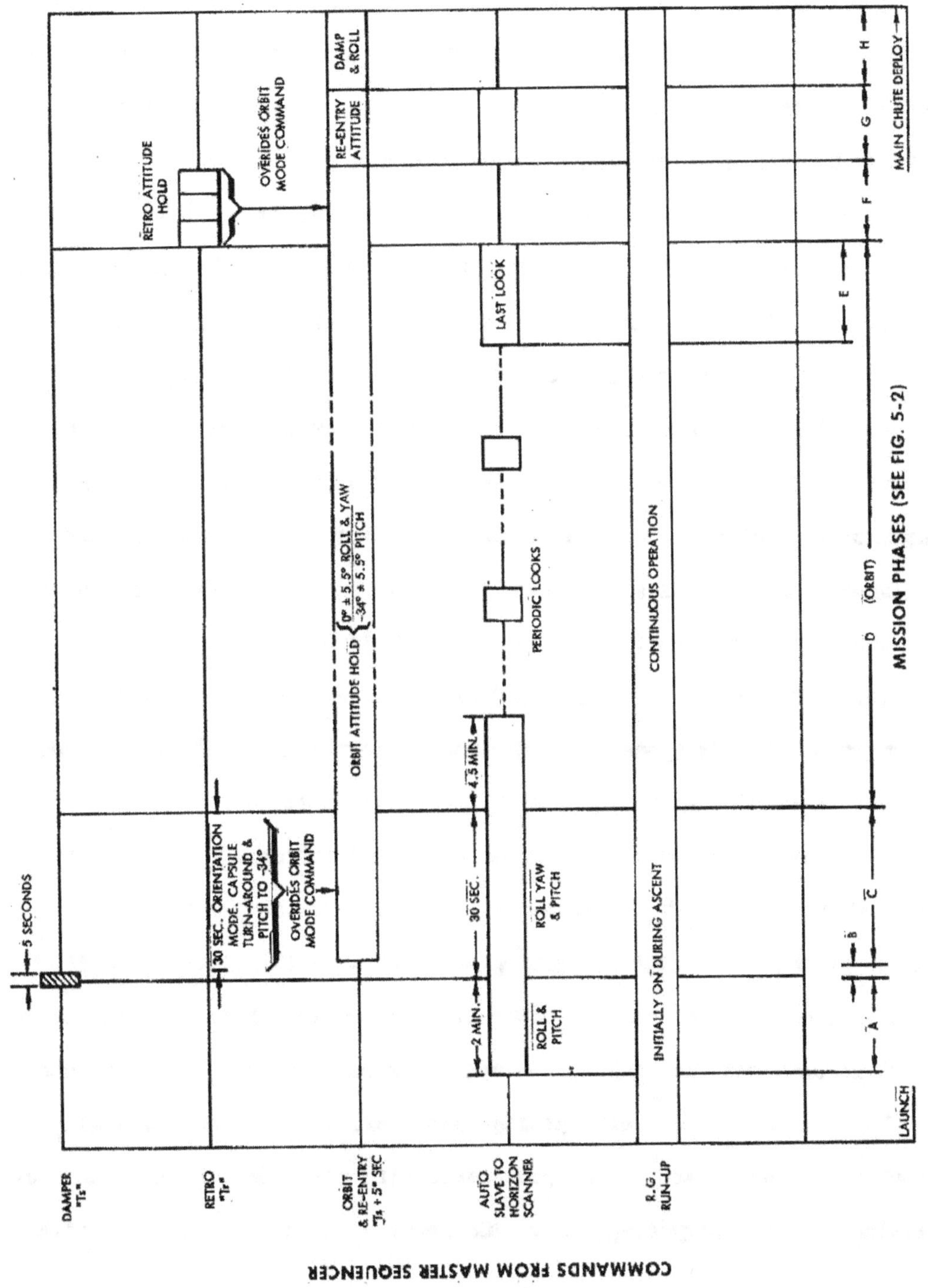

IFM18-126

Figure 5-3. A.S.C.S. Mission Profile

may change any one or all three of the spacecraft attitudes maintained by the ASCS by changing the space reference plane or planes of the attitude gyros. To maintain the new reference plane or planes, the Horizon Scanner slaving command must be stopped by placing the gyro switch in the FREE position. New reference planes may be established by the astronaut while the ASCS is in operation by placing the gyro switch in the FREE position and placing pitch torquing switch in the OFF position. Also manually turning off the ASCS fuel in the axis or axes affected, utilizing manual control to position the spacecraft, and then caging and uncaging the gyros. The ASCS may then be returned to fully automatic operation in all three axes with the exception of Horizon Scanner slaving. To utilize Horizon Scanner slaving, the spacecraft attitudes must be within the observation range of the scanners and the gyro switch must be placed in the NORMAL position.

The ASCS receives Tr signal and maintains spacecraft in high torque retrograde attitude (phase F). Horizon Scanner slaving is discontinued at this time. Thirty seconds after retrograde attitude command, the retro rockets are fired. During the period of retrograde rocket firing the ASCS utilizes high torque action to hold the spacecraft within one degree of the ideal angles. Retrograde rocket firing command and ASCS high torque switching command occur simultaneously. Rocket firing is completed in 20 seconds and the high torque switching command is held for 23 seconds.

Upon completion of retro package jettison, the ASCS automatically pitches the spacecraft to the post-retro fire attitude (phase G) in preparation for re-entry drag. The ASCS returns to orientation mode with

constant scanner operation to accurately maintain the re-entry attitude.

Finally, when re-entry is sensed by the .05g accelerometer switch, the eighth and last phase (H) of the ASCS performance starts with the turning off of the attitude gyro power. During this period the ASCS initiates and maintains a constant roll rate of 10° to 12° per second to minimize touchdown dispersion. Rate damping is provided to stabilize the re-entry trajectory. ASCS operation in this phase continues until main chute deployment, at which time all ASCS power is removed.

Pilot-override provisions permit interruptions of the "normal" sequence by manual or fly-by-wire control manipulation and return to the "normal" ASCS MODE. Thus to a significant degree the astronaut is the intelligent "back-up" for the ASCS. Full utilization of this reliability augmentation principle has led to gyro caging and other switching features which are intended to make the spacecraft manually controllable. The following table lists the switch and control positions necessary to achieve the four basic modes of control after attaining orbit. Variations of the various modes can be obtained by further switch manipulation.

CONTROL MODE	SWITCH POSITIONS			"T" HANDLES POSITIONS	
	SELECT-NORM	FBW-AUX. DAMP	LOW & HIGH- LOW ONLY	AUTO "T"	MANUAL "T"
AUTOMATIC	NORMAL	Either	Either	PUSH ON	PULL OFF
FLY-BY-WIRE	SELECT	FBW	As Needed	PUSH ON	PULL OFF
DIRECT (No Damping)	NORMAL	Either	Either	PULL OFF	PUSH ON
DIRECT (With Damping)	SELECT	AUX. DAMP	Either	PUSH ON	PUSH ON

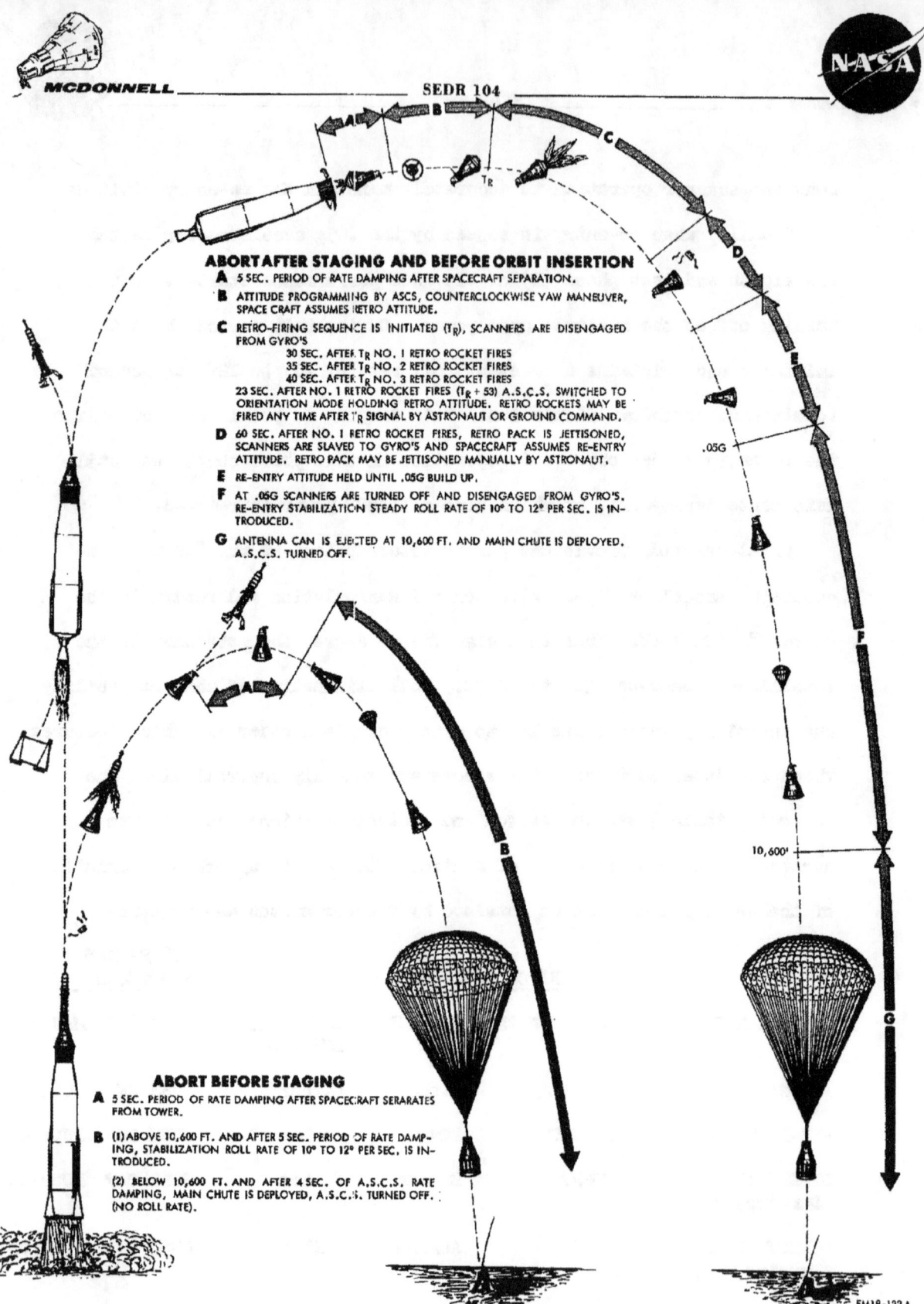

Figure 5-4, A.S.C.S. Emergency Operation.

5-6. ABORT SEQUENCING

In general, abort sequencing (See Figure 5-4) is programmed to correspond to the safest procedures at all times. The possible abort situations can be divided into three types, namely (1) abort, before tower separation when ASCS rate damping is required; (2) abort after tower separation but before the trajectory is truly orbital; and (3) abort from orbit. The following paragraphs, 5-7 through 5-9, discuss ASCS sequencing in each of the abort conditions.

5-7. ABORT BEFORE TOWER SEPARATION

If an abort mission is started during the period when the booster and sustainer engines are burning, the ASCS is utilized for rate damping only **after** the following external operations have been achieved.

(1) Booster and sustainer engines cut-off.

(2) Spacecraft separation from adapter.

(3) Escape tower rocket firing.

(4) Retro rocket separation from spacecraft.

(5) Timed arrival at approximate peak of trajectory.

(6) Separation of escape tower from spacecraft.

Upon completion of the later operation, the ASCS is commanded to provide rate damping, using the rate gyros which are continuously energized during the normal ascent and "abort trajectory" flight. A constant roll rate of $10°$ to $12°$ per second is employed. Rate damping ceases upon deployment of the main chute.

SEDR 104

5-8. ABORT AFTER TOWER SEPARATION

The first operation is engine cut-off. This is followed immediately by spacecraft separation, posigrade firing, and the normal mission post-separation signal sequence to the ASCS. The effect is immediate damping of any tendency to tumble. After 5 seconds of rate damping, the automatic sequence commands spacecraft turn around and an attitude angle of 34 degrees. Then either the astronaut or ground command must initiate retrograde sequencing. Upon achieving the proper roll, pitch and yaw angles within rather wide "permission" bounds (See Paragraph 5-5, Page 5-6), the ASCS enables rapid-sequence retro rocket firing to proceed.

NOTE

ASCS "permission interlock" during retro fire can be over-ridden at any time by the astronaut.

After retrograde operation, the abort mission in this case proceeds as in the normal mission post-retrograde sequence (except for the difference in trajectory time and distance intervals.)

5-9. ABORT FROM ORBIT

Whenever an abort from orbit is initiated, the normal automatic or manual retrograde operations will apply. However, if manual retrograde operations are utilized the pre-retrograde period of gyro slaving to the Horizon Scanners ("last look") will be eliminated.

5-10. SYSTEM OPERATION

Overall system operation is best explained by Figure 5-5. The Amplifier Calibrator receives inputs from sensors on the left side of the page and generates outputs to Display and Reaction Control devices on

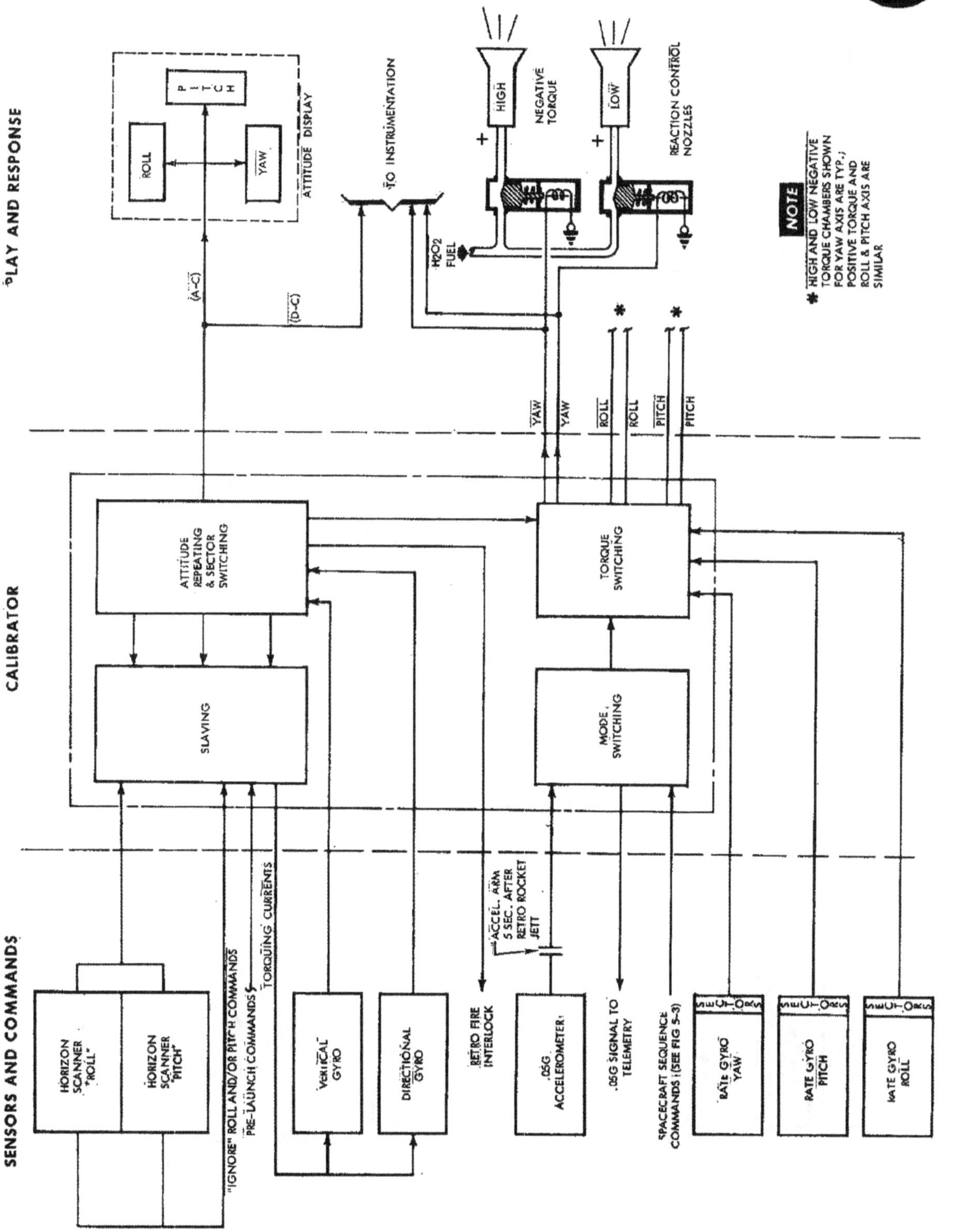

MCDONNELL — SEDR 104

Figure 5-5 A.S.C.S. Block Diagram

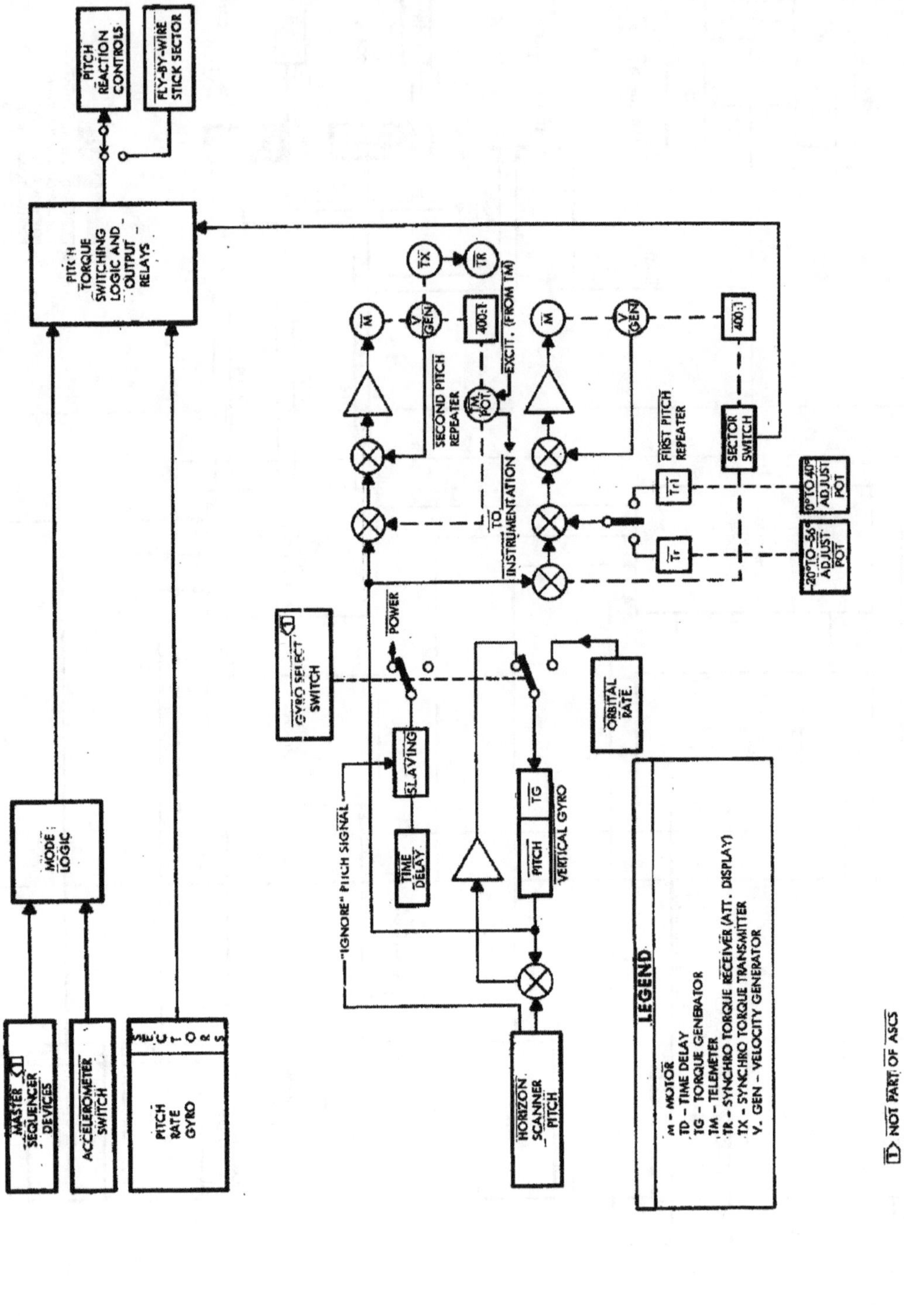

Figure 5-6. Pitch Axis Block Diagram

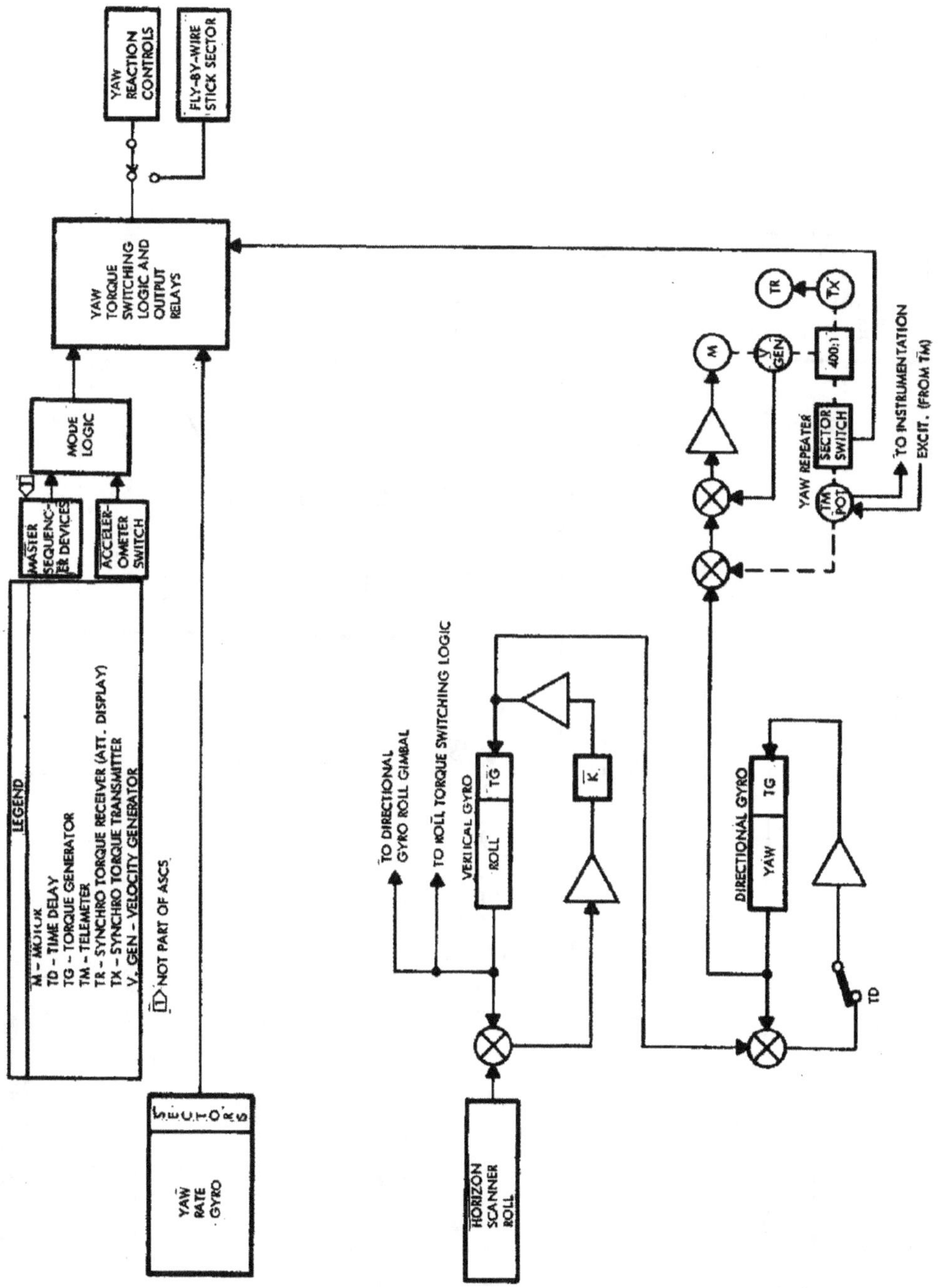

Figure 5-7 Yaw Axis Block Diagram

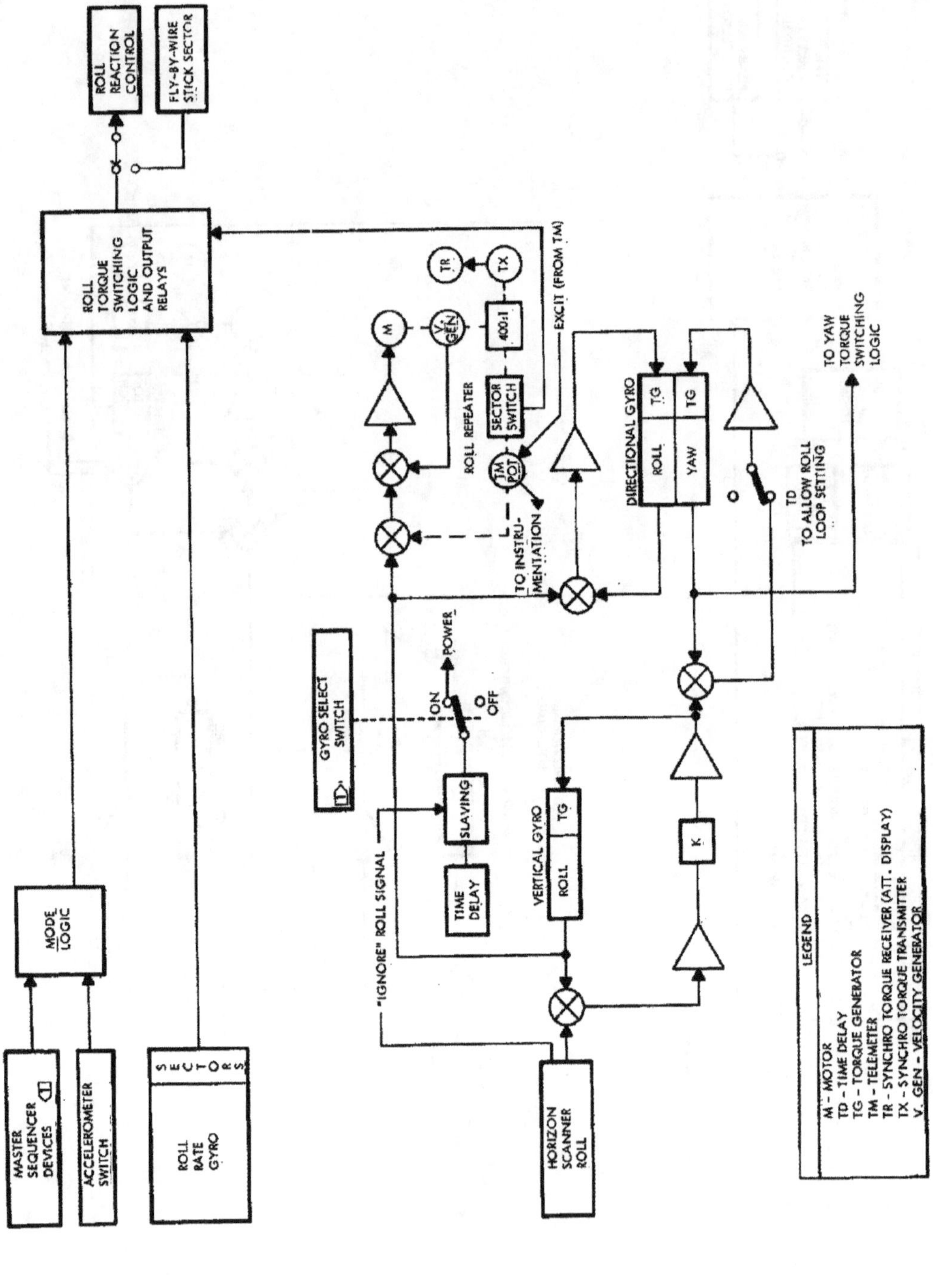

Figure 5-8, Roll Axis Block Diagram

the right. The four basic operations are slaving, repeating, mode switching and torque switching. Data flow pertaining to the individual Yaw, Roll and Pitch channels is illustrated in Figures 5-6, 5-7 and 5-8. In general, these diagrams are straightforward and require no explanation. However, the method utilized in deriving Directional information is unique to a degree and warrants the following discussion.

The pitch gimbal (vertical gyro) is precessed continuously during the orbital phase of the normal mission, so that the spacecraft "local vertical" reference revolves 360 degrees during each orbital cycle. The gyro slaving principles which permit Directional (yaw) information to be derived are as follows: After initial slaving and settling of the roll and pitch loops, the ASCS controls the spacecraft to the command pitch attitude, and to level roll attitude. Initially, after separation and spacecraft turn around, some yaw error (as great as 10 degrees) may exist due to directional drift during boost. Since the Roll gimbal of the vertical gyro is the inner gimbal, yaw misalignment of the spacecraft causes the Roll gimbal output to contain an error component due to the constant orbital (pitch) angular rate. Thus a comparison of the Roll Horizon Scanner and vertical gyro roll indications will provide an error signal producing a roll gimbal torquing rate. This torquing rate which is a direct function of yaw error is used to slave the yaw gimbal of the directional gyro.

Another area that warrants discussion is that of torque switching, i.e., the thrust output of the Reaction Control System in conjunction with the various modes of ASCS operation.

Figure 5-10 serves as an introduction to the torque switching behavior of the ASCS. For maximum conservation of control fuel, the behavior varies according to the ASCS mode appropriate at a given time. A so-called "phase-plane" plot of angular rate vs. angle is shown in the lower right corner of Figure 5-10 adjacent to a typical Pitch time-history for the "Orbit" mode. Current ASCS design permits a plus or minus 5.5 degree oscillation about the nominal orbital attitude, which in turn is referenced to the Horizon Scanner's sensed "Horizontal". The oscillation is nonsinusoidal because of the discontinuous torque program; pitch rate is a square wave, and pitch angle a sawtooth, both having a characteristic period of 240 seconds. Portrayed on the phase-plane, the "Orbit" mode oscillation is a gentle drift from -5.5 degrees relative pitch angle to +5.5 degrees relative pitch (-39.5° to -28.5° degrees, referenced to true horizontal). This drift lasts for approximately one-half-period of two minutes. When the error becomes +5.5 degrees, a low torque pulse causes the angular rate to reverse, where upon the second half-period is spent drifting slowly through zero to -5.5 degrees error.

As another example of ASCS torque-switching, Figure 5-10 shows the "Retrograde-Hold" torque logic phase-plane diagram. In this case high-torque nozzles are utilized instead of the low-torque nozzles which were adequate to control the gentle orbit oscillation. A series of attitude gyro repeater sector switches and rate-gyro pickoff sector switches are used to define steplike boundaries within the phase-plane. A typical contour is shown to illustrate the motion resulting from a large disturbance torque while in this mode. When the spacecraft motion results in a pitch rate value above the right-hand stair step, high negative torque is

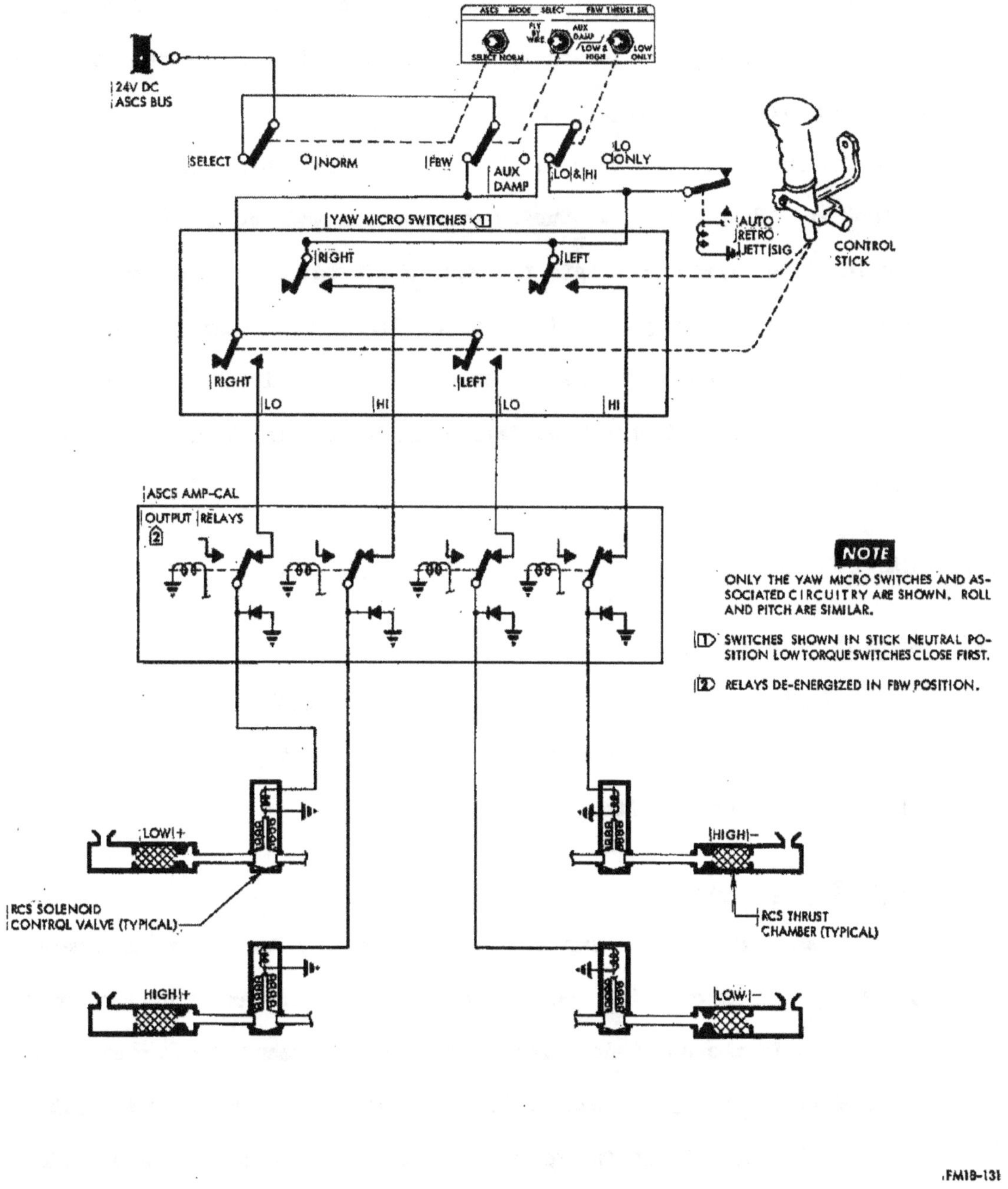

Figure 5-9. A.S.C.S. Fly By Wire Control

applied until the spacecraft attains a negative rate and rotates into the "no-torque" region. The inverse occurs if the retro rocket thrust eccentricity or other disturbances force the spacecraft into a situation calling for positive thrust. The net effect of the torque-switching logic shown is to maintain rapid and reliable control during the important operation of retrograde firing.

Other modes of operation requiring torque switching logic are "Orientation" and "Rate Damper". During orientation mode both high and low torquing is utilized to rotate the spacecraft to new preset attitudes. Both high and low torque is also applied during rate damper mode but only rate gyro signals are needed as a basis for switching logic. In this case, torque switching boundaries are horizontal lines on the phase-plane.

5-11. SYSTEM UNITS

5-12. AMPLIFIER CALIBRATOR

The Amplifier Calibrator unit can be "functionally" divided into four sections. These functional sections are slaving, repeating, mode switching and torque switching.

5-13. ATTITUDE GYRO SLAVING

This section contains amplifiers and summing networks which accept roll and pitch information from the Horizon Scanners and generate currents to torquers in the attitude gyros. Thus, upon command from an external timing device, the Gyros Roll, Pitch and Yaw gimbals are aligned with corresponding directions in, or perpendicular to the orbit plane. (Ref. Para. 5-10).

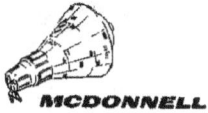

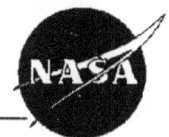

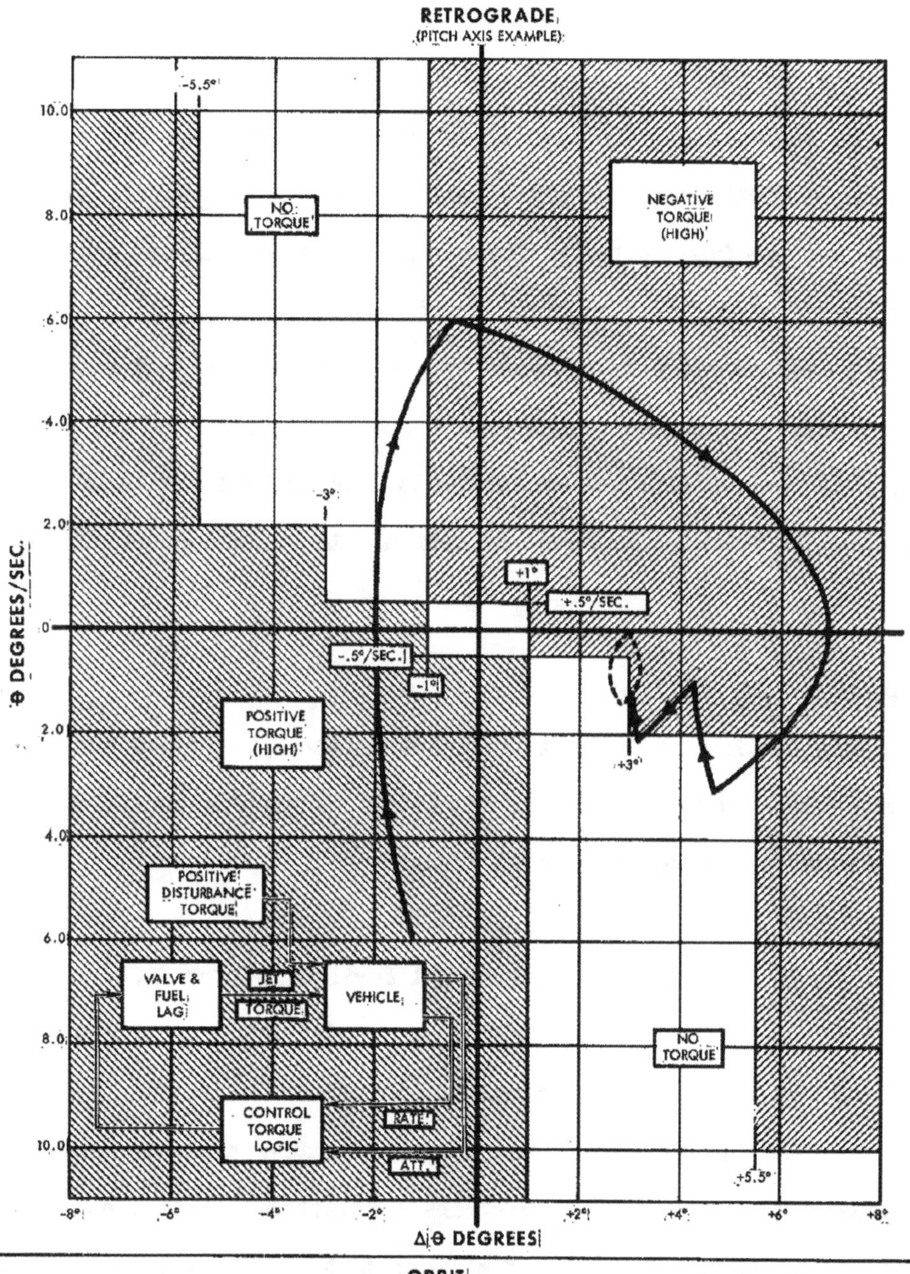

ORBIT
(PITCH AXIS EXAMPLE)

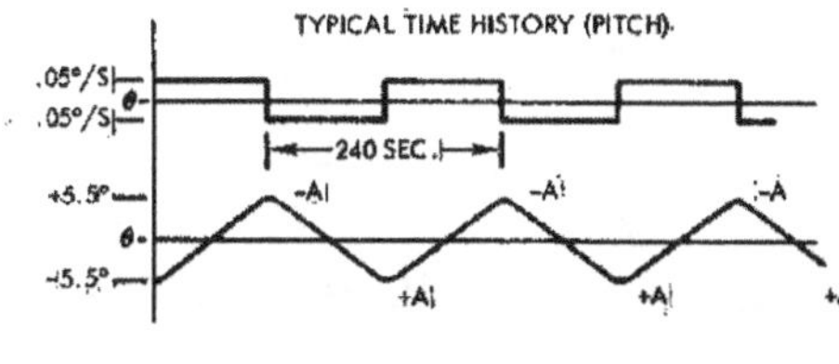

PUSLE	DURATION IN SEC.
±A	0.2
±B	0.147
±C	0.252
±D	0.437
±E	0.755

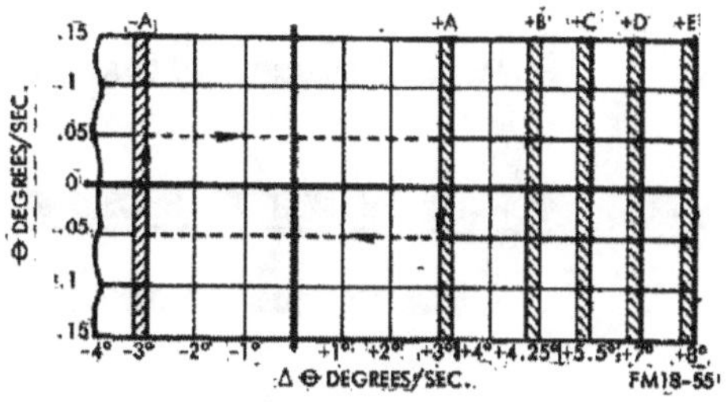

Figure 5-10. Torque Logic Phase Plane Diagram

SEDR 104

5-14. REPEATER SECTION

The repeater section is a group of servo-mechanisms (four in present design, including two for pitch angle repeating). <u>Attitude</u> gyro outputs, which are received at the calibrator in proporational or "analog" form, are amplified and used to drive shafts which serve as roll, pitch and yaw signal sources for both internal (torquing switching) and external (display and telemetry) purposes. The on-off reaction control of the spacecraft makes it desirable to use conductive sectors on the shafts of three of the repeaters. The sectors serve as attitude-level references for torque switching.

5-15. MODE SWITCHING SECTION

This section of the calibrator establishes the proper attitude angle bias, torque switching status, and interlock signals corresponding to the ASCS mode commanded by external devices.

NOTE

The sum of all such external devices is, for ASCS design purposes a "master sequencer" which coordinates all automatic functions.

The mode-switching section uses compact, solid-state switching circuits. Although these circuits contain many transistors, diodes, and other electrical components, they are of a class that is not critically dependent upon reference voltage or temperature levels.

5-16. TORQUE SWITCHING SECTION

The torque switching section contains transistor and diode circuits

SEDR 104

similar to those in the mode-switching section. Torque switching circuits receive the step-function outputs of the attitude gyro repeaters, plus the outputs of the rate gyros. The latter (rate) signals come from sector switches replacing the usual proportional rate gyro pickoffs. Using these step-wise indications of attitude and rate conditions, along with the mode switching section output defining the current phase of the mission "decisions" are made which result in energizing of the appropriate Reaction Control Valves.

5-17. ACCELEROMETER SWITCH

The acceleration switch is a hermetically sealed instrument. The basic mechanism consists of a centrally located mass supported by a cantilever spring. The mass is damped by the viscous shear action of the fluid which fills the case. Switch actuation is caused by the displacement of the mass element. An Acceleration force of .05g, in the axis normal to and in the direction away from the base, is required to close the circuit. Mechanical stops are provided to restrain the mechanism and to protect against damage when subjected to excessive acceleration.

5-18. ATTITUDE GYROS

The function of the attitude gyros (vertical and directional) is to determine attitude angles between a set of fixed axes in the moving spacecraft and the reference axes which are fixed in the orbital plane but which are moving with the local vertical. Both attitude gyros are "free" gyroscopes with slaving capability. A means is incorporated for caging and for obtaining electrical signal (synchros) outputs which define the attitude of the gyros with respect to two mutually perpendicular axes. The attitude gyros possess unrestricted mechanical freedom in the outer axis and ± 83° (minimum) of mechanical freedom in the inner axis. It is noted that the degree of gyro freedom does not necessarily reflect the attitudes permissible by manually steering the spacecraft in orbit. Due to limitations in the Horizon Scanner system and the repeater section of the Amplifier Calibrator, manual control of the spacecraft should be limited to ± 30° in roll and yaw and within ± 30° of the -14.5° pitch attitude. However, barring equipment malfunction, exceeding these limits will not prejudice the success of a mission. If these limits are exceeded, it is recommended that the gyro switch be placed in the FREE position. Input power requirements are 115 volt 400 cps single phase (gyro motor), and 26 volt, 400 cps (synchro and torque motor).

5-19. ATTITUDE AND RATE INDICATOR

The Attitude and Rate Indicator is mounted on the upper portion of the Main Instrument Panel. The indicator provides visual indications of spacecraft Rate and Attitude in the Yaw, Pitch and Roll planes. The attitude indicators are driven by the attitude gyro synchro outputs (through the Amplifier Calibrator). The attitude indicators are calibrated to indicate capsule attitude within a range of ± 180° except for Yaw which shall indicate 0°, 80°, and 270° in a clockwise direction. The rate portion of the indicator is driven by the miniature rate transducers. The range of rate indication is 0 to ± 6°/second for all

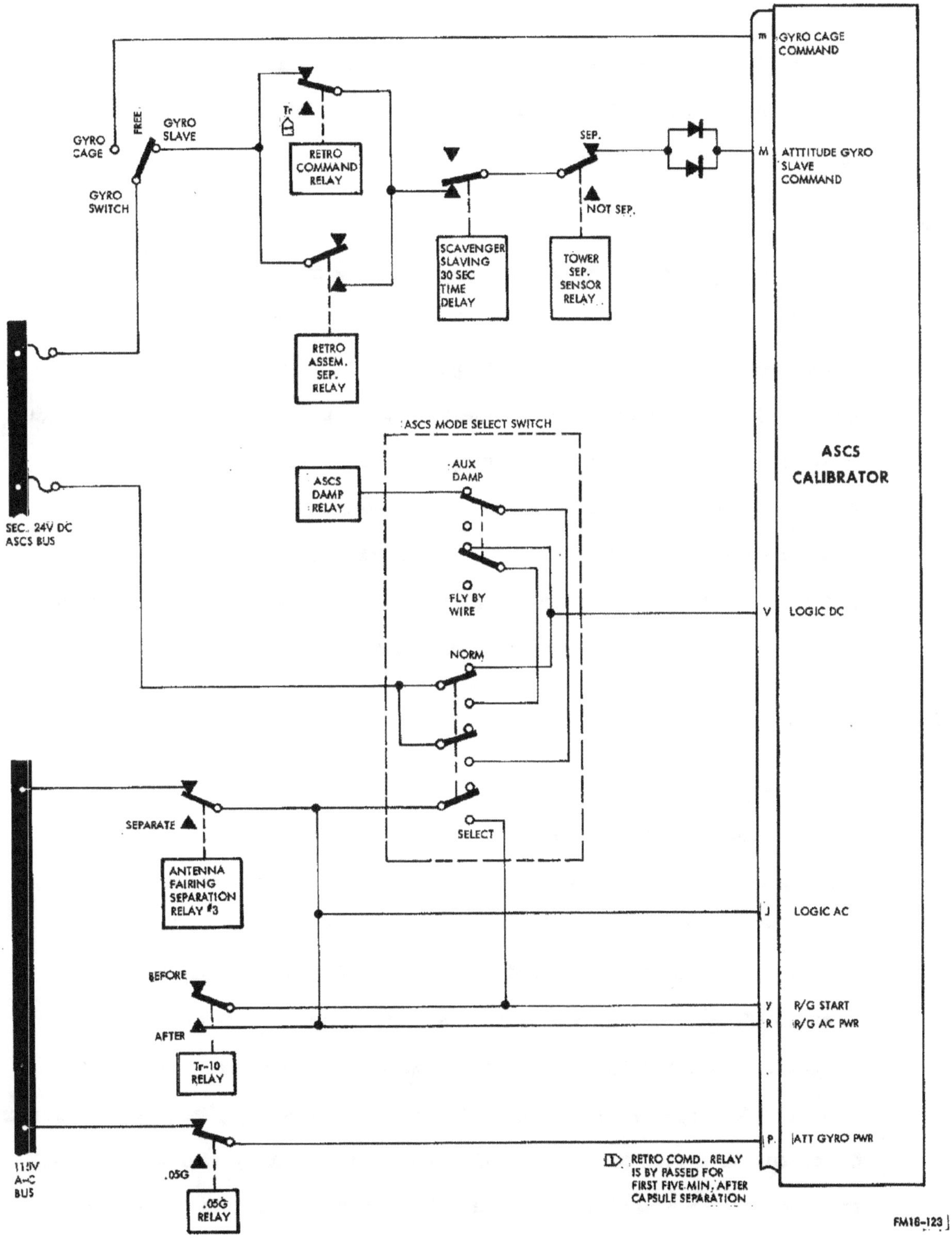

SEDR 104

Figure 5-11 Power Distribution Diagram

three indicators. The roll rate indicator has the additional capability of being externally switched to a range of 0 to 15°/sec. in order to monitor re-entry roll rate.

5-20. RATE GYROS

The rate gyros perform electrical circuit switching functions at specific rates of angular velocity about an axis perpendicular to the base of each unit, referred to as the "input axis". Rate gyros are used in the pitch, roll and yaw axes, respectively. Each rate gyro consists of a high speed rotor, mounted in a gimbal ring, in such a manner that it is free to precess about one axis only (the output axis) which is perpendicular to the spin axis of the rotor. The output signals are generated by the motion of wipers, attached to the gimbal ring, moving across the contacts of sector switches. Input power requirements are met by 115 volts, 400 cps.

5-21. REACTION CONTROL SYSTEM

5-22. SYSTEM DESCRIPTION

The Reaction Control System is used for spacecraft yaw, pitch and roll control. The system utilizes nitrogen gas pressure to expel hydrogen peroxide from a bladder into the thrust chamber catalyst beds. The right angle thrust chambers produce thrust by decomposing 90% hydrogen peroxide (H_2O_2). The system is divided into three different portions. One portion is for automatic control (ASCS), one for manual control, and a reserve fuel system to be utilized during the re-entry phase of the mission. The automatic system and the manual system are basically similar with the exception of the method used to control the flow of fuel to the thrust

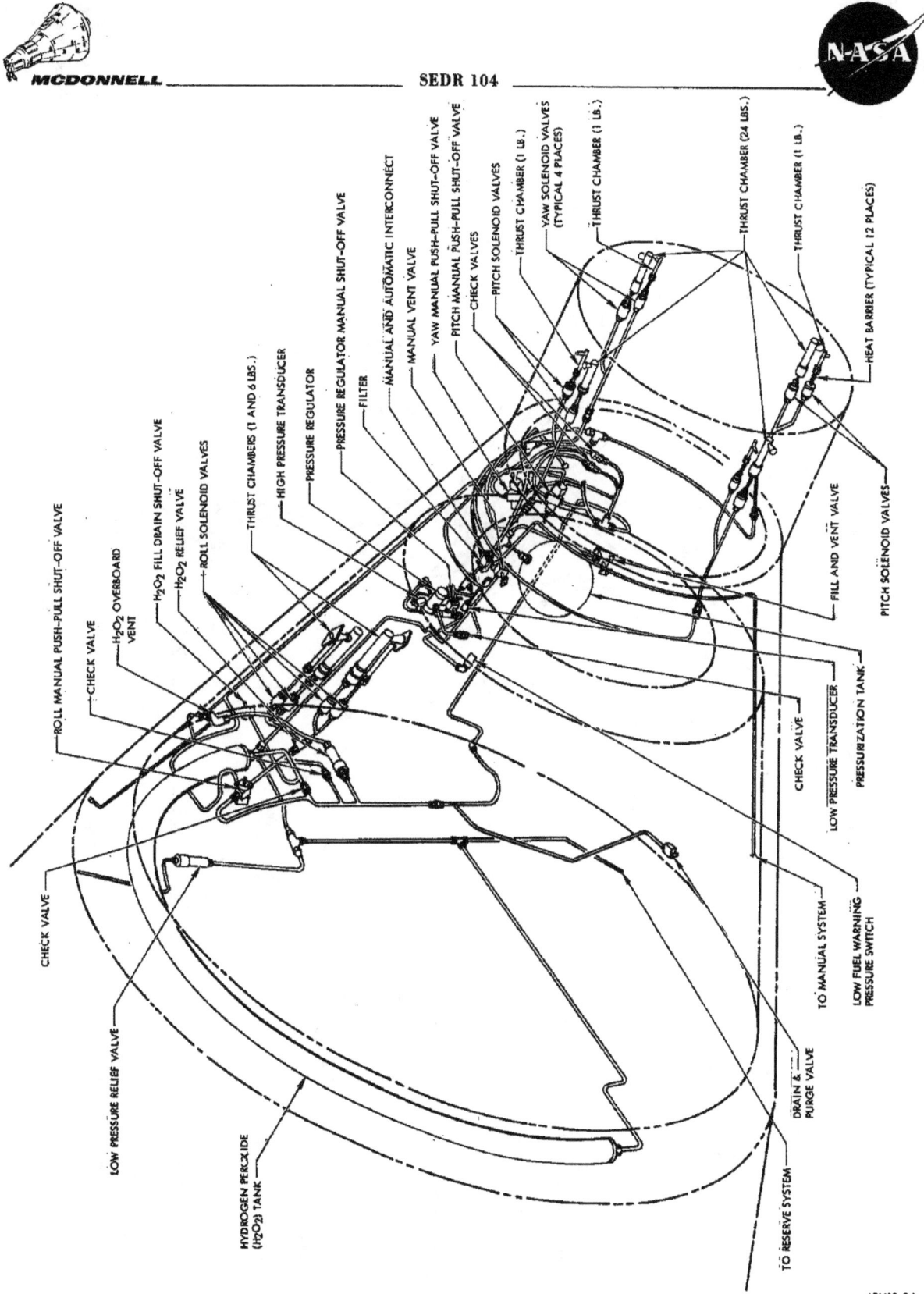

Figure 5-12 Automatic R.C.S. Installation

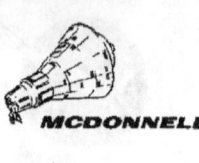

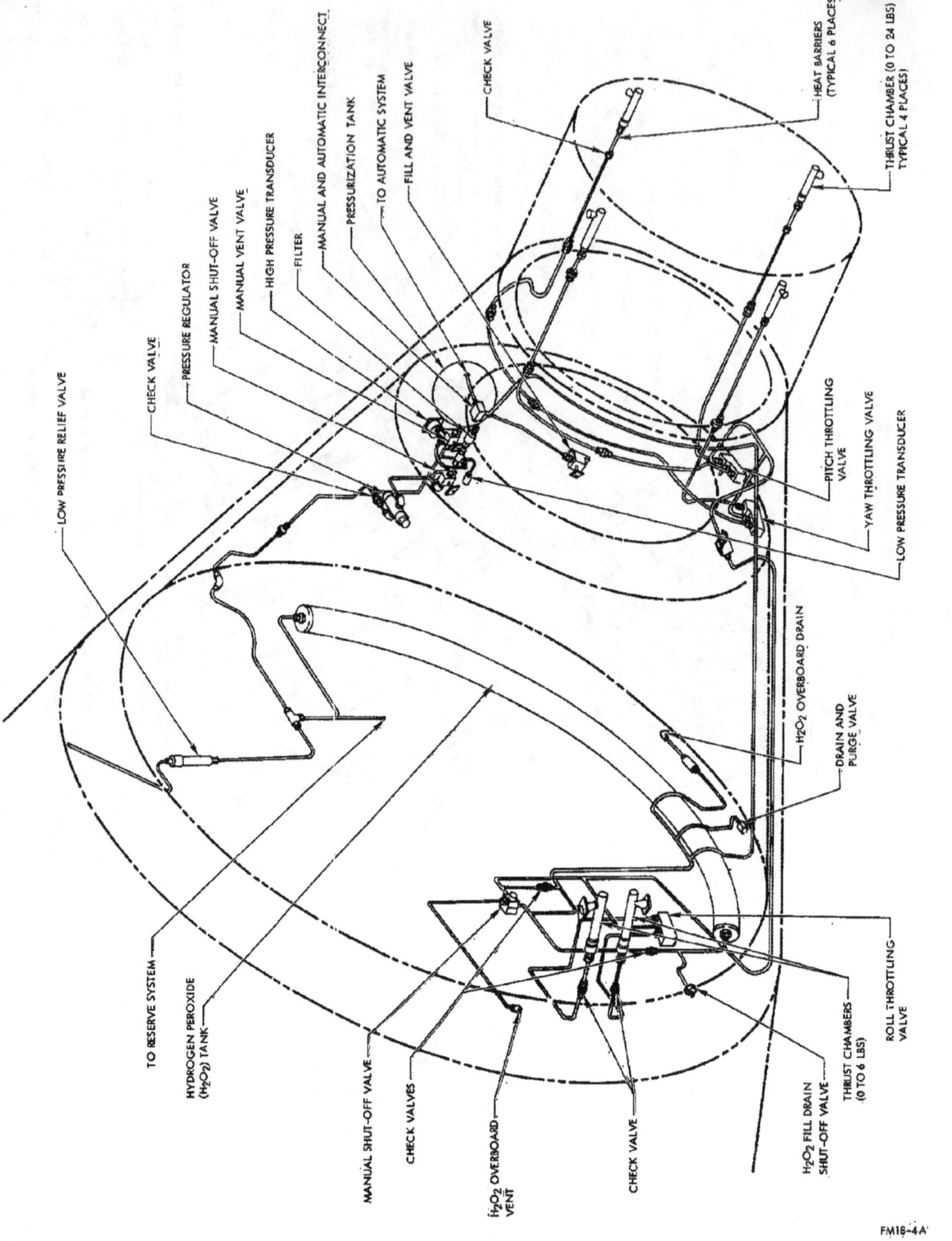

Figure 5-13 Manual R.C.S. Installation

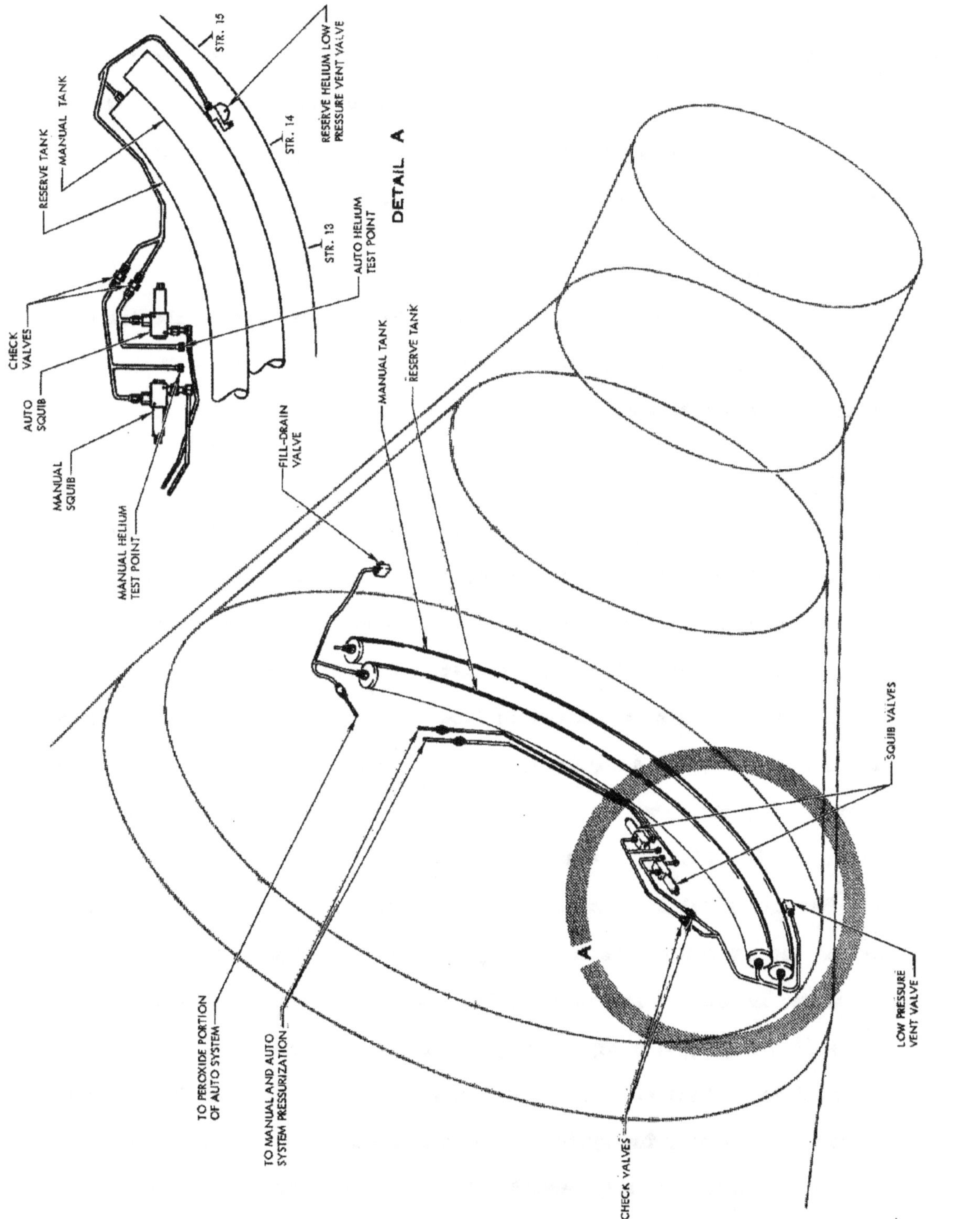

SEDR 104

Figure 5-14 Reserve RCS Installation

chambers. In the automatic system, the flow of fuel to the thrust chambers is controlled by electrically actuated solenoid valves. The flow of fuel to the thrust chambers of the manual system is controlled by mechanically actuated proportional valves.

5-23. **SYSTEM OPERATION**

The following paragraphs 5-24 through 5-26, briefly describe the operation of the automatic, manual and reserve systems. Figure 5-17 should be followed closely in conjunction with the following text.

5-24. **AUTOMATIC SYSTEM**

The automatic system can be divided into two major parts; the pressurization portion and the fuel portion. The pressurization sphere, located within the astronaut's compartment, is serviced with nitrogen to a pressure of 3000 psi. When the regulator shutoff needle valve is opened, the pressurizing gas is allowed to flow into the pressure regulator where the output pressure is regulated to $480 \, {}^{+10}_{-8}$ psi. From the regulator, the pressurizing gas flows through a checkvalve and enters the fuel tank where it surrounds a bladder containing the fuel.

Two electrical transducers are installed in the pressurization system. One of the transducers is located upstream of the regulator shutoff needle valve and is used to indicate the fuel quantity. The pressurizing gas is allowed to expand to occupy a larger volume as the fuel is expelled from the fuel tank bladder. This expansion results in a decrease in pressure within the pressurizing system. The decrease in pressure is sensed by the electrical transducer and converted into a voltage which operates a control

fuel quantity indicator located on the astronaut's main instrument panel. The control fuel quantity indicator is calibrated to show percent of fuel remaining.

The second electrical transducer is located upstream of the fuel tank and is used to indicate the pressure on the fuel bladder. The voltage from this transducer is displayed on the H_2O_2 pressure indicator on the astronaut's main instrument panel. The H_2O_2 pressure indicator is calibrated to indicate pressure from 400 to 700 psi.

A pressure switch located in the pressurization system, upstream of the pressure regulator, is used to illuminate a fuel quantity warning light on the astronaut's main instrument panel. The pressure switch is set to actuate at a pre-determined low fuel level.

A pressure relief valve is also incorporated into the pressurization system and is set to relieve excessive system pressure. The physical location of the various components in the automatic pressurization system can be seen in Figure 5-12.

The fuel portion of the automatic system begins with the half toroidal fuel tank. Located inside the tank is a flexible bladder containing approximately 32 pounds of liquid 90% H_2O_2. The pressurizing gas surrounding the outside of the bladder forces the fuel out of the bladder through a perforated transfer tube located inside the bladder. The fuel flows through a checkvalve and into the downstream lines.

Located in each axis, yaw, pitch and roll, is a manually actuated shutoff valve. These valves are linked to pull handles located on the

SEDR 104

astronaut's left hand console. By actuating these shutoff valves, the fuel supply can be shutoff to each individual thrust axis in the event a solenoid malfunction should occur. The ability to isolate each axis individually allows the automatic system to operate normally on the remaining axes, requiring the astronaut to back-up manually, only the malfunctioning axis.

Downstream of the manual shutoff valves are the electrically controlled solenoid valves. Upon receiving a signal from the ASCS or fly-by-wire control system, the appropriate solenoid valve opens. Fuel enters the solenoid valve and passes into the corresponding thrust chamber where it decomposes and produces the desired thrust. There are 1 lb. and 24 lb. thrust chambers, providing low and high thrust, about each of the following axes: yaw left, yaw right, pitch up and pitch down. There are 1 lb. and 6 lb. thrust chambers, providing low and high thrust, about the following axis: roll CW and roll CCW. Check valves, located between the manual shutoff valves and the solenoid valves, allow for thermal expansion of fuel trapped downstream of the manual shutoff valves. A relief valve in the fuel system relieves excessive pressure. See Figure 5-12 for the physical location of the automatic fuel components.

5-25. **MANUAL SYSTEM**

The manual system pressurization system is identical to that used in the automatic system with one exception. The manual pressurization system contains no pressure switch and no fuel quantity warning light.

The fuel portion of the manual system begins with the half toroidal fuel tank. Located inside the tank is a flexible bladder containing

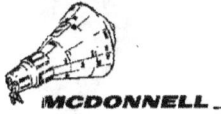

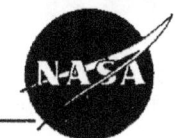

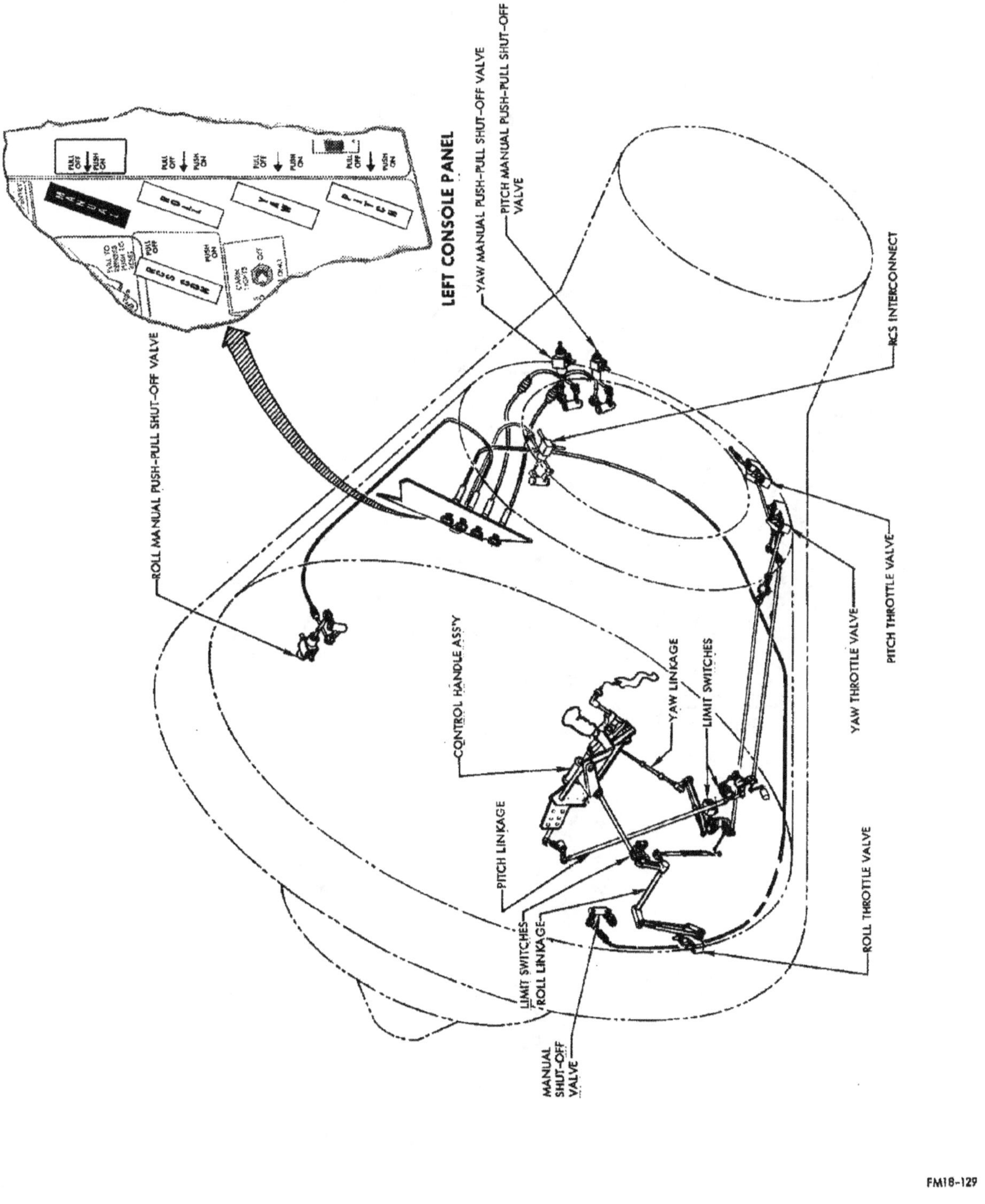

FM18-129

Figure 5-15. R.C.S. Control Linkage

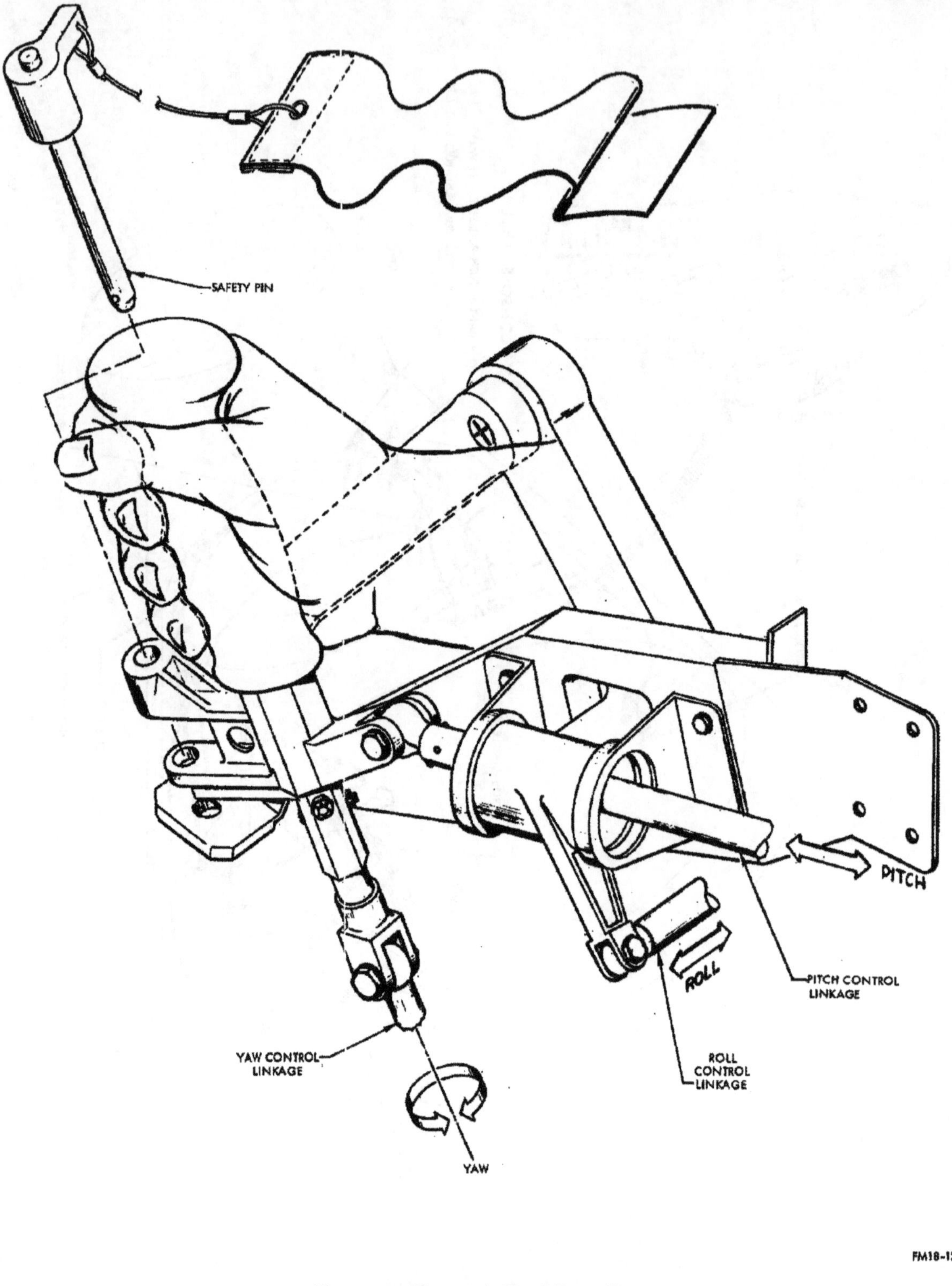

Figure 5-16/ Three Axis Hand Controller.

approximately 23.4 pounds of liquid 90% H_2O_2. The pressurizing gas surrounding the outside of the bladder forces the fuel out of the bladder through a perforated transfer tube located inside the bladder. The fuel flows through a checkvalve and into the downstream lines.

Located downstream of the manual system checkvalve and upstream of the proportional control throttle valves, is a mechanically actuated shutoff valve. The shutoff valve is linked mechanically to a pull handle located on the astronaut's left console. When the pull handle is pulled out, the fuel supply to the proportional control throttle valves is shut off. From the manual system shutoff valve, fuel flows to the proportional control throttle valves. The throttle valves, see Figure 5-19, are linked mechanically to the astronaut's hand controller, see Figure 5-16. Thrust output is directly proportional to hand controller displacement. Spring cartridges located in the hand controller linkage provide control handle "feel" and return the handle to neutral. Shear pins, located in the throttle valve bellcranks, are designed to shear with increased effort on the part of the astronaut. The ability to shear the mechanical linkage on each axis prevents the jamming of one throttle valve from disabling the entire system.

From the throttle valves, fuel flows through a checkvalve and enters the thrust chamber. The checkvalves installed downstream of the throttle valves require positive pressure for the fuel to flow. This requirement eliminates the erratic thrust that would occur if the fuel downstream of the throttle valve were allowed to drain through the thrust chamber after the throttle valve had returned to neutral.

A maximum thrust of 24 lbs. is available on the following axes: yaw right, yaw left, pitch up and pitch down. A maximum thrust of 6 lbs., is available on the following axis: roll CW and roll CCW.

A checkvalve located between the manual shutoff valve and the throttle valves allows for thermal expansion of fuel trapped downstream of the manual shutoff valve. A relief valve installed in the manual fuel system relieves excessive pressure. The physical location of the various components of the manual system can be seen in Figure 5-13.

The automatic and the manual system can be interconnected by a mechanically actuated interconnect valve. The interconnect valve is mechanically linked to a pull handle on the astronaut's left console. By opening the interconnect valve, the fuel of either system may be used in the other system.

5-26. **RESERVE SYSTEM**

The reserve system consists primarily of a fuel tank, similar in construction to the fuel tank used in the automatic and manual systems, with a capacity of 15 lbs. of 90% H_2O_2. The reserve fuel tank is connected to the automatic portion of the reaction control system and is designed to be used during the re-entry portion of the mission.

In order to utilize the reserve fuel, it is necessary to supply a pressurizing gas to surround the reserve fuel bladder in a manner similar to that used in the automatic and manual fuel tanks. Pressuring gas is made available to the reserve fuel bladder from either the automatic pres-

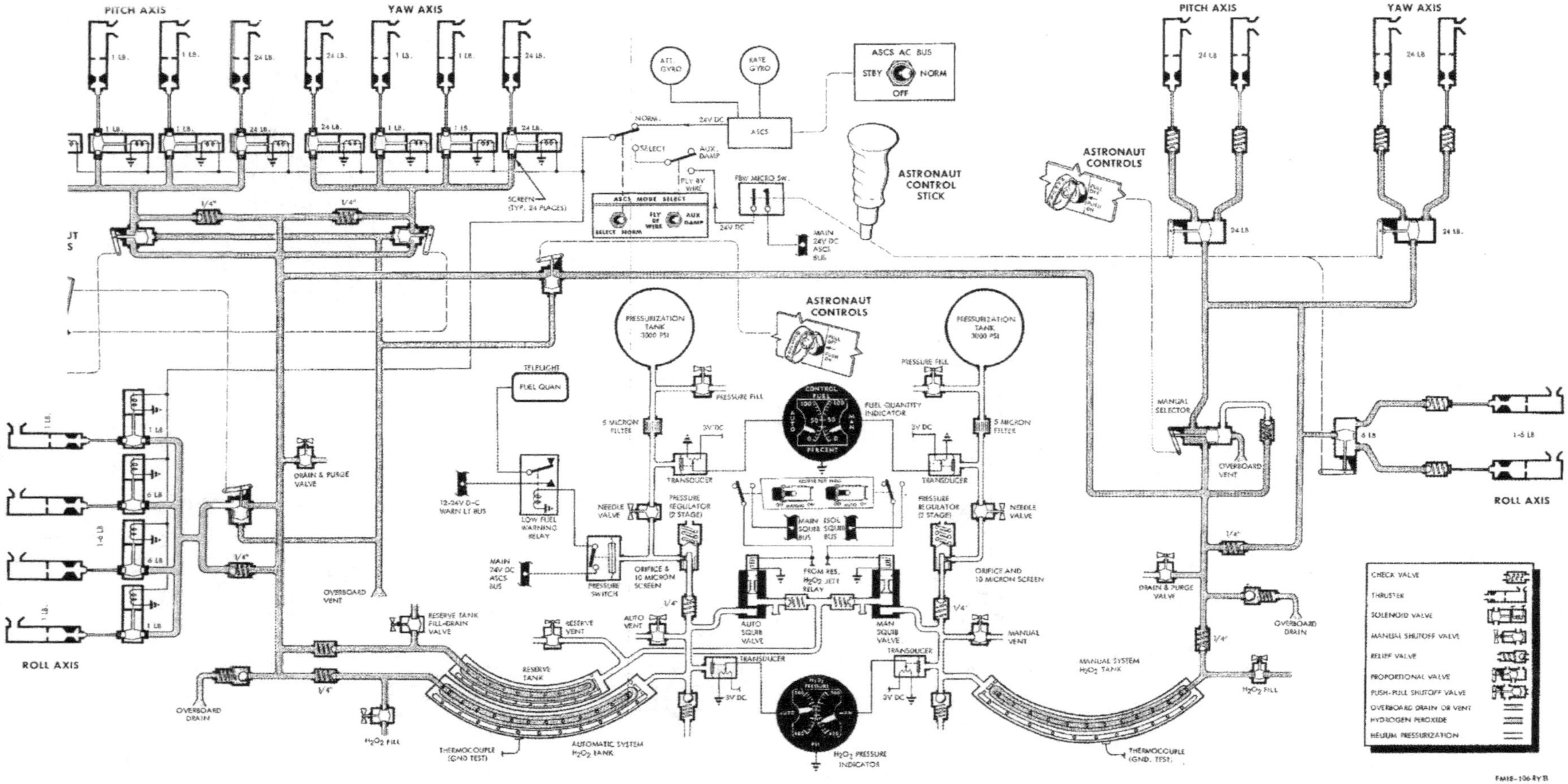

Figure 5-17 Reaction Control System

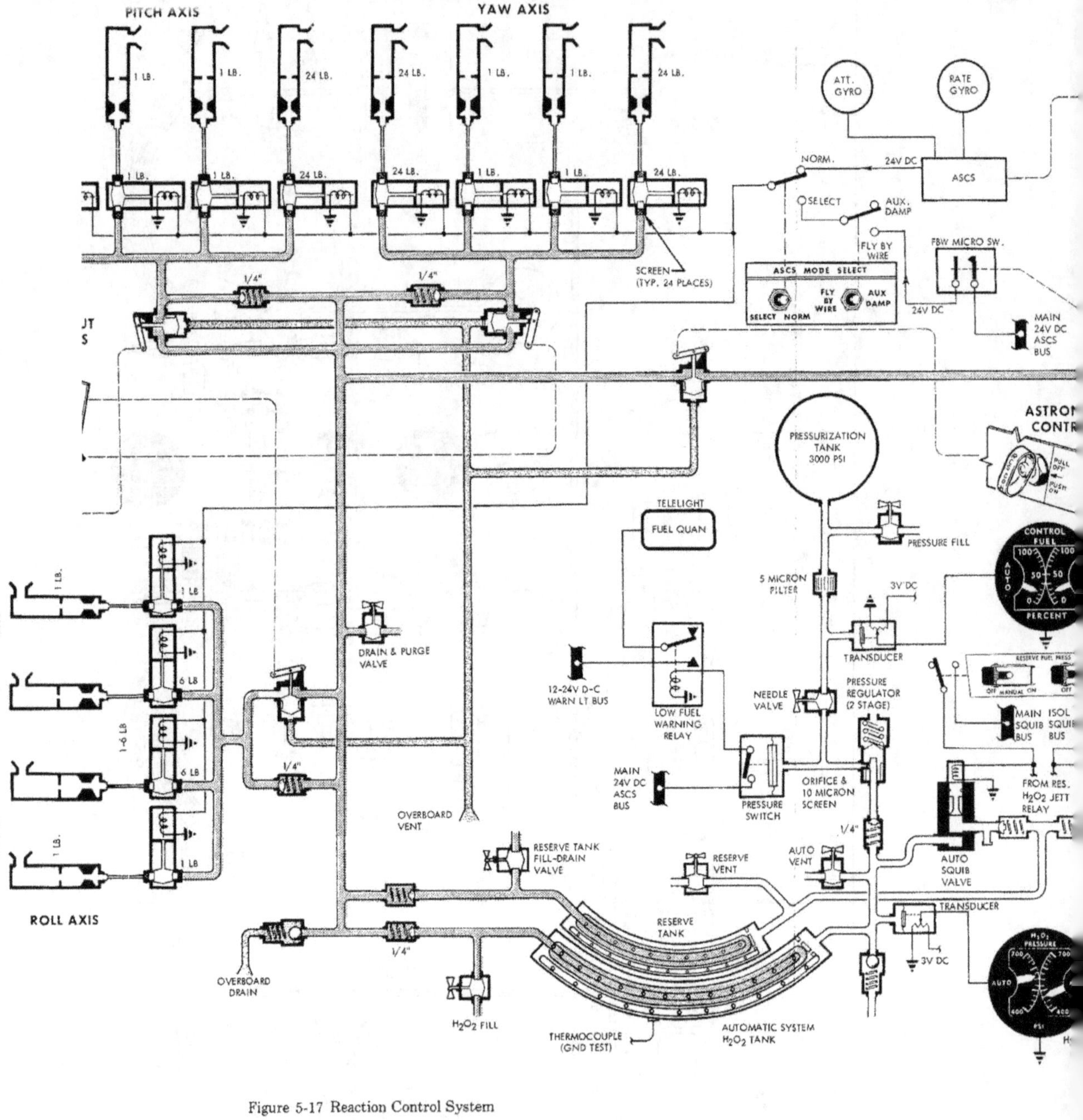

Figure 5-17 Reaction Control System

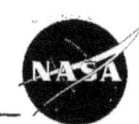

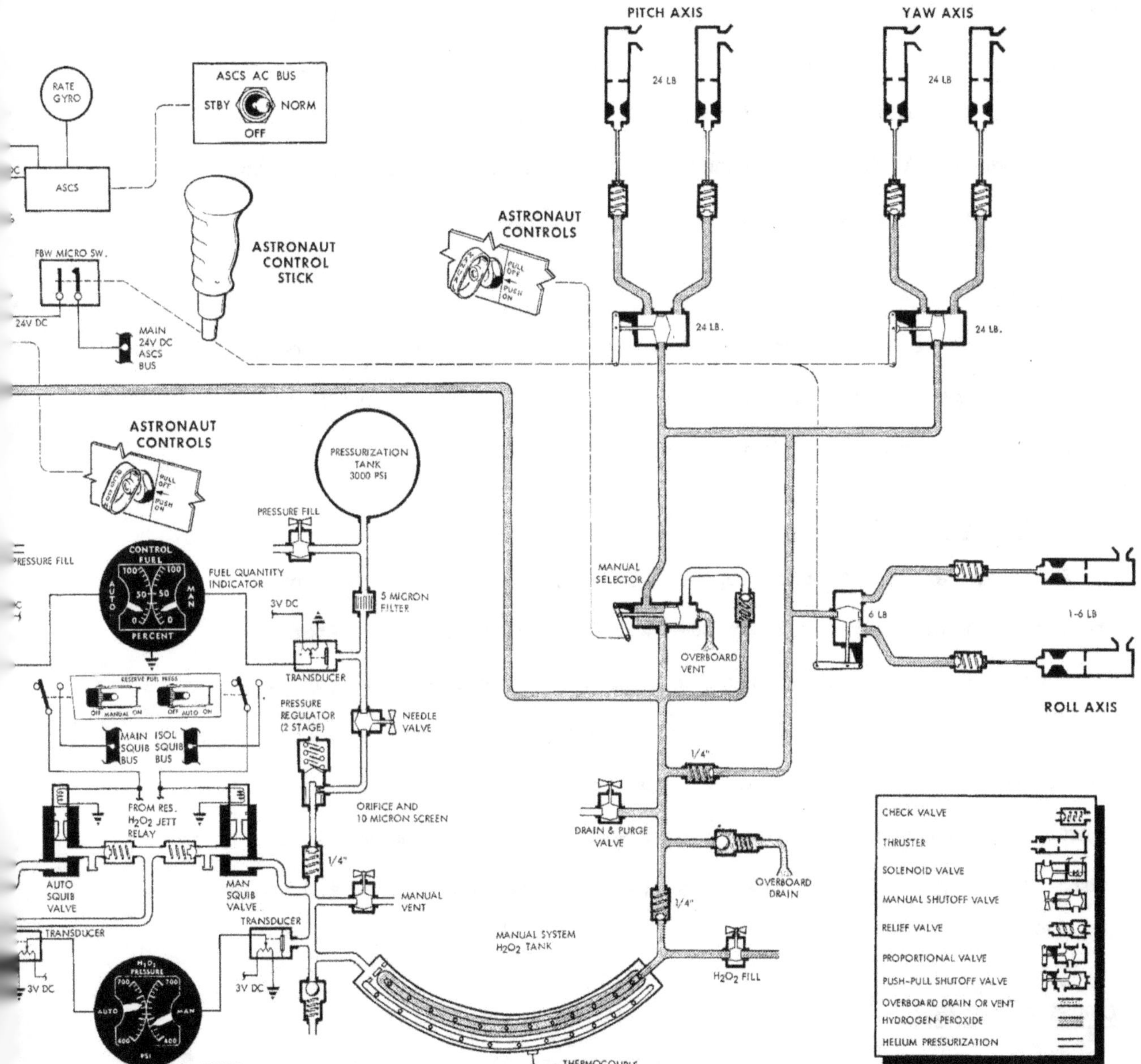
PITCH AXIS
YAW AXIS
24 LB
24 LB
RATE GYRO
ASCS AC BUS
STBY
NORM
OFF
ASCS
ASTRONAUT CONTROLS
ASTRONAUT CONTROL STICK
FBW MICRO SW.
24V DC
MAIN 24V DC ASCS BUS
24 LB.
24 LB.
ASTRONAUT CONTROLS
PRESSURIZATION TANK 3000 PSI
PRESSURE FILL
PRESSURE FILL
CONTROL FUEL
AUTO
MAN
PERCENT
FUEL QUANTITY INDICATOR
5 MICRON FILTER
3V DC
TRANSDUCER
MANUAL SELECTOR
OVERBOARD VENT
6 LB
1-6 LB
ROLL AXIS
RESERVE FUEL PRESS
MAIN SQUIB BUS
ISOL SQUIB BUS
PRESSURE REGULATOR (2 STAGE)
NEEDLE VALVE
1/4"
FROM RES. H2O2 JETT RELAY
ORIFICE AND 10 MICRON SCREEN
DRAIN & PURGE VALVE
1/4"
AUTO SQUIB VALVE
MAN SQUIB VALVE
MANUAL VENT
OVERBOARD DRAIN
1/4"
TRANSDUCER
TRANSDUCER
MANUAL SYSTEM H2O2 TANK
H2O2 FILL
3V DC
3V DC
H2O2 PRESSURE
H2O2 PRESSURE INDICATOR
THERMOCOUPLE (GND. TEST)
CHECK VALVE
THRUSTER
SOLENOID VALVE
MANUAL SHUTOFF VALVE
RELIEF VALVE
PROPORTIONAL VALVE
PUSH-PULL SHUTOFF VALVE
OVERBOARD DRAIN OR VENT
HYDROGEN PEROXIDE
HELIUM PRESSURIZATION

FM18-106 RYB

surization system, the manual pressurization system or both. Pressurizing gas is made available by firing a squib valve connecting the reserve system to the appropriate pressurization system. Firing of the squib valves is accomplished by astronaut actuation of either of two toggle switches provided on the left console of the instrument panel. In addition, the squib valves are both fired automatically at antenna fairing separation during re-entry if they have not been fired previously.

5-27. SYSTEM UNITS

Due to the general nature of the system components, a discussion of each is considered unnecessary. However, two items (propellant fuel and thrust chambers) do warrant brief explanations.

5-28. PROPELLANT FUEL (H_2O_2)

Hydrogen peroxide is a clear, colorless liquid soluble in all proportions in water and most substances which are miscible with water. Hydrogen peroxide when catalytically decomposed releases water vapor, oxygen gas and heat. H_2O_2 decomposition when properly contained and controlled is capable of producing usable thrust. One pound of H_2O_2 solution (90%) when properly decomposed will produce approximately 60 cubic feet of gas. Hydrogen peroxide (90%) freezes at 11.3°F, and boils at 268°F.

5-29. THRUST CHAMBERS

The thrust chamber assemblies (See Figure 5-18) consist of a stainless steel chamber that contains a distribution disc followed by a catalyst bed and then a nozzle. The catalyst bed contains a stack of nickel screen

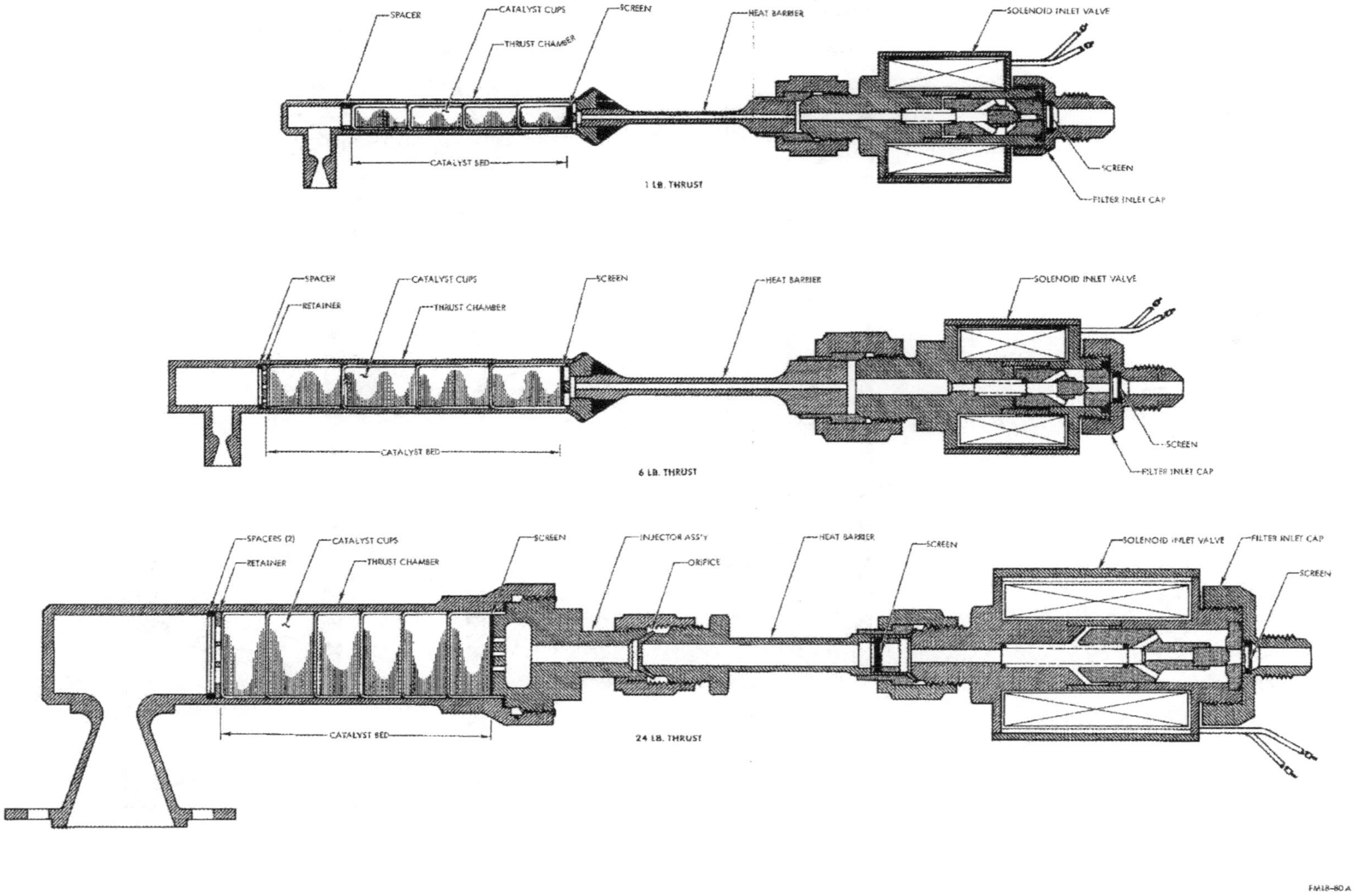

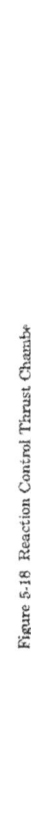

Figure 5-18 Reaction Control Thrust Chamber

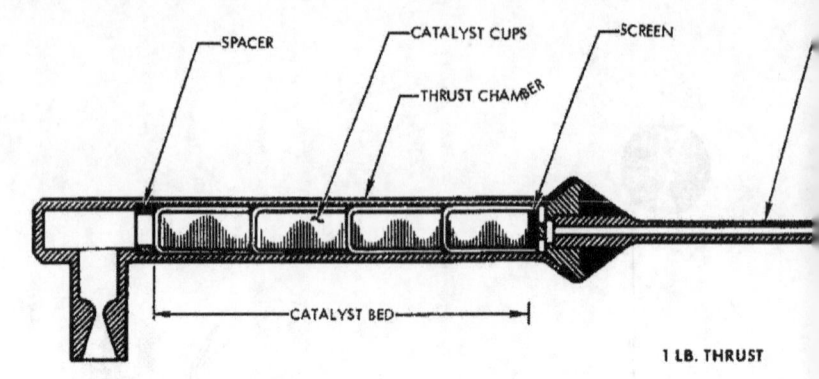
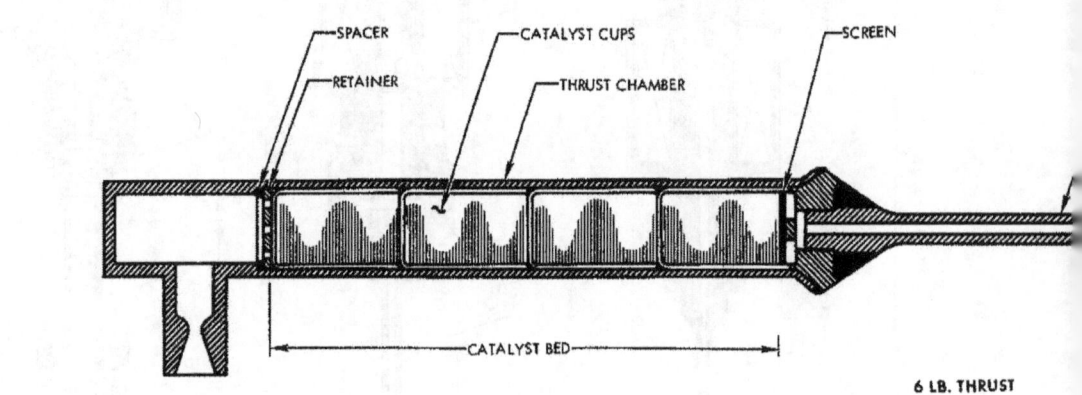
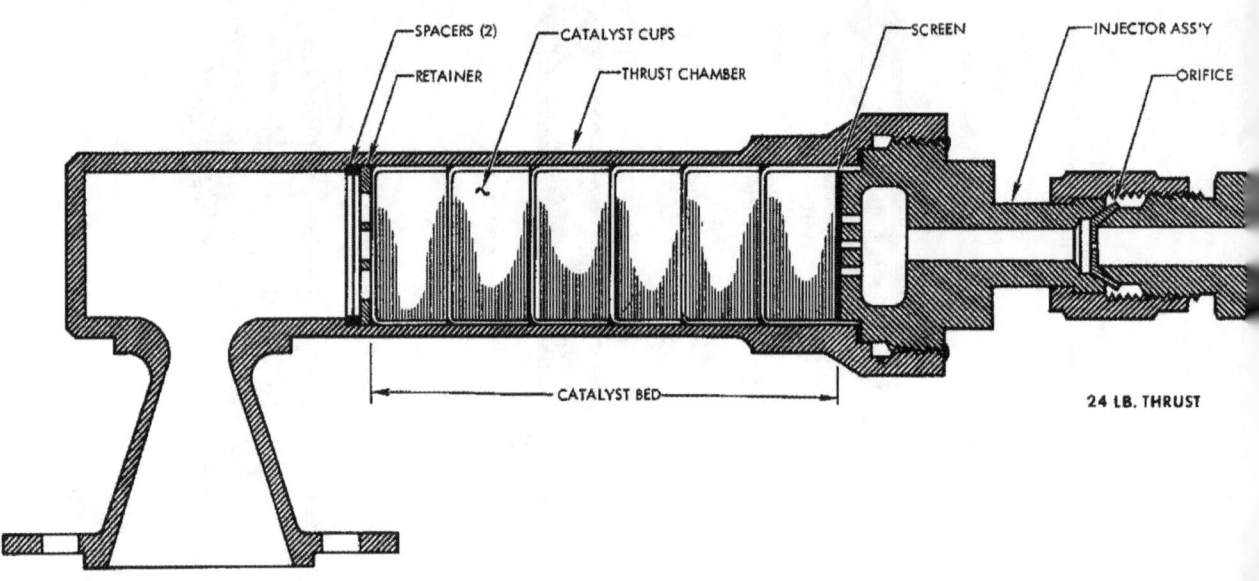

Figure 5-18 Reaction Control Thrust Chamber

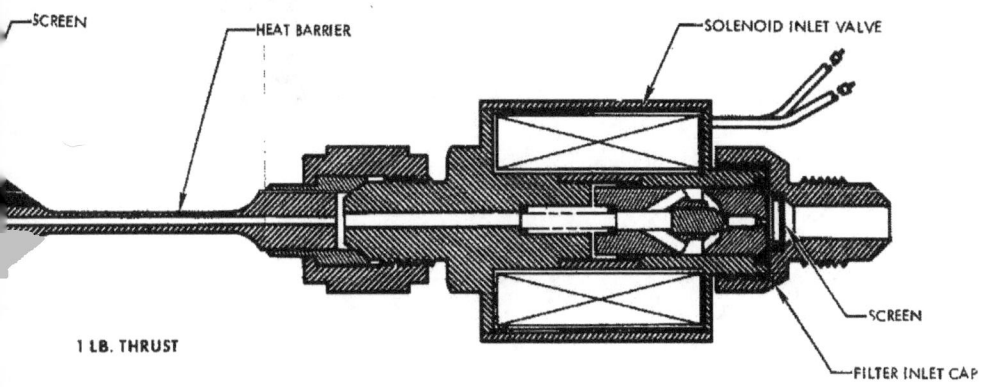

1 LB. THRUST

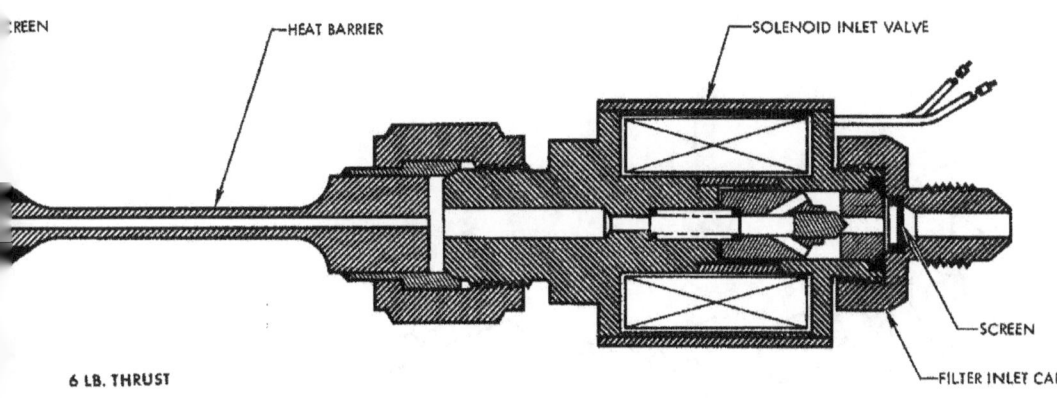

6 LB. THRUST

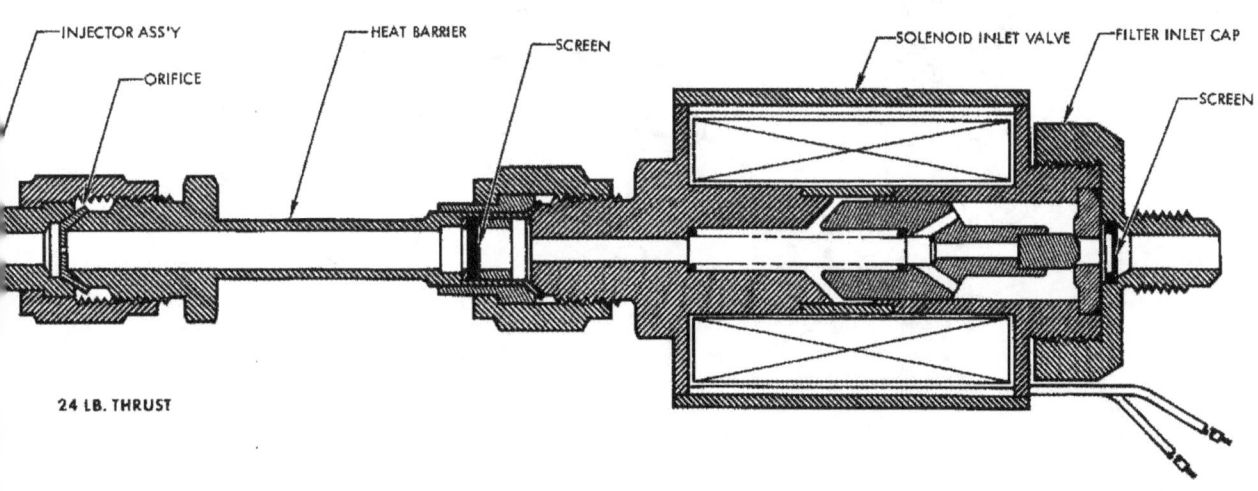

24 LB. THRUST

FM18-80A

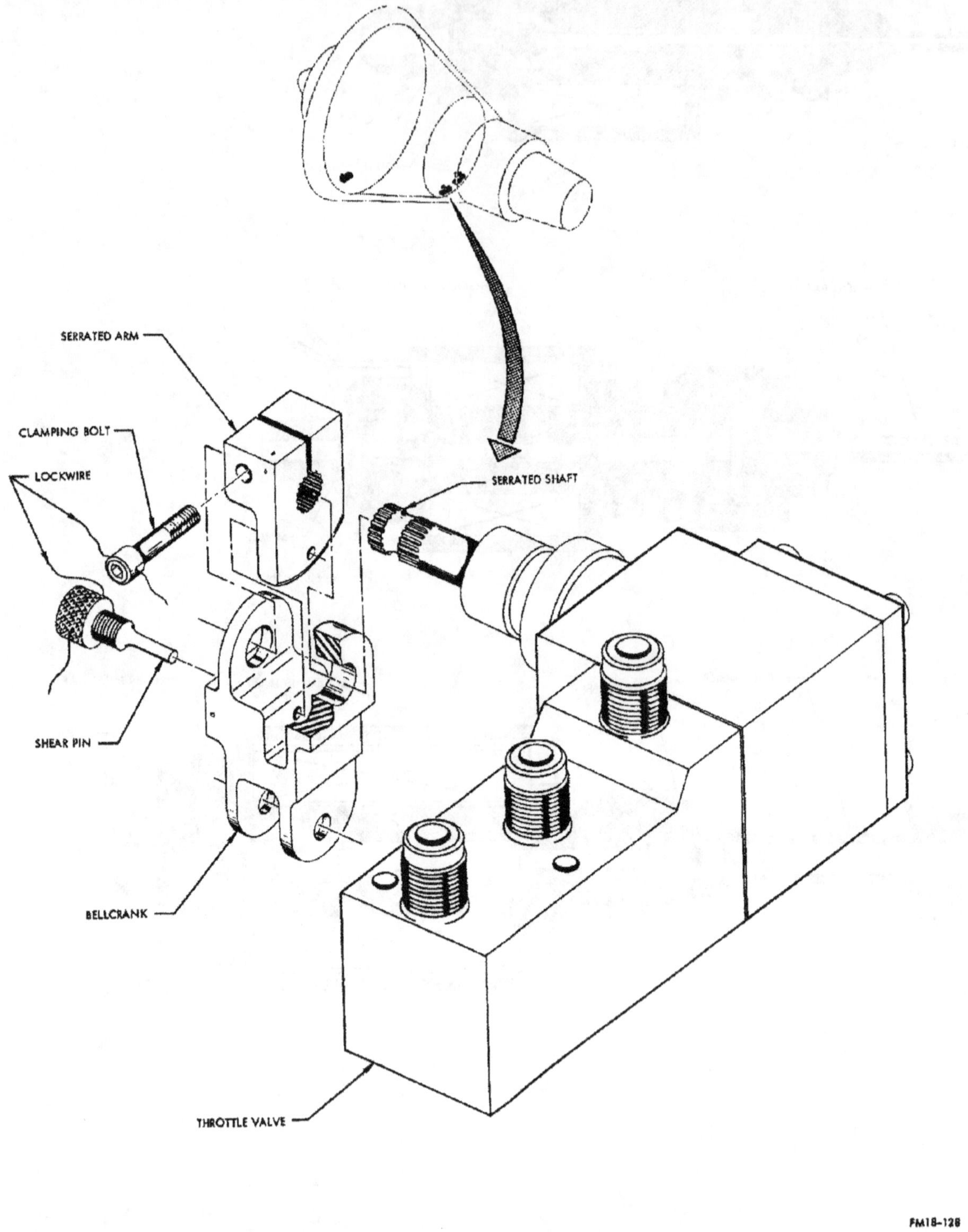

Figure 5-19 Manual R.C.S. Throttle Valve and Bellcrank

SEDR 104

wafers. The screen gauge resembles common household screen. The screen is plated with a silver-gold coating that enhances the catalytic properties of the nickel. The open area between the catalyst bed and the right angle nozzle forms a short plenum chamber to smooth out the flow prior to reaching the nozzle throat.

H_2O_2 enters the thrust chamber upon actuation of the solenoid valve. The stainless steel plate distributes the flow and presents the catalyst bed with a uniform input. Upon entering the first stage of the catalyst bed, a violent reaction takes place. Expanding gases rush through the remainder of the catalyst bed resulting in a thrust output, in the right angle nozzle. The majority of the decomposition (and most violent) takes place within the first two catalyst cups. Temperatures of approximately $1400°F$ can be expected in this area. The remainder of the catalyst cups are to assure a complete decomposition process and to prevent any liquid form of H_2O_2 from reaching the nozzle.

5-30. HORIZON SCANNER SYSTEM

5-31. SYSTEM DESCRIPTION

The Horizon Scanner System incorporates two identical scanning units. The purpose of the Horizon Scanner System is to provide a roll and pitch reference during the orbital phase of the normal mission. The scanners produce an output signal that slaves the ASCS attitude gyros to the proper angles upon command from an external programmer.

5-32. BASIC CONSTRUCTION

Figure 5-20 is a photograph of a Horizon Scanner Unit. All major com-

Figure 5-20 Wide Angle Horizon Scanner

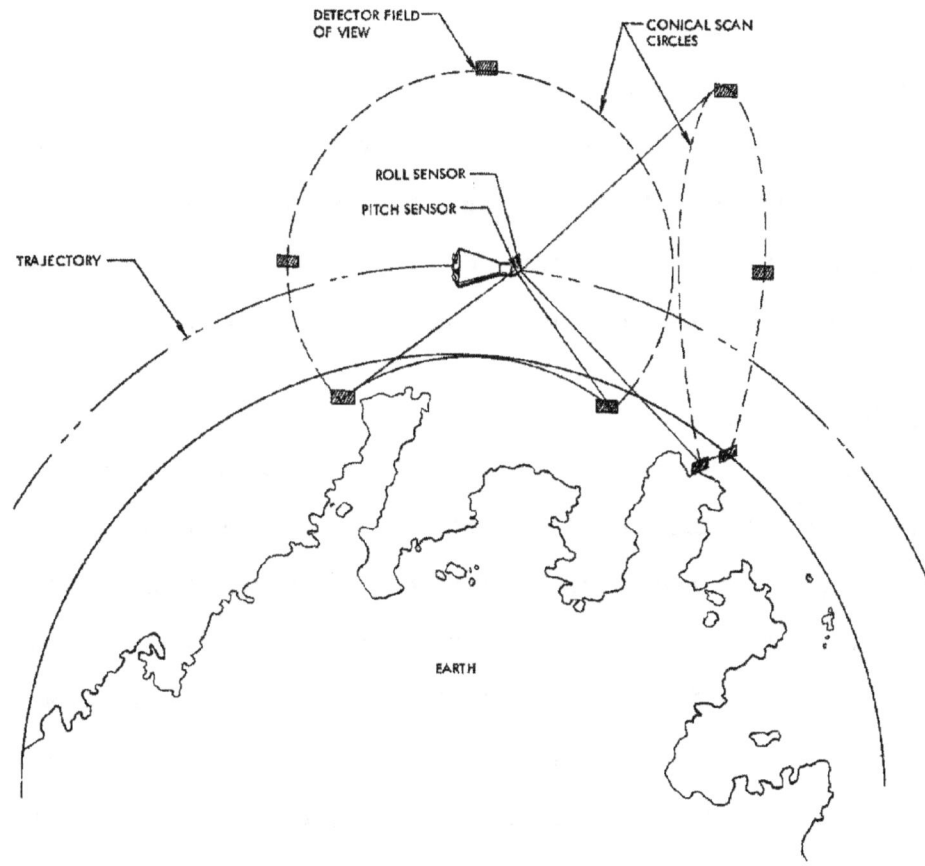

Figure 5-21 Horizon Sensor Scan Pattern

ponents and subassemblies are mounted from the large circular plate and include the scanning prism assembly, prism drive system, infrared detector, electronics, synchronous switches, electrical connector and cover. The circular plate is flange mounted so that the scanning prism compartment projects into the space outside of the vehicle. The electronic system is completely transistorized and the various functional sections are fabricated on separate printed circuit boards. Three of these printed circuit boards are enclosed in the shielded housing fastened to the circular plate. For rapid servicing the four posts with attached boards can be replaced as a single unit, or individual boards can be replaced as required.

5-33. SPECIAL FEATURES

The Horizon Scanner has a number of special features. It is compact in size (6 5/32" long x 5 7/8" diameter overall) and light in weight (3.02 lbs). The scanner is equipped with a centrifugally - activated shutter. The shutter prevents solar radiation from dwelling upon the detector and resulting in probable damage during those periods when the scanning prism is not rotating. Another feature is a special circuit which can be used to disconnect the error signals from the vehicle reaction devices during those periods when the presence of the sun in the scan path or the loss of horizon would result in erroneous error signals. The final feature of significance is that only a single power source providing 110 volts, 400 cycles and 3.2 VA for each scanner is required to operate the entire system. The highly regulated power supply in the system eliminates the need for the bulky batteries usually required to bias the infrared detector.

SEDR 104

5-34. SYSTEM OPERATION

Operation of the Horizon Scanners depends upon infrared radiation received from the earth as compared to the essentially zero radiation from space. These differences in radiation levels provide a sharp radiation discontinuity at the horizon. The Scanner system uses this discontinuity for both day and night vertical reference sensing. When the spacecraft is oriented so that the earth is present in its scanning path, there will in general be two points where the scan intersects the earth's horizon (See Figure 5-21). The scanner detects the thermal discontinuity, or change in radiation level, between the earth and space at the two horizon points. The Scanner then bisects the included angle from itself to the Horizon points, compares the direction of the bisector with that of a fixed reference in the spacecraft and generates linear error signals proportional to the angle between the bisector and the fixed reference. As previously stated, these error signals (roll and pitch) are used to slave the ASCS attitude gyros.

Figure 5-22 shows a simple block diagram of the Horizon Scanner. The following discussion entails a brief explanation of the functioning of each block as related to the overall operation.

5-35. RADIATION GRADIENT AT HORIZON

There is a large difference in the radiation which the detector receives as it scans across the boundary between space and the upper atmosphere (troposphere). This change is approximately equal to that from black bodies at $0°K$ and $200°K$ respectively, and the radiance difference is approximately 3.003 watts/cm^2 - steradian. The location of this gradient is sharply defined, and it is much larger than any others that can be encountered during the scan cycle.

5-36. CORRECTION FOR REFLECTED SOLAR RADIATION

Sharp radiation gradients do exist because of reflected solar radiation. Such gradients are found at cloud edges, topographical irregularities on the earth's surface and the terminator line between night and day. These radiation changes can be filtered out so that the horizon gradient is the only one that is detected by the system. Selective filtering can be accomplished since most of the reflected solar radiation falls in the spectral region between 0.2 and 2.0 microns, while the radiation emitted by the earth and troposphere is at wavelengths longer than 5 microns. The filtering is accomplished by a germanium prism and field lens in front of the detector. As a filter, germanium sharply cuts off all radiation at wavelengths shorter than 1.8 microns while transmitting very uniformly radiation from 1.8 to 20 microns. The use of this filter removes over 90 percent of the reflected solar radiation. Signal clipping techniques in the electronics remove any residual effects.

5-37. SCANNING AND RADIATION DETECTION

Details of the scanning prism assembly can be seen in Figure 5-20. The infrared detector is fixed to the center of the circular plate and its field of view extends through the circular opening in the center of the scanning assembly. The detector field of view is 2° by 8° and the presence of the scanning prism has the effect of deflecting it 55° from the normal. Thus the apex angle of the scanning cone is 110°. In opera-

tion, the drive system rotates the scanning prism and the detector field scans the field of view through the conical pattern described previously. Different amounts of radiation strike the detector during various portions of the scan cycle, and the amplitude of the detector output changes accordingly. The detector output signal is processed by the electronics system and the error signal produced is available at the electrical connector.

5-38. SYNCHRONIZING GENERATOR

Closely associated with the prism drive system is the reference signal generator. The output of this generator is a square wave signal at a frequency of 30 cycles per second. This signal is the fixed reference against which the detector horizon signals are compared. The reference signal is triggered by the interaction between a magnetic pickup coil and a semi-circular steel vane. The vane is imbedded in a slot cut into the surface of the scanning prism assembly gear. A pickup is mounted so that the end of its magnetized core comes close to the surface of the vane. As the scanning prism assembly turns, the ends of the vane pass by the end of the magnetized pickup coil core, generating the reference pulse. A subsequent electronic network converts the pulse to a phase locked 30 cycle square wave. The use of this signal will be considered later in this section.

5-39. SUN SHUTTER

The sun shutter consists of a pair of spring loaded metal slides which fit into opposed transverse slots through the tube section of the scanning mirror assembly. When the scanning mirror assembly is not rotating spring tension pulls the two slides together and the detector field is obstructed. When the scanning mirror is turning, the centrifugal force on the slides is sufficient to open the shutter.

5-40. INFRARED ARRANGEMENT

Infrared radiation from the field of view strikes the infrared detector and produces the electrical signal which is processed by the electronics system. The infrared detector is a thermistor bolometer with its active element immersed in the germanium lens.

The active element is a rectangular flake of thermistor material and is connected in a bridge circuit with a similar compensating flake which is shielded from radiation. The two flakes are oppositely biased and their junction is connected to the input of the preamplifier which follows. By immersing the active element in the rear surface of the germanium lens the overall detectivity can be increased by a factor of about 3.5 over an unimmersed detector having the same field of view. The material in the thermistor flake has a high negative temperature coefficient of resistance. That is, when the temperature of the material is raised, the flake resistance decreases. Since the surface of the thermistor flake is blackened, it absorbs impinging radiation and its resistance is decreased. When the shutter is closed, both flakes in the detector bridge are at the same temperature. Since both flakes have the same linear characteristics, their resistances are the same. Gradual variations in ambient temperature change the resistances of both flakes by equal amounts and the voltage of their junction remains the same. When the shutter opens, incoming radiation is focused on the active element; the compensating

element is shielded from outside radiation. The temperature of the active element is changed, and its resistance becomes different from that of the compensating element. As a result, there is a voltage change at the junction of the two flakes and this change is connected to the electronics system. As the scanning prism turns and causes the detector field of view to cross the horizon, there is a sharp change in the radiation level striking the detector. The result of the radiation changes during a complete scan cycle is the generation of an approximate squarewave signal at a frequency of 30 cycles per second.

The electronics system is physically arranged so that functionally related parts are located close to each other. The electronics system is divided into eight major circuits, located on individual printed circuit boards. In some cases, the requirements of compact and economical construction have resulted in two or three sub-circuits being located on one board. Thus, the functionally related booster amplifier, signal centering circuit and phase inverter-limiter are located on one board which the block diagram (Figure 5-22) shows as divided by dotted lines. Although the power supply and reference generator circuits are not closely related in function they are both located on the same printed circuit board.

The paragraphs that follow describe the functions of the major sections and sub-sections of the electronics system. The description is made with reference to the waveforms generated by system operation, and these are shown in Figure 5-23. Functional description will be made at the level of the major circuits and sub-circuits shown in the block diagram Figure 5-22.

5-41. IMMERSED DETECTOR

The radiation falling upon the detector determines the waveshape of the detector output signal which is to be processed by the electronics system. The radiation characteristics are determined by the scanning cycle described previously, and are shown as the first waveform, WF-1, of Figure 5-23. The waveform shows that earth radiation is higher than space radiation and that there is an abrupt shift from one level to the other as the detector scans across the horizon. The change in radiation requires 200 microseconds to take place because the detector is of finite size, and this time is required for a complete shift of radiation level across the entire surface of the detector. Thirty complete cycles of radiation change take place in one second.

WF-2 shows the detector output signal which results from the radiation changes taking place at the detector. This signal resembles the radiation signal with the exception that the shift between the two levels takes a longer time. The reason for this is that approximately 2 milliseconds is required for the active detector flake to reach the half-level of its now stabilized output. The detector output signal has an amplitude in the order of 2 millivolts.

5-42. PRE-AMPLIFIER AND BOOSTER AMPLIFIER CIRCUITS

The junction of the two thermistor flakes is direct-coupled to the input of the pre-amplifier. The pre-amplifier has a voltage gain of 400 at 30 cycles per second. Direct-coupling is used between pre-amplifier stages to provide good low-frequency response and to prevent phase shift. Negative feedback is used within the pre-amplifier to provide stable gain,

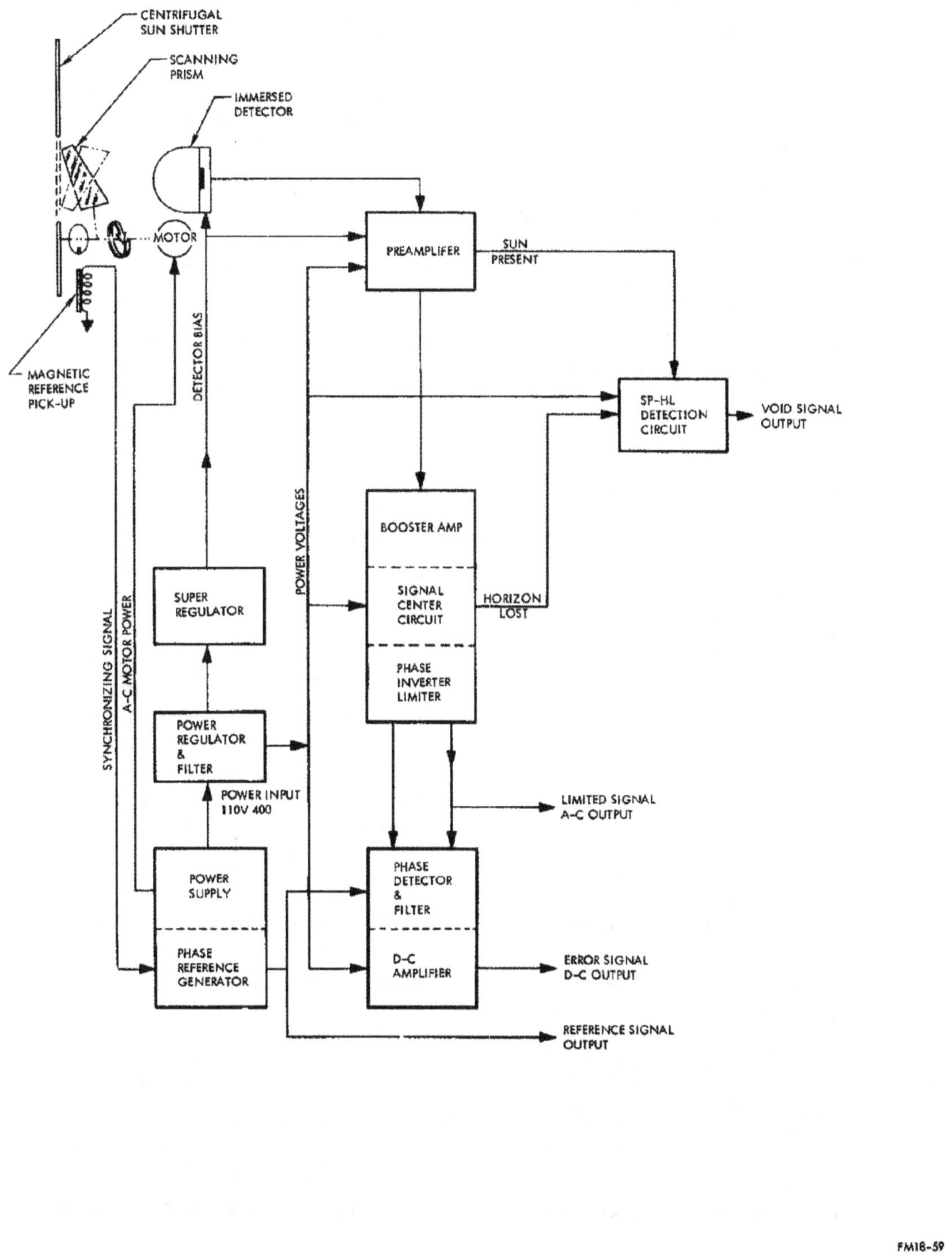

Figure 5-22 Horizon Scanner Block Diagram

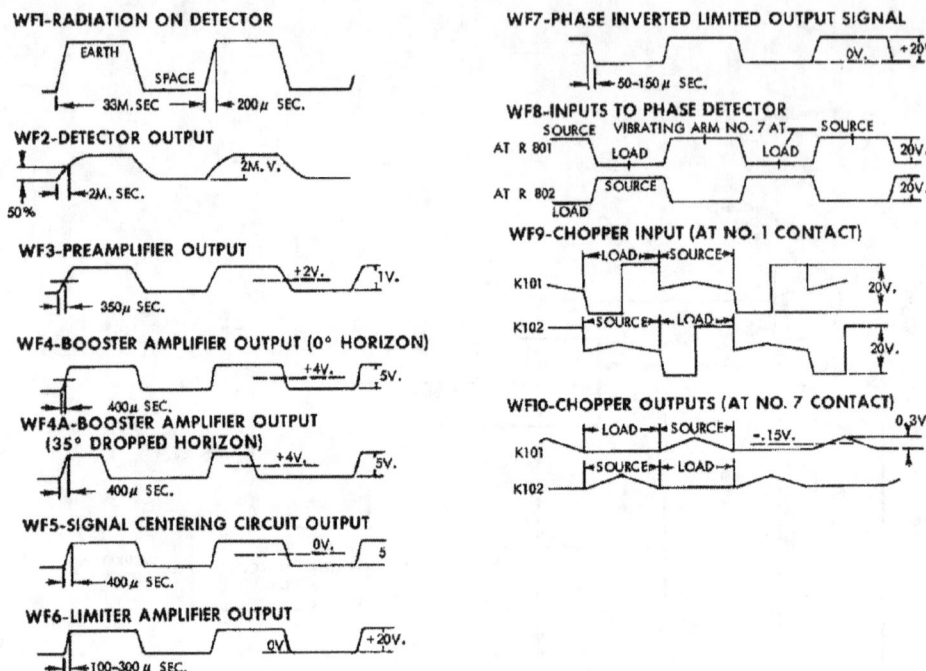

INTERNAL SIGNAL WAVEFORMS

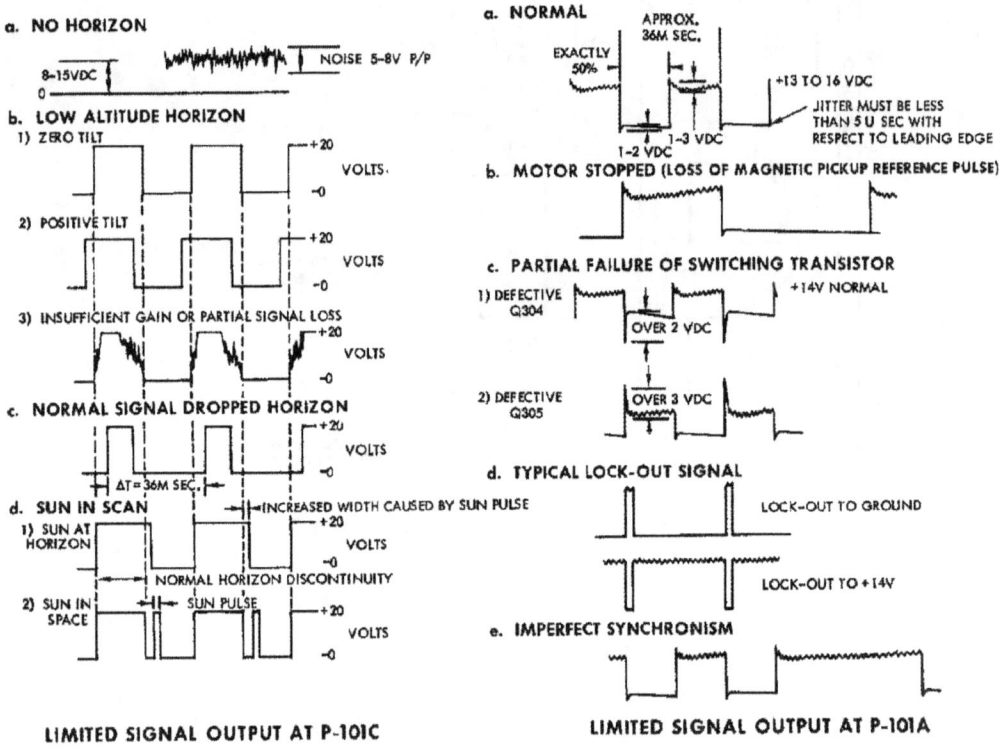

LIMITED SIGNAL OUTPUT AT P-101C LIMITED SIGNAL OUTPUT AT P-101A

Figure 5-23 Horizon Scanner Wave Forms

and the RC coupling network in the feedback loop provides a high-frequency boost to compensate for the long detector time constant. WF-3 shows the effect of this boost. The rise time of the waveform has been reduced to approximately 350 microseconds at the half-level point. The booster amplifier provides an additional voltage gain of 5 to 30 cycles per second. WF-4 shows the output of the booster amplifier. The peak-to-peak signal amplitude is in the order of 5 volts.

5-43. SIGNAL CENTERING CIRCUIT

In the signal processing considered previously there has been no particular interest in the voltage level of the average signal. However, this average signal level is important in system operation. The reason for this is that the error signal must be determined only by the phase angle between the horizon and the fixed reference in the vehicle, and amplitude variations in the signal should have no effect. Amplitude variations will take place because of changes in earth temperature at different parts of the trajectory or orbit. When these amplitude variations are combined with the rise-time characteristic of the detector there is a difference in phase between different portions of the leading edge and the fixed reference signal. Error signals would also be affected by amplitude changes due to changes in amplifier gain and supply voltage. Limiting can be used to eliminate the amplitude variations but the limiting slice must be taken at a point of minimum phase variation. These variations are greatest at the peaks of the wave and least at the center. Using an RC circuit to couple the signal to the limiter would balance equal areas of the signal waveform above and below ground.

Changes in the angle of horizon depression would cause a shift in the d-c level of the signal. Hence, a signal centering circuit is employed before the limiters to assure that the same center slice is sample for phase shift under all conditions. The signal centering circuit consists of two diodes connected back-to-back as d-c restorers. The diodes conduct on opposite peaks and thus permit the associated capacitors to charge up to opposite peak values of the signal. The two levels are then summed in a resistive divider network, and the sum is sampled by tapping the divider. An emitter follower couples this signal which is shown in WF-5 to the limiter circuit.

5-44. LIMITING AND PHASE SPLITTING CIRCUITS

The signal next enters the first of a pair of cascaded feedback amplifiers, each of which acts as a limiter and phase inverter. The amplifier consists of a grounded emitter stage which performs the phase inversion, and an emitter follower. The feedback ratio is about 50:1 and the overall gain of the section is about 30. The output swing is 10 volts each side of the fixed 10-volt level. A low output impedance is maintained during the time when the emitter follower is in cutoff by feeding the signal from the collector of the grounded emitter stage directly through a shunt diode. The first section of the feedback amplifier is fed by the signal centering circuit. Its output signal is the "limited signal output" and shown in WF-6. Part of this signal is fed to another limited amplifier substantially the same as the first, where it undergoes a second inversion to become a mirror image of the output of the first section. The output of the limiter section is thus the dual signal shown in WF-8. While either signal carries

SEDR 104

the signal information, the presence of the image signal will be found useful in cancelling out undesirable ripple components in the rectified signal.

5-45. PHASE DETECTOR

A pair of symmetrical, limited signals enter the detector section (WF-8). From these the detector derives a d-c signal which is proportional to the phase difference between the reference pulse and the midpoint of the two horizon intercepts. The phase sensitive rectifier consists of two SPDT polarized relays, or choppers, driven in phase opposition by the reference signal. These are designated K101 and K102 (WF9). The use of two choppers provides the advantages of full wave rectification, notably low ripple.

To understand the action of the synchronous rectifier, it is essential to know the relative phasing of the drive and horizon signals. Since the two choppers are driven 180° out of phase, the arm of one connects its capacitor to the source while the arm of the other connects its capacitor to the load. Switch-over takes place when the radial sector of the scanning beam crosses the vertical reference mark of the sensor and switching back occurs 180° later. Thus, when it is connected to the source, the capacitor receives part of the sky pulse and part of the earth pulse. The capacitor is charged positively during the switched in portion of the earth pulse and negatively during the switched in portion of the sky pulse. If the sensor horizontal is parallel to the horizon, each capacitor is negatively charged an amount equal to the positive charge. Therefore, the net charge is zero.

If the sensor tilts with respect to the horizon the amount that each capacitor charges positively is not equal to the amount it charges negatively. The net charge is, therefore, no longer zero. The net signal at the input of the d-c amplifier is thus positive for positive tilt of the sensor (cw as viewed from the sensor along the scan axis) and negative for a tilt in the opposite direction. WF-9 and WF-10 indicate respectively the voltage of each chopper and the uncombined output.

5-46. D-C AMPLIFIER

The output of the phase detector is combined and filtered in an R-C network at the input of the d-c amplifier. The signal at this point varies approximately 100 mv per degree of tilt and the average level is - 0.15 volts. The amplifier input is at high impedance to maintain a low ripple factor. With a gain of approximately three, the output of the amplifier is 286 mv per degree tilt of the sensor, reversing polarity at zero tilt. Part of the output is fed back to the emitter of the input stage. The balanced circuit configuration minimizes the output drift with temperature fluctuations.

5-47. SP-HL DETECTION CIRCUIT

There are two conditions under which unwanted error signals are generated, namely, when the sun appears in the scan and when the horizon is lost. Signals produced under these conditions trigger a logic circuit which indicates by its output that the sun is present or the horizon is lost (hence the designation SP-HL circuit). This output can be used to disconnect the d-c error output from the vehicle guidance system. The

SEDR 104

effect of sun presence is shown graphically in the waveforms of Figure 5-23. The sun pulse introduces an unsymmetrical element into the signal train, and the horizon information derived from it is likely to be false. The presence of the sun's radiation is perceived at the detector. The sun radiation is hundreds of times greater than that of the earth. The stars and other bodies produce negligible signals. When a sun pulse occurs, the second stage of the pre-amplifier puts out a negative pulse with a peak amplitude of three to four volts. This pulse causes the Void Signal Circuit to produce an output. When a horizon is present in a normal scan a signal of 5 of 6 volts from the signal centering circuit suffices to keep the void circuit amplifier shutoff. The absence of the signal when the horizon is lost has the same effect as a sun pulse -- it causes the void amplifier to conduct with a consequent output current of 4 ma into a load of 2000 ohms or less.

5-48. PHASE REFERENCE SIGNAL CIRCUIT

A phase reference signal is produced by the scanning system whenever it passes through its highest point with respect to the Sensor. The reference signal is generated in the scanning system. It consists of two equally spaced pulses, one positive and one negative, for each revolution of the scanning system. These pulses trigger the bistable multivibrator. The two-level detector section is in synchronism with the scan cycle. The output of the reference generator under various operating conditions is shown in Figure 5-23.

5-49. **POWER SUPPLY**

All the power required to operate the sensor is derived from the 110 volt, 400 cps line by a built-in power supply. Input to the supply is through the transformer. The primary of this transformer is tapped like an auto-transformer to provide low voltage a-c to operate the scanning motor. The transformer secondary output is full wave rectified to produce, -30, +30 and +16 volts d-c with respect to ground. The +30 and -30 volt outputs are fed to the Regulator. The +16 volts supplies the reference generator, and void signal circuits. Part of the transformer secondary voltage is rectified separately to produce unregulated power for use in the reference generator and the void signal output current amplifier.

5-50. **POWER REGULATOR AND SUPER REGULATOR CIRCUITS**

The Regulator circuits convert the outputs of the power supply into regulated voltages for use in the sensor. Most of the voltages are regulated by cascaded zener diodes which maintain a substantially constant voltage across their terminals by an effect similar to break-down in a gas discharge regulator. The regulator also contains a symmetrical arrangement of transistors connected as emitter followers. Since the base potential of each transistor is fixed by zener action, the output voltage is accordingly regulated with reasonably low noise. This output is filtered and further regulated in the Super Regulator circuit to provide the highly regulated voltage required by the detector and pre-amplifier. This voltage is extremely stable and its noise content is essentially transistor noise. The zener diodes used in these circuits are 1/4 watt units which regulate within 5%.

SECTION VI

SEQUENCE SYSTEM, LAUNCH RETROGRADE OR ABORT

TABLE OF CONTENTS

TITLE	PAGE
Normal Mission Sequence	6-3
Escape System	6-14

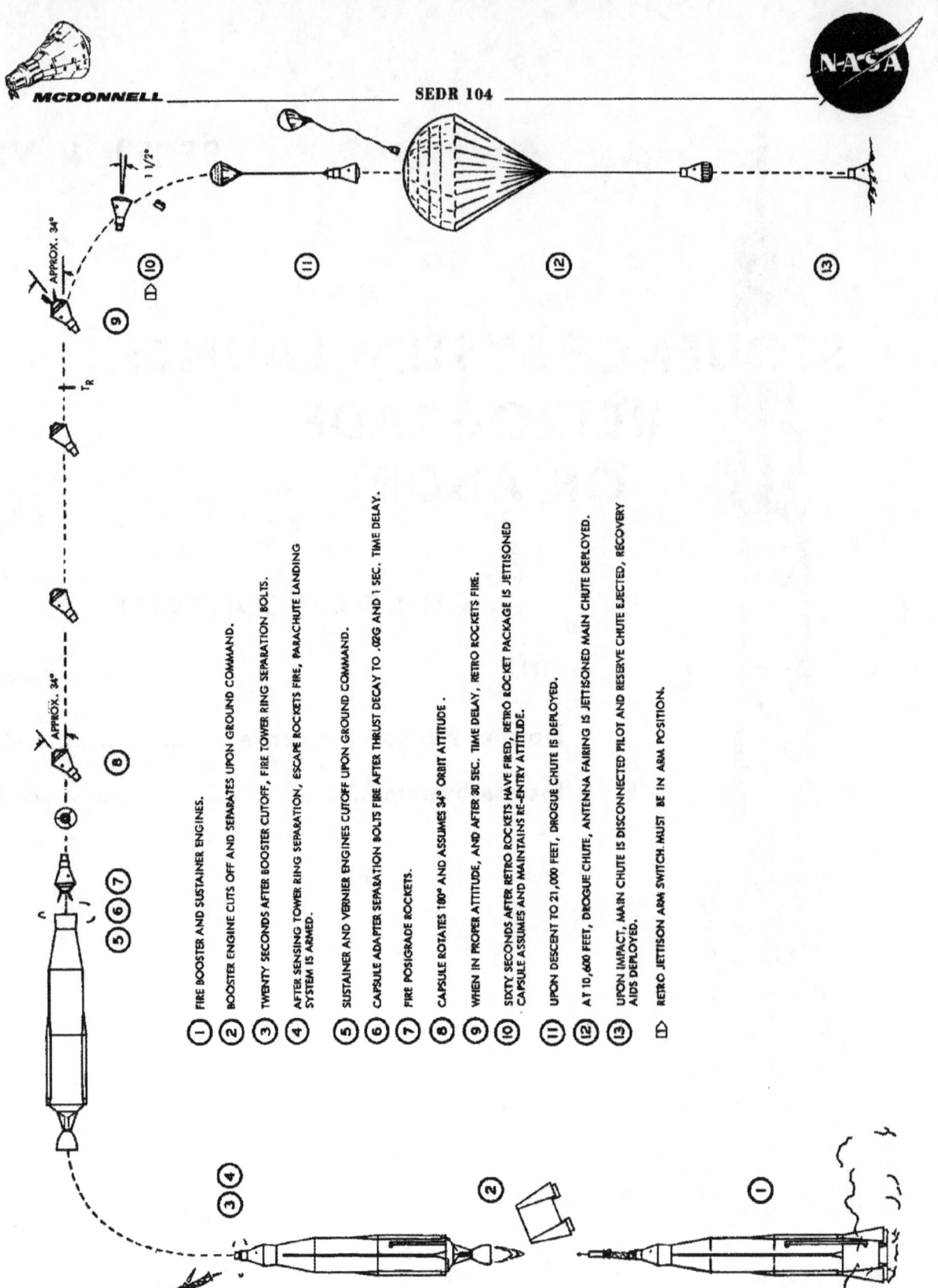

Figure 6-1 Normal Mission Sequence

SEDR 104

VI. SEQUENCE SYSTEM, LAUNCH THROUGH RETROGRADE OR ABORT

6-1. NORMAL MISSION SEQUENCE

6-2. LAUNCH THROUGH STAGING

6-3. DESCRIPTION

The launch through staging sequence establishes basic references at time of launch and then remains inactive until staging. At staging, the missiles's booster engine separates, resulting in the escape tower bolts being fired after a twenty second time delay. The escape rockets are fired immediately after tower bolt detonation and subsequently the landing system becomes armed.

6-4. OPERATION

The sequence system is initiated by two 28 V d-c signals from the missile which occur at 2 inches after liftoff (See Figure 6-2). This is known as time zero reference and energizes a Time Zero Latching Relay in the No. 3 Launch and Orbit Relay Box located within the spacecraft. An astronaut controlled back-up switch is provided in the event the 28 V signals from the missile do not reach the spacecraft. These same signals are also sent to the Maximum Altitude Sensor and the Satellite Clock. The signal to the Maximum Altitude Sensor results in establishing the function of time liftoff versus the time an abort may occur. At approximately 135 seconds missile staging will occur whereby the mechanical separation of the booster engine will cause the loss of spacecraft power to the Booster Engine Separation Sensor Relay. Through this de-energized relay, power will be applied to the Tower Jettison 20 Second Time Delay

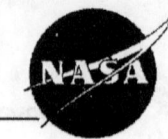

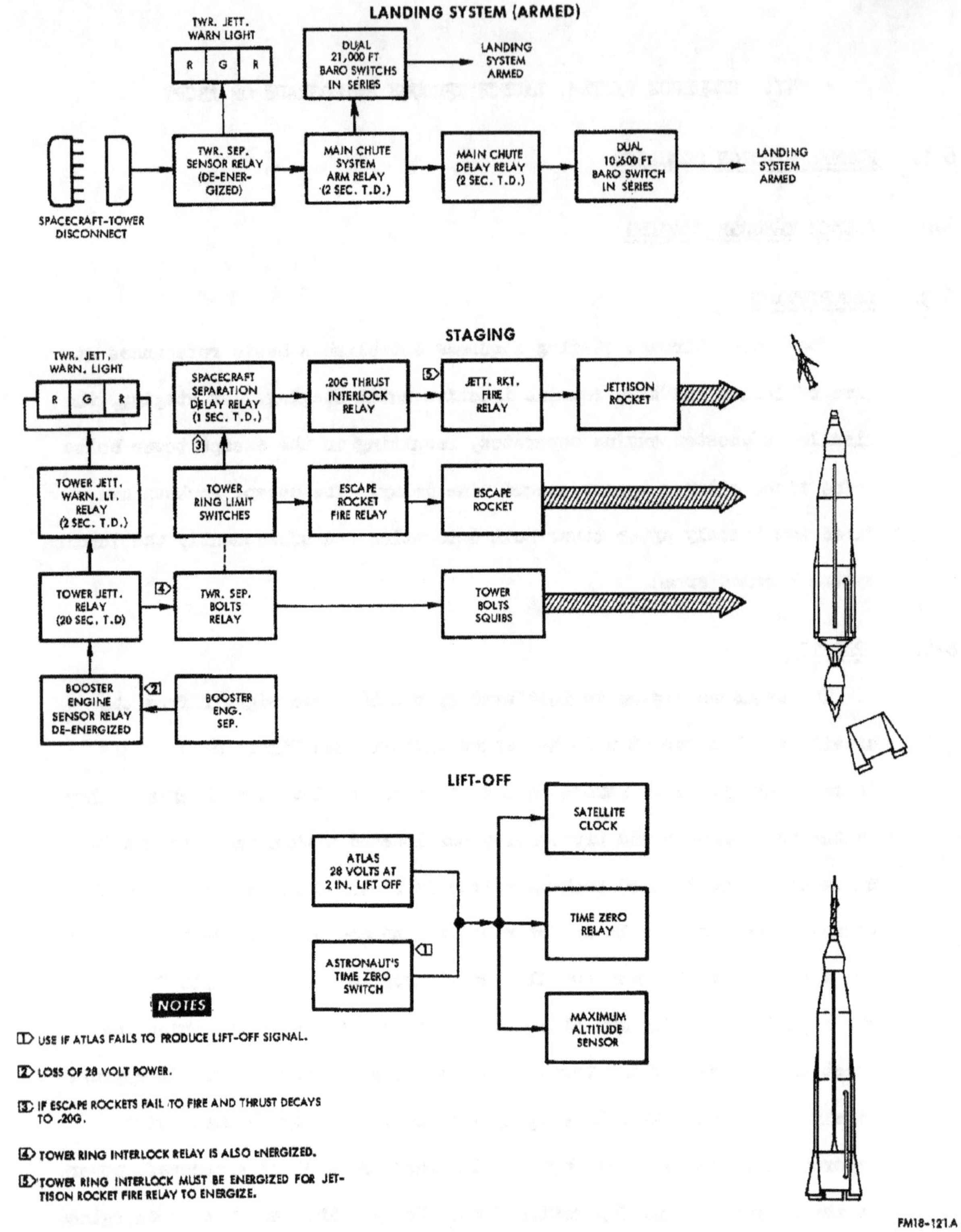

LANDING SYSTEM (ARMED)
TWR. JETT. WARN LIGHT
R G R
DUAL 21,000 FT BARO SWITCHS IN SERIES
LANDING SYSTEM ARMED
SPACECRAFT-TOWER DISCONNECT
TWR. SEP. SENSOR RELAY (DE-ENERGIZED)
MAIN CHUTE SYSTEM ARM RELAY (2 SEC. T.D.)
MAIN CHUTE DELAY RELAY (2 SEC. T.D.)
DUAL 10,600 FT BARO SWITCH IN SERIES
LANDING SYSTEM ARMED
STAGING
TWR. JETT. WARN. LIGHT
R G R
SPACECRAFT SEPARATION DELAY RELAY (1 SEC. T.D.)
.20G THRUST INTERLOCK RELAY
5
JETT. RKT. FIRE RELAY
JETTISON ROCKET
3
TOWER JETT. WARN. LT. RELAY (2 SEC. T.D.)
TOWER RING LIMIT SWITCHES
ESCAPE ROCKET FIRE RELAY
ESCAPE ROCKET
TOWER JETT. RELAY (20 SEC. T.D)
4
TWR. SEP. BOLTS RELAY
TOWER BOLTS SQUIBS
BOOSTER ENGINE SENSOR RELAY DE-ENERGIZED
2
BOOSTER ENG. SEP.
LIFT-OFF
ATLAS 28 VOLTS AT 2 IN. LIFT OFF
1
ASTRONAUT'S TIME ZERO SWITCH
SATELLITE CLOCK
TIME ZERO RELAY
MAXIMUM ALTITUDE SENSOR
NOTES
1 USE IF ATLAS FAILS TO PRODUCE LIFT-OFF SIGNAL.
2 LOSS OF 28 VOLT POWER.
3 IF ESCAPE ROCKETS FAIL TO FIRE AND THRUST DECAYS TO .20G.
4 TOWER RING INTERLOCK RELAY IS ALSO ENERGIZED.
5 TOWER RING INTERLOCK MUST BE ENERGIZED FOR JETTISON ROCKET FIRE RELAY TO ENERGIZE.
FM18-121A

Figure 6-2 Launch Through Staging Block Diagram

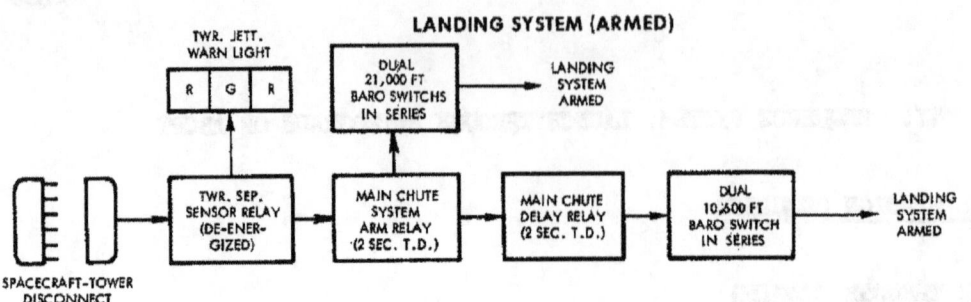

Relay. After 20 seconds, power will be applied through the contacts of the Tower Jettison 20 Second Time Delay Relay to energize the Tower Separation Bolts Power Relay and the Tower Jettison Warning Light Relay (2 sec. T.D.). The Tower Separation Bolts Power Relay is armed by both the Main and Isolated DC Squib Bus through the Squib Arm Switch. When energized, the Isolated Squib Bus power fires two of the five squibs (2 bolts) and Main Squib Bus power fires three of the five squibs (3 bolts. The Tower Ring Interlock Relay will also be energized when the Tower Separation Bolts Relay is energized. As the three segmented Tower Clamp Ring separates, the three Tower Clamp Ring Limit Switches return to the normal position and allow Isolated and Main Squib Bus power through their contacts. The Isolated Squib Bus power energizes both the Emergency Escape Rocket Fire Relay and the Emergency Jettison Rocket Fire Relay, while the Main Squib Bus power energizes both the Escape Rocket Fire Relay and the Jettison Rocket Fire Relay. As the contacts of the Emergency Jettison and Jettison Rocket Fire Relays are connected in parallel, either relay will fire all squibs of the Jettison Rocket from the two different power sources. The Emergency Escape and Escape Rocket Fire Relays are connected in the identical same manner and will fire all squibs of the Escape Rocket from both power sources.
Power to energize the Emergency Jettison and Jettison Rocket Fire Relay is routed through the .2g Thrust Relay which is energized through the Spacecraft 1 Second Time Delay and Thrust Cutoff Sensor. The .2g sensor plus the 1 Second Time Delay Relay allows the Tower Clamp Ring to separate and the Escape Rockets to fire separating the tower from the spacecraft with the Jettison Rocket unfired. When this is accomplished, two electrical disconnects between the tower and the spacecraft are separated and remove power from the

three Tower Separation Sensor Relays. Through the de-energized No. 1 Tower Separation Sensor Relay, the green JETT TOWER light on the tele-light panel illuminates. When the tower and the spacecraft separate, the No. 1 and No. 2 Tower Separation Relays are de-energized, allowing power to energize the No. 1 and No. 2 Main Chute System Arm 2 Second Time Delay Relays. After 2 seconds delay, the Main Chute Relays arm, the 21,000 Foot Baroswitches and the Main Chute Delay 2 Second Time Delay Relays. After two seconds, the 10,600 Foot Baroswitches are armed. The power circuit will hold at these two points until the spacecraft descends down through the 21,000 foot range, at which time the landing sequence is initiated. Refer to Section VII of this manual.

6-5. SECOND STAGING

6-6. DESCRIPTION

Second staging is initiated by sustainer engine cutoff at which time the Spacecraft Adapter Bolts are fired providing acceleration has decayed to .20g. The three Posigrade Rockets and the four explosive electrical disconnects are fired immediately after the bolts are detonated and result in spacecraft separation. Spacecraft separation is sensed and initiates five seconds of rate damping which is followed by orbit orientation in which the spacecraft rotates 180° degrees and settles into a 34° orbit attitude.

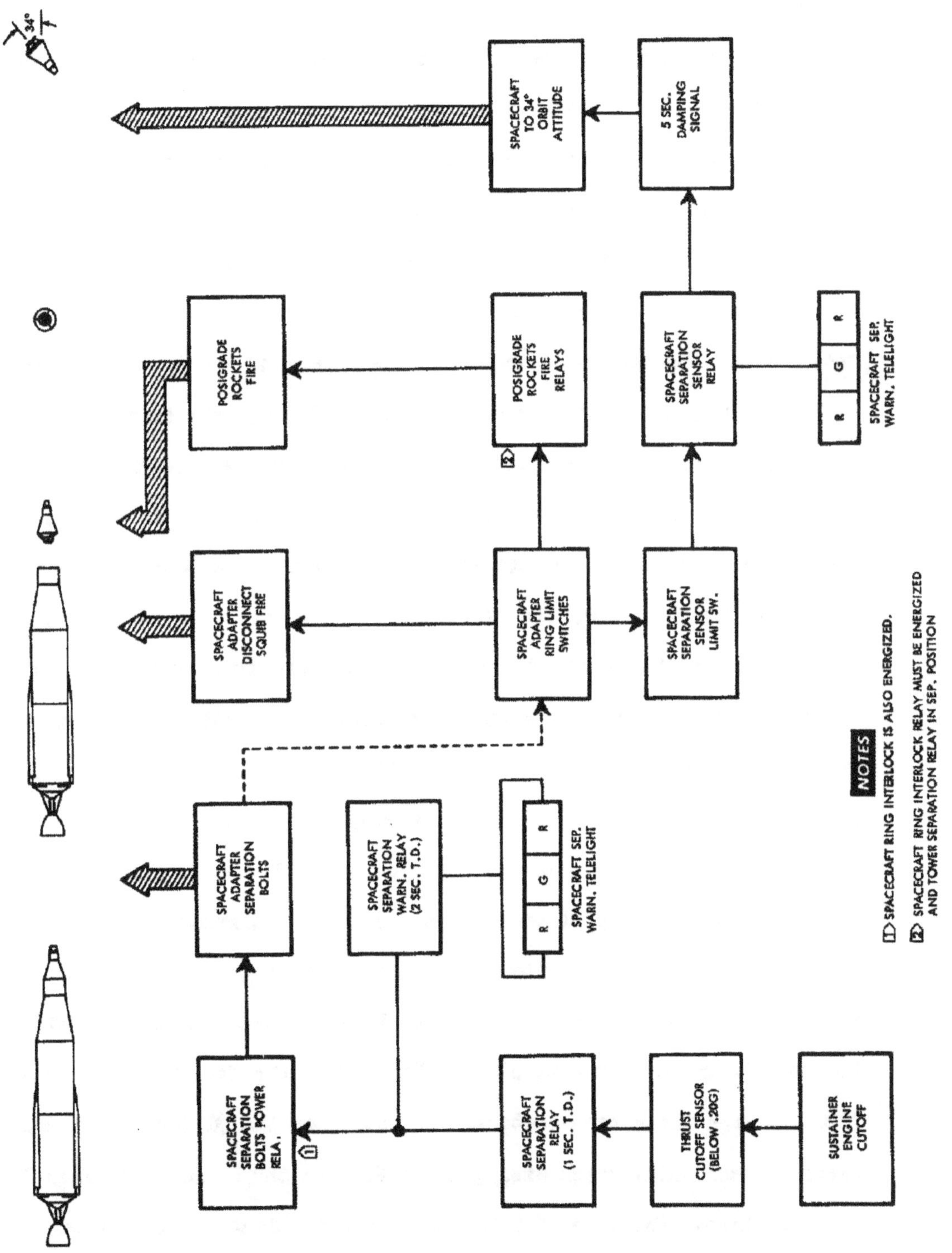

SEDR 104

Figure 6-3 Second Staging Block Diagram

6-7. OPERATION

At approximately 285 seconds after launch, second staging will occur (See Figure 6-3). At this time a 28 volt d-c signal from the missile will energize the Sustainer Engine Cutoff Relay. When the thrust drops below .20g, the .20g Switch in the Thrust Cutoff Sensor closes. Power is then supplied through the Spacecraft Separation 1 Second Time Delay and Sustainer Engine Cutoff Relays after the 1 Second Time Delay to energize the Spacecraft Separation Bolts Power and the Spacecraft Separation Warning 2 Second Time Delay Relays. Through the energized contacts, power from the Main and Isolated Bus fire the Spacecraft Separation Bolts. Also, Main Bus power is supplied through the Warning Light Time Delay Relay to illuminate the red SEP CAPSULE Sequence Light on the Left Hand Console. When the tri-segmented Spacecraft-to-Adapter Clamp Ring separates, it allows the Spacecraft Adapter Ring Limit Switch to close supplying power to energize the Posigrade Rocket Fire, Emergency Posigrade Rocket Fire and the Spacecraft Adapter Disconnect Squib Fire Relays. The Main and Isolated Busses supply power through the energized contacts of these relays to fire the Posigrade Rockets and the four Spacecraft Adapter Explosive Disconnects. The Posigrade Rockets create sufficient thrust to separate the spacecraft from the adapter. This allows the three Spacecraft Separation Limit Switches which are attached to the retrograde straps to close. Power from the Isolated Squib Bus flows through the closed contacts and activates the #1 Spacecraft Separation Sensor Relay which extinguishes the red SEP CAPSULE Sequence Light and illuminates the green SEP CAPSULE Sequence Light. Power through the

SEDR 104

#1 Spacecraft Separation Sensor Relay is applied to the Five Second Time Delay Damping Signal Relay. Activation of the Damping Signal Relay actuates the Orbit Orientation Relay bringing the spacecraft to a 34° (blunt end up) orbit attitude.

6-8. RE-ENTRY

6-9. DESCRIPTION

In order for the spacecraft to impact at a designated area, the re-entry sequence must be initiated approximately 3000 nautical miles up range of the touchdown point. The method of initiating normal re-entry sequence is by the closing of the Retrograde Firing Signal Switch within the Satellite Clock. The switch may be activated by the run-out of time pre-set into the clock prior to launch for a calculated re-entry time. Timing starts at booster liftoff. The time may also be pre-set by the astronaut or by ground command when necessary. The sequence may be started directly by using ground command transmitters and the Spacecraft Command Receiver. The final method is for the astronaut to manually start the sequence by pressing the Retro Sequence Button. The last two methods by-pass the Satellite Clock. A brief resume of the sequence starts with the closing of the Retrograde Firing Signal Switch, which energizes the Retro Sequence Fire 30 Second Time Delay Relay. After the spacecraft has attained the proper attitude and the time delay has run out, the three Retro Rockets will fire 5 seconds apart. When the Retro Rockets fire and the Auto Retro Jettison Switch is in the "ARM" position, the Retro Rocket Assembly Jettison 60 Second Time Delay Relay is energized and at the run-out of the 60 second time delay the Retro Package is jettisoned. The Retro Package separation is sensed and results in the separation of the three Retro Package Electrical Umbilicals.

6-10. OPERATION

The Satellite Clock Retro Fire Switch is armed by Main Squib Bus power through the Spacecraft Separation Relay contacts (See Figure 6-4). With the Retro Delay Switch in the "NORM" position and the Retro Fire Switch closed by any of the three previously mentioned methods, the Retro Rocket Sequence Indicator Relay is activated illuminating the green RETRO SEQ Indicator Light. With the Retro Delay Switch in the "INST" position, power is supplied directly to the contact of the Attitude Permission Relay No. 1 by-passing the 30 second delay. The astronaut may manually start the retro sequence by pressing the Retro Sequence Switch on the left hand console which will energize the Emergency Retro Sequence Relay and allow the normal sequence to be followed. The Retro Interlock Switch in the ASCS Calibrator closes allowing power to flow through the Emergency Retro Fire No. 1 Relay and energize the No. 1 and No. 2 Attitude Permission Relays when the spacecraft is in the proper position for retro rocket firing. The red RETRO ATT Telelight is switched on when the Retro Rocket Sequence Indicator Relay energizes at the start of the Retro Sequence.

Normally, the Attitude Permission Relay extinguishes the red RETRO ATT Telelight, illuminating the green light and energizing the Retro Signal Latch and the Retro Rocket Fire Relays. Power from the Isolated Squib Bus is now routed through the Retro Rocket Fire Relays to the

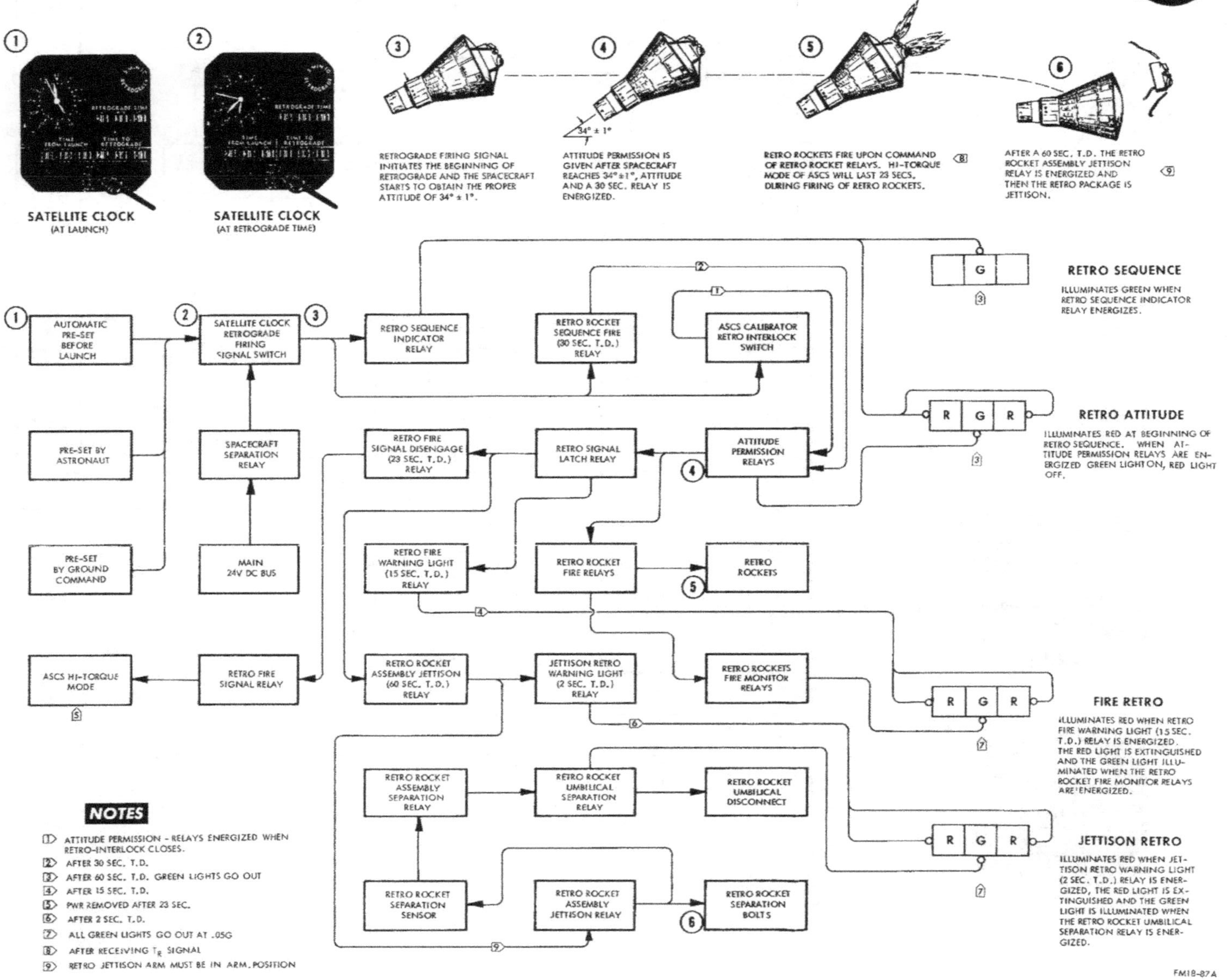

Figure 6-4 Normal Re-entry Sequence

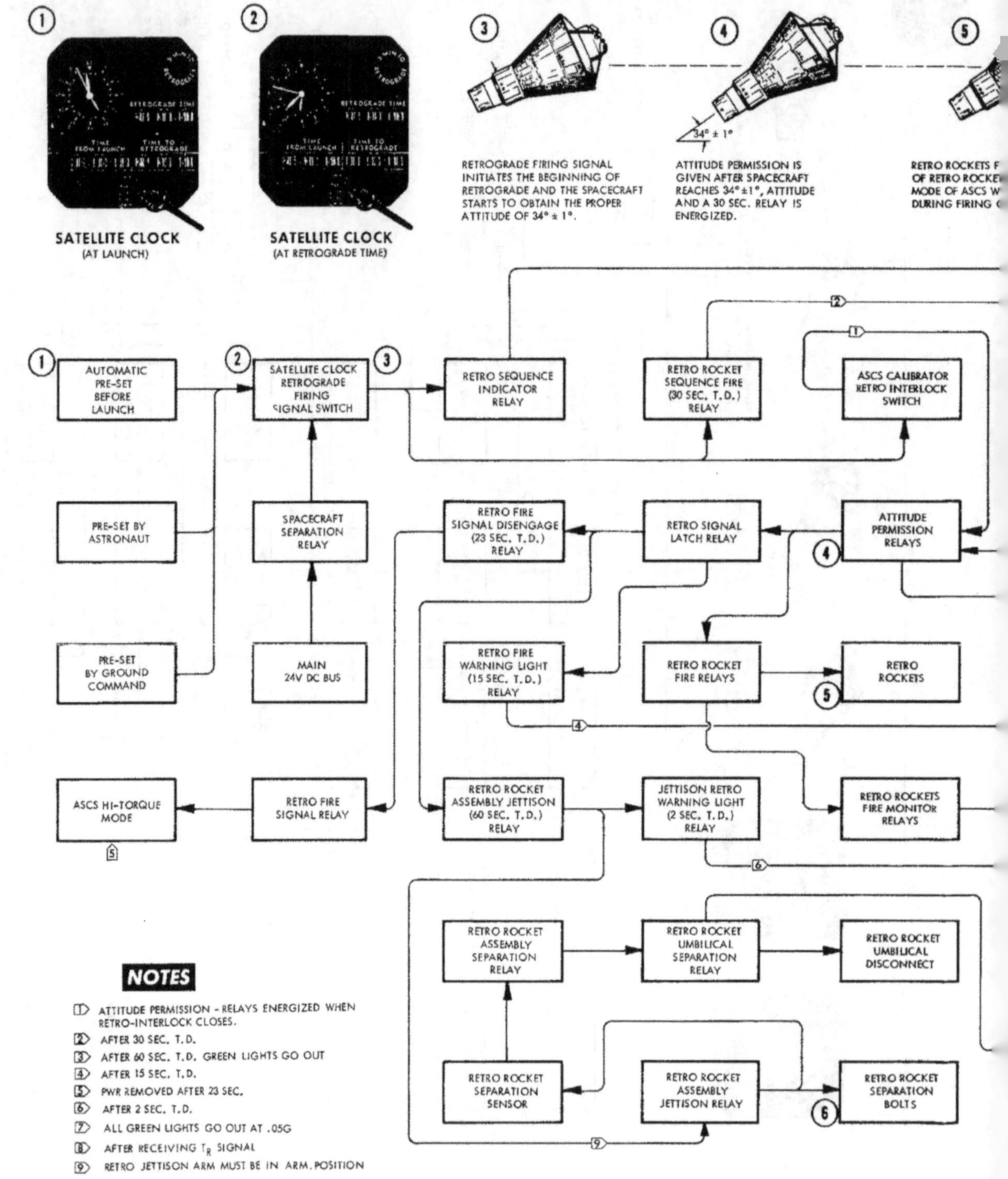

① **SATELLITE CLOCK** (AT LAUNCH)

② **SATELLITE CLOCK** (AT RETROGRADE TIME)

③ RETROGRADE FIRING SIGNAL INITIATES THE BEGINNING OF RETROGRADE AND THE SPACECRAFT STARTS TO OBTAIN THE PROPER ATTITUDE OF 34° ± 1°.

④ ATTITUDE PERMISSION IS GIVEN AFTER SPACECRAFT REACHES 34° ± 1°, ATTITUDE AND A 30 SEC. RELAY IS ENERGIZED.

⑤ RETRO ROCKETS F OF RETRO ROCKE MODE OF ASCS W DURING FIRING C

NOTES

1. ATTITUDE PERMISSION - RELAYS ENERGIZED WHEN RETRO-INTERLOCK CLOSES.
2. AFTER 30 SEC. T.D.
3. AFTER 60 SEC. T.D. GREEN LIGHTS GO OUT
4. AFTER 15 SEC. T.D.
5. PWR REMOVED AFTER 23 SEC.
6. AFTER 2 SEC. T.D.
7. ALL GREEN LIGHTS GO OUT AT .05G
8. AFTER RECEIVING T_R SIGNAL
9. RETRO JETTISON ARM MUST BE IN ARM. POSITION

Figure 6-4 Normal Re-Entry Sequence

SEDR 104

④ ATTITUDE PERMISSION IS GIVEN AFTER SPACECRAFT REACHES 34°±1°, ATTITUDE AND A 30 SEC. RELAY IS ENERGIZED.

⑤ RETRO ROCKETS FIRE UPON COMMAND OF RETRO ROCKET RELAYS. HI-TORQUE MODE OF ASCS WILL LAST 23 SECS. DURING FIRING OF RETRO ROCKETS.

⑥ AFTER A 60 SEC. T.D. THE RETRO ROCKET ASSEMBLY JETTISON RELAY IS ENERGIZED AND THEN THE RETRO PACKAGE IS JETTISON.

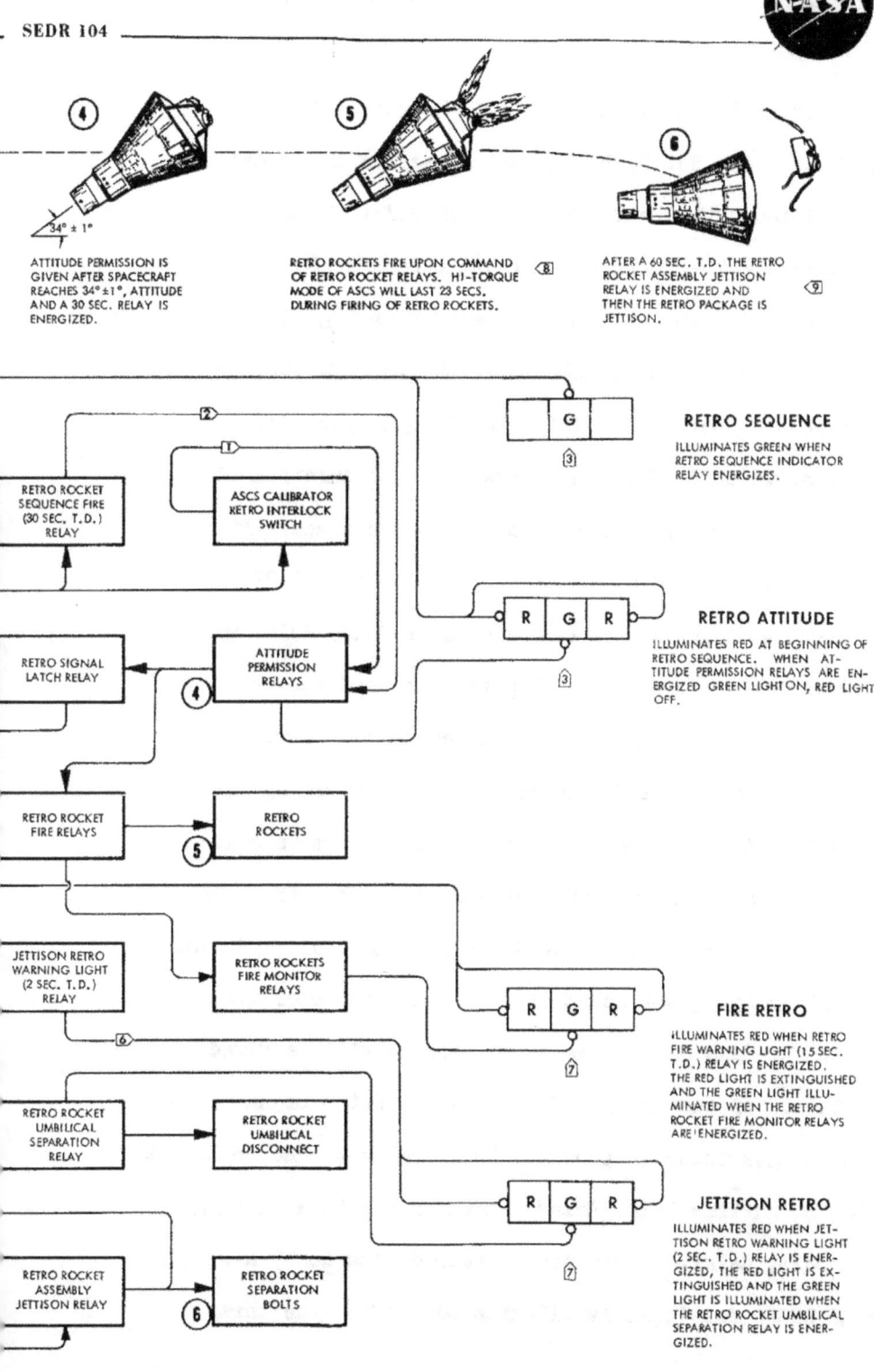

RETRO SEQUENCE
ILLUMINATES GREEN WHEN RETRO SEQUENCE INDICATOR RELAY ENERGIZES.

RETRO ATTITUDE
ILLUMINATES RED AT BEGINNING OF RETRO SEQUENCE. WHEN ATTITUDE PERMISSION RELAYS ARE ENERGIZED GREEN LIGHT ON, RED LIGHT OFF.

FIRE RETRO
ILLUMINATES RED WHEN RETRO FIRE WARNING LIGHT (1.5 SEC. T.D.) RELAY IS ENERGIZED. THE RED LIGHT IS EXTINGUISHED AND THE GREEN LIGHT ILLUMINATED WHEN THE RETRO ROCKET FIRE MONITOR RELAYS ARE ENERGIZED.

JETTISON RETRO
ILLUMINATES RED WHEN JETTISON RETRO WARNING LIGHT (2 SEC. T.D.) RELAY IS ENERGIZED, THE RED LIGHT IS EXTINGUISHED AND THE GREEN LIGHT IS ILLUMINATED WHEN THE RETRO ROCKET UMBILICAL SEPARATION RELAY IS ENERGIZED.

FM18-87A

Retro Rockets in turn firing first the Left (No. 1) Rocket, after a 5 second time delay the Bottom (No. 2) Rocket and 5 seconds later the Right (No. 3) Rocket. Through the Retro Signal Latch Relay, a circuit will be completed to energize the Retro Fire Signal Disengage 23 Second Time Delay Relay. The power to the coil of the Retro Fire Signal Disengage Relay allows the circuit to be completed to the ASCS Calibrator resulting in high-torque RCS operation. This high-torque operation will last for 23 seconds, which is 3 seconds more than the duration of total retro rocket firing. The high-torque mode holds the spacecraft in the 34° attitude while the rockets are firing. At the end of 23 seconds, the Retro Fire Signal Disengage 23 Second Time Delay Relay will energize removing power from the Retro Fire Signal Relay and thus removing the high-torque signal. With the Attitude Switch in the "AUTO" position, the astronaut may press the Retro Fire Switch energizing the No. 1 Emergency Retro Fire Relay allowing Isolated Bus power to energize the No. 2 Emergency Retro Relay, firing the Retro Rockets when the spacecraft assumes the proper attitude. With the Retro Attitude Switch in the "BY-PASS" position, and pressing the Retro Fire Switch which energizes the Attitude Permission By-Pass Relay, the Retro Rockets will be fired regardless of the attitude of the spacecraft. When the Retro Signal Latch is energized, Main Squib Bus power is supplied to energize the Retro Rocket Assembly Jettison 60 Second Time Delay Relay and the Retro Fire Warning Light 15 Second Time Delay Relay. After the 15 second time delay has run out, the red FIRE RETRO Telelight is illuminated. After the three

SEDR 104

Retro Fire Monitor Relays are energized, the red FIRE RETRO Telelight goes out and the green light is illuminated. At the end of the 60 second time delay, the Retro Rocket Assembly Jettison 60 Second Time Delay Relay allows power to energize the Retro Rocket Assembly Jettison Relay and the Jettison Retro Warning Light 2 Second Time Delay Relay. As the 2 second time delay is expired, the red JETT RETRO Telelight is illuminated. The Retro Rocket Assembly Jettison Relay directs Main and Isolated Bus power to the two squibs of the Retro Rocket Assembly Jettison Bolt. The bolt will fracture and the package will drop free of the spacecraft, being assisted by a coil spring installed between the heat shield and Retro Package Assembly. The dropping of the Retro Package Assembly from the spacecraft will allow the three Retro Rocket Assembly Separation Sensors (single pole limit switches) to return to their normal position energizing the Retro Rocket Assembly Separation Relay. This will allow the Retro Rocket Assembly Umbilical Separation Relay to energize, firing the six squibs of the three Retro Rocket Package Umbilicals and jettison the electrical umbilical plugs milliseconds after dropping of the Retro Package. When the Retro Rocket Assembly Umbilical Separation Relay is activated, it removes power from the red JETT RETRO Telelight and illuminates the green light. When the Retro Rocket Assembly Separation Relay is energized, it activates the Accelerometer Arm 5 Second Time Delay Relay. At the end of the 5 second time delay, the relay functions supplying a ground for Main Bus power which operates the .05g Relay. The energizing of the No. 1 and No. 2 .05g Retro Sequence Drop, .05g Jettison Assembly Sequence Drop, and .05g Retro Telelight Power Drop through

the contacts of the .05g Relay removes power from the various relays and switches involved in the retro sequence as well as extinguishing the retro telelights.

6-11. ESCAPE SYSTEM

6-12. GENERAL DESCRIPTION

The escape system primarily consists of a tower assembly designed to provide a safe means of abort between pre-launch and staging. By utilizing the Posigrade Rocket System, escapes may still be initiated after booster staging and throughout sustainer operation until orbit. The tower assembly consists of a 10 foot, tubular steel structure with a 4 foot Escape Rocket mounted to its tapered end. A segmented clamp ring with 3 explosives bolts secures the base of the tower to the recovery compartment upper flange. Attached to the Escape Rocket Nozzle Adapter Plate is a Jettison Rocket which is used to jettison the tower assembly after the Escape Rocket has been fired for an abort; however, under normal launch conditions the Escape Rocket is fired to accomplish tower separation at time of booster engine separation.

6-13. ESCAPE BEFORE LIFTOFF BEFORE UMBILICAL DISCONNECT

Only one ground controlled signal will energize the Mayday Relays. This signal is a direct hardline from the block house Abort Switch to the Mayday Relays. The sequence (Ref. Figure 6-5) to energize the Mayday Relays is as follows: The Abort Switch is activated energizing the 50 Micro-Second Time Delay Relay. 28 volts is applied through the 50

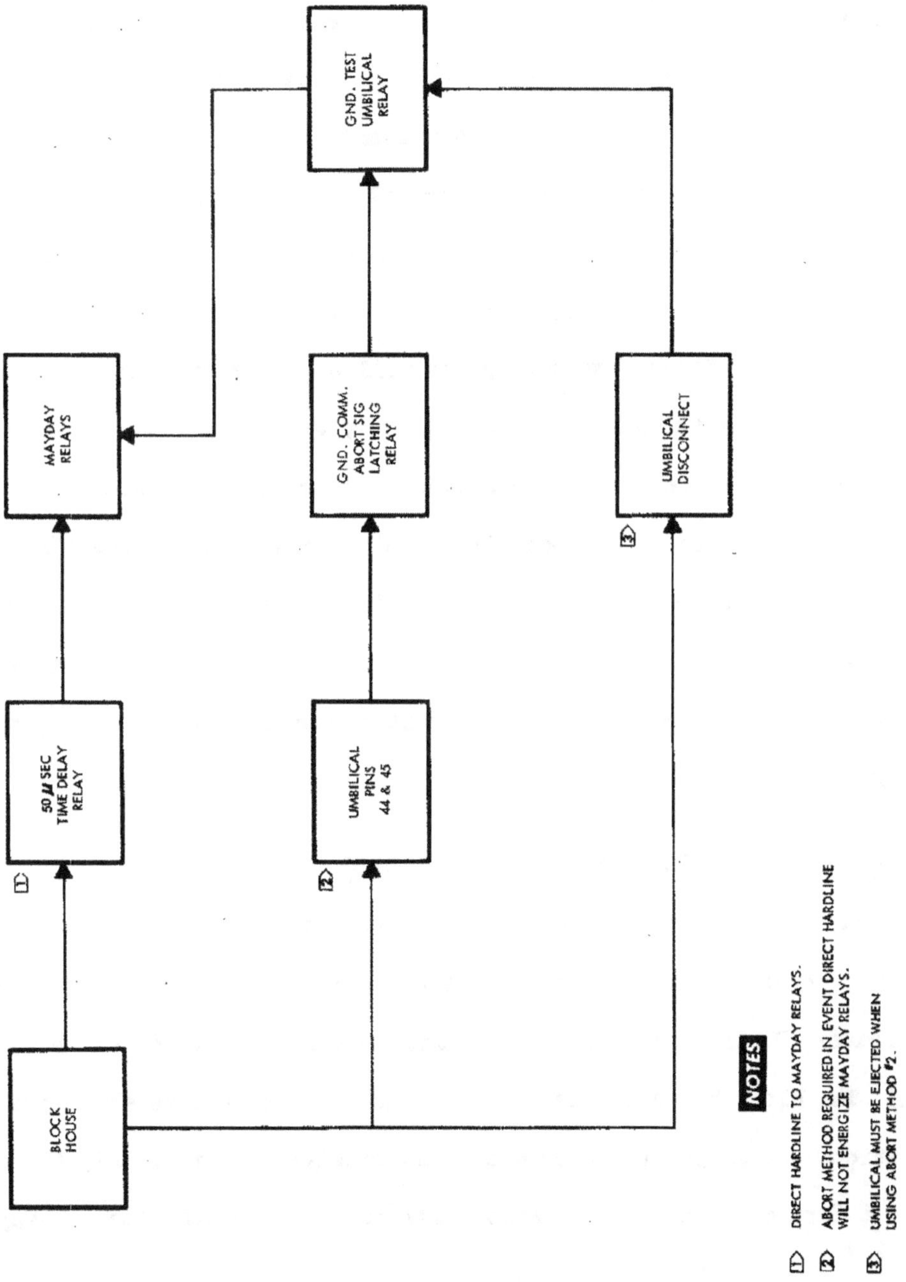

MCDONNELL — SEDR 104

Figure 6-5 Escape Before Liftoff Before Umbilical Disconnect

Micro-Second Time Delay Relay contacts and through the missile to the Mayday Relays. In the event the spacecraft must be aborted on the launch pad and the missile is unable to transmit the hardline abort signal, there is one other method which may be selected. Umbilical Pins 44 and 45 are abort wired and transmit 28 V power from the block house to the spacecraft's Ground Command Abort Signal Latching Relay energizing and locking in the relay. Through this energized relay spacecraft 28 V Squib Arm Bus power is transmitted to the pole of the Ground Test Umbilical Relay; however, power will not continue through this relay until the relay is de-energized. The only way the relay may be de-energized is by ejecting the umbilical. Therefore, if this abort method is required to be used, it would be necessary for the block house conducter to select the Abort Switch (power to pins 44 and 45) and within milliseconds thereafter the umbilical is ejected. After the Mayday Relays are energized, the escape sequence is the same as explained in paragraph 6-16.

6-14. ESCAPE BEFORE LIFTOFF AFTER UMBILICAL DISCONNECT

During countdown, there will be approximately 50 to 90 seconds between time of spacecrafts unbilical eject and time zero, which is two inches liftoff. During this period, the three available methods of abort are: (1) The block house to missile hardline abort signal as explained in the previous paragraph; (2) Ground command receiver abort signal; (3) Astronaut's Abort Handle. These three methods all result in energizing the Mayday Relays.

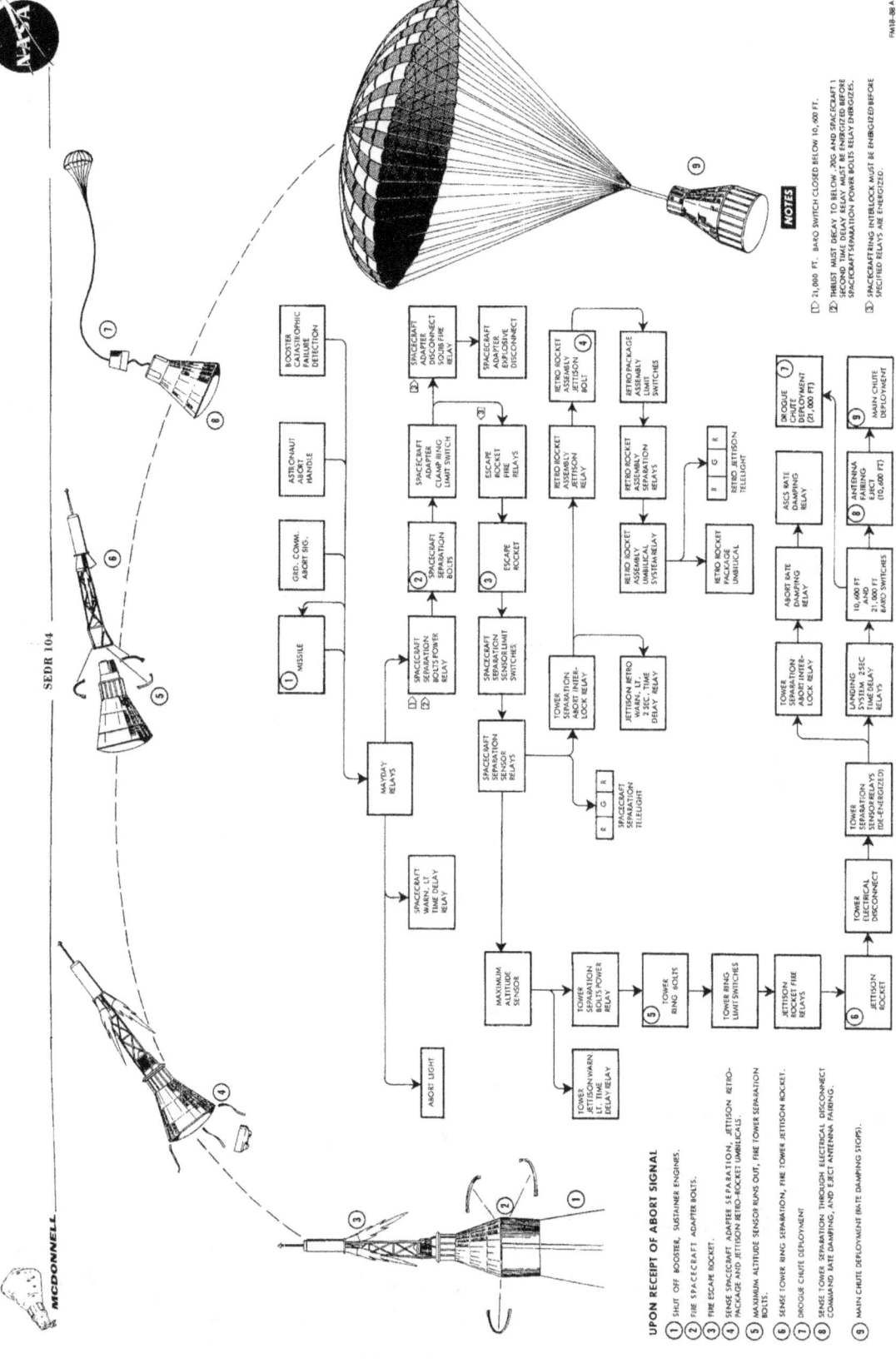

Figure 6-6 Escape Before Tower Separation

MCDONNELL — SEDR 104

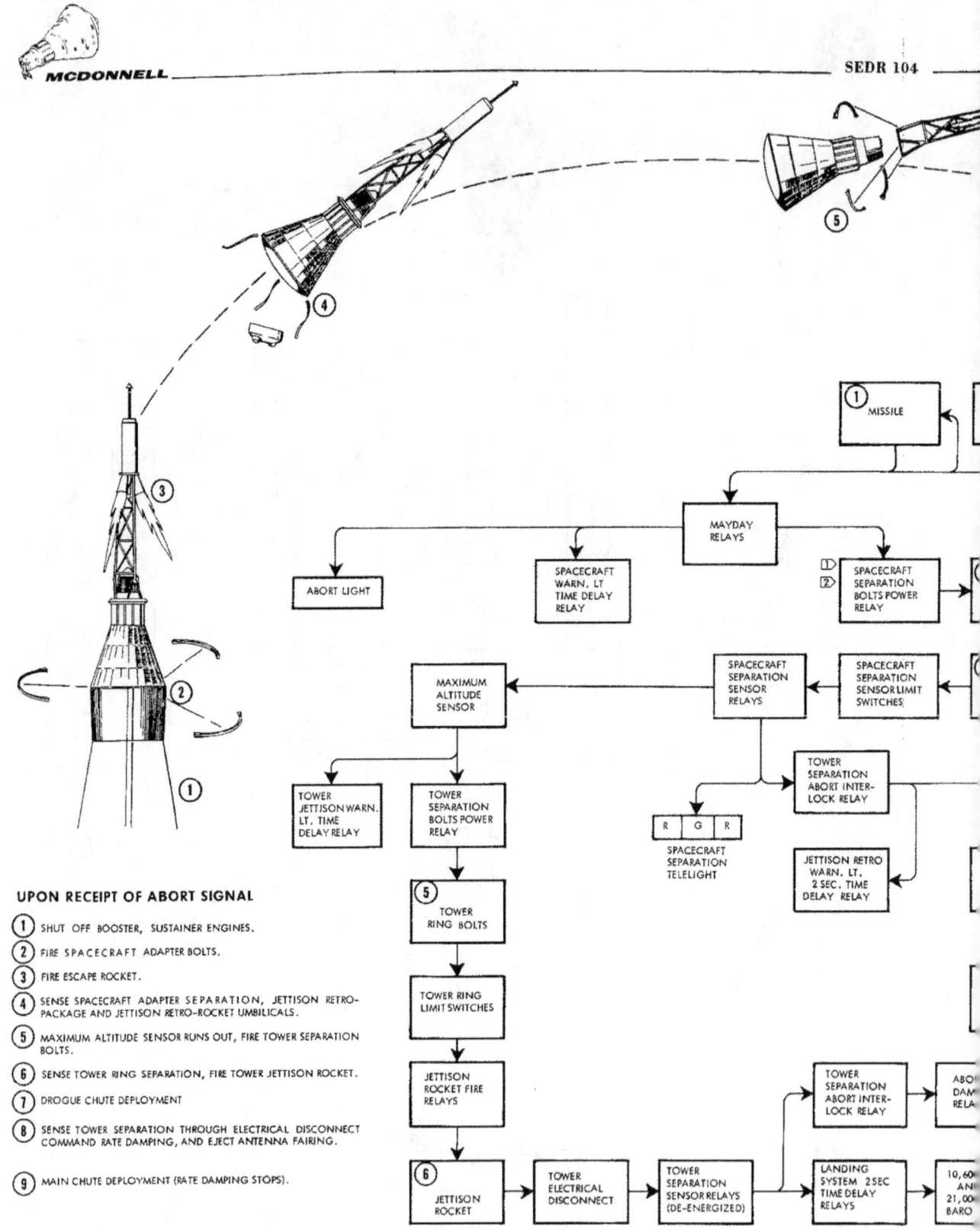

UPON RECEIPT OF ABORT SIGNAL

1. SHUT OFF BOOSTER, SUSTAINER ENGINES.
2. FIRE SPACECRAFT ADAPTER BOLTS.
3. FIRE ESCAPE ROCKET.
4. SENSE SPACECRAFT ADAPTER SEPARATION, JETTISON RETROPACKAGE AND JETTISON RETRO-ROCKET UMBILICALS.
5. MAXIMUM ALTITUDE SENSOR RUNS OUT, FIRE TOWER SEPARATION BOLTS.
6. SENSE TOWER RING SEPARATION, FIRE TOWER JETTISON ROCKET.
7. DROGUE CHUTE DEPLOYMENT
8. SENSE TOWER SEPARATION THROUGH ELECTRICAL DISCONNECT COMMAND RATE DAMPING, AND EJECT ANTENNA FAIRING.
9. MAIN CHUTE DEPLOYMENT (RATE DAMPING STOPS).

Figure 6-6 Escape Before Tower Separation

SEDR 104

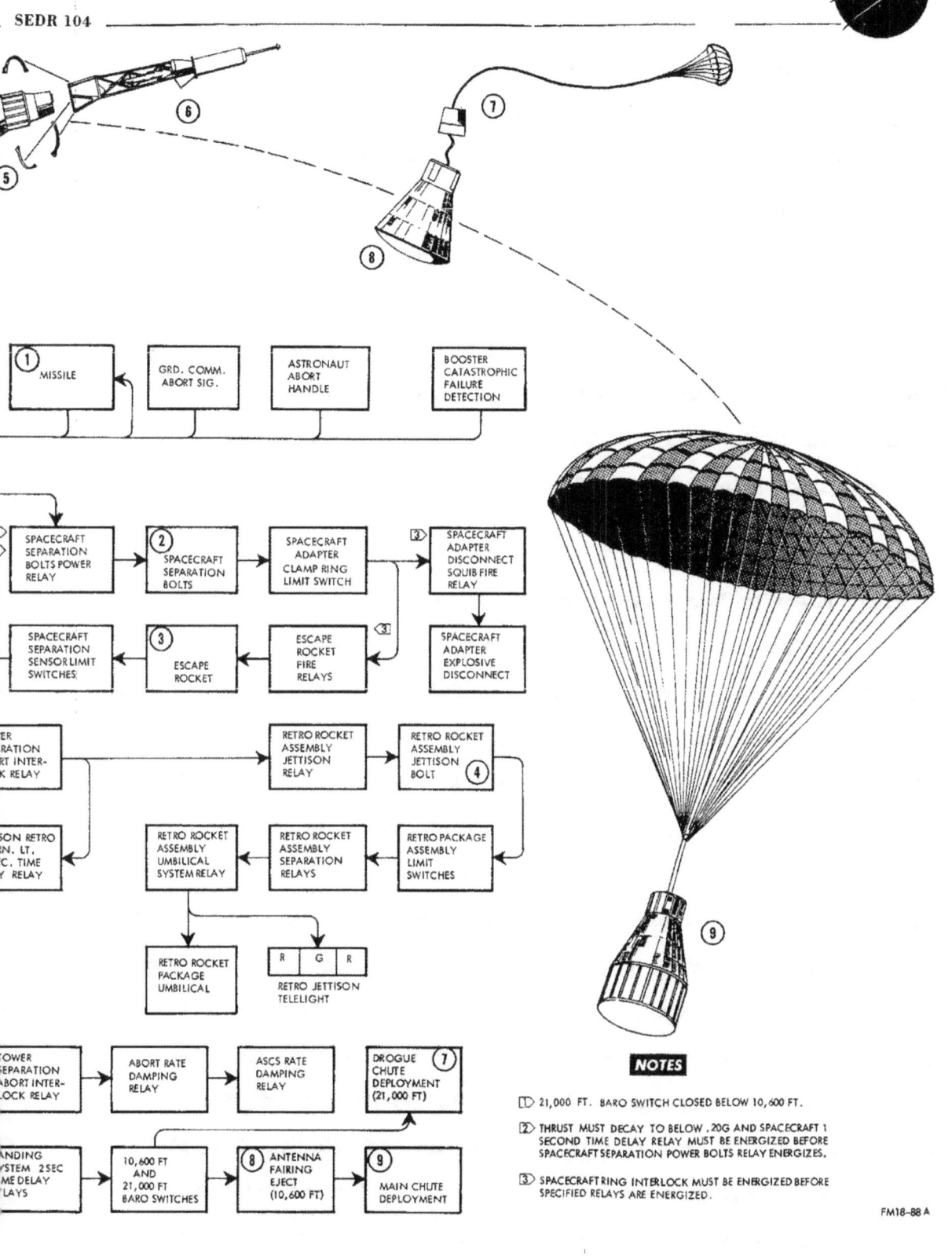

NOTES

1. 21,000 FT. BARO SWITCH CLOSED BELOW 10,600 FT.

2. THRUST MUST DECAY TO BELOW .20G AND SPACECRAFT 1 SECOND TIME DELAY RELAY MUST BE ENERGIZED BEFORE SPACECRAFT SEPARATION POWER BOLTS RELAY ENERGIZES.

3. SPACECRAFT RING INTERLOCK MUST BE ENERGIZED BEFORE SPECIFIED RELAYS ARE ENERGIZED.

FM18-88 A

SEDR 104

6-15. ESCAPE AFTER LIFTOFF BEFORE TOWER SEPARATION

After liftoff, (Time Zero), there are three methods by which an abort may be initiated. They are: (1) Ground command receiver abort signal and (2) Astronaut Abort Handle, both of which were possible methods in the previous paragraph, (3) The Booster Catastrophic Failure Detection System. This third method has been noneffective in the two previous paragraphs due to the #1 Time Zero Relay being de-energized. However, the #1 Time Zero Relay is energized two inches after liftoff and completes a circuit to the Mayday Relays if the Catastrophic Failure Detection Relay is de-energized by loss of power from the missile.

6-16. OPERATION

When the Mayday Relays are energized, the abort sequence (See Figure 6-6) will occur as follows: The abort Light on the left hand console will illuminate, the Spacecraft Separation Bolts Power Relay is energized and the Spacecraft Separation Warning Light Time Delay Relay is energized. The Abort Relay in the Maximum Altitude Sensor is energized after Spacecraft Adapter separation. The Maximum Altitude Sensor computes the time delay required for the spacecraft to reach a safe dynamic pressure before jettisoning the Escape Tower. The Spacecraft Separation Bolts Squibs will be fired, releasing the Spacecraft-Adapter Clamp Ring and allowing the three limit switches to return to their normal positions energizing the Emergency Escape Rocket Fire Relay, the Escape Rocket Fire Relay and the Spacecraft Adapter Disconnect Squib Fire Relay, firing the Escape Rocket and the four Spacecraft Adapter Explosive Disconnect Squibs. The Escape Rocket's 56,000 pounds of thrust will separate the spacecraft from the missile and carry it away from the sustainer at a small angle. The Spacecraft Separation Sensor

SEDR 104

Limit Switches also energize Spacecraft Separation Sensor Relays, which turns on the green SEP CAPSULE Telelight, and also energizes the Tower Separation Abort Interlock Latching Relay. The Abort Interlock Relay energizes the Retro Rocket Assembly Jettison Relay and fires the two squibs of the Retro Rocket Assembly Jettison Bolt. The bolt will fracture and the package will drop free of the spacecraft, being assisted by a coil spring installed between the Heat Shield and Retro Package Assembly for this purpose. When the spacecraft reaches a maximum altitude, contacts in the Maximum Altitude Sensor will close and energize the Tower Separation Bolts Power Relay firing the bolts. As the three tower bolts are fractured, the segmented Tower Clamp Ring separates allowing the three Tower Ring Limit Switches to return to their normal position energizing the Emergency Jettison and Jettison Rocket Fire Relays. Through these relays and their parallel contacts Main and Isolated Bus power will fire the squibs of the Jettison Rocket. The tower will be jettisoned clear of the spacecraft resulting in separating the two Tower to Spacecraft Electrical Disconnects. The separation of either disconnect will de-energize the Tower Separation Sensor Relays energizing the Abort Rate Damping Relay through the contacts of the Tower Separation Abort Interlock Relay. This relay will send a signal to the ASCS commanding rate damping until time of Main Chute deployment. De-energizing the Tower Separation Relays will also start two 2 second timers in the recovery sequence. The first timer will arm the 21,000 Foot Baroswitches and two seconds later the second timer arms the 10,600 Ft. Baroswitches.

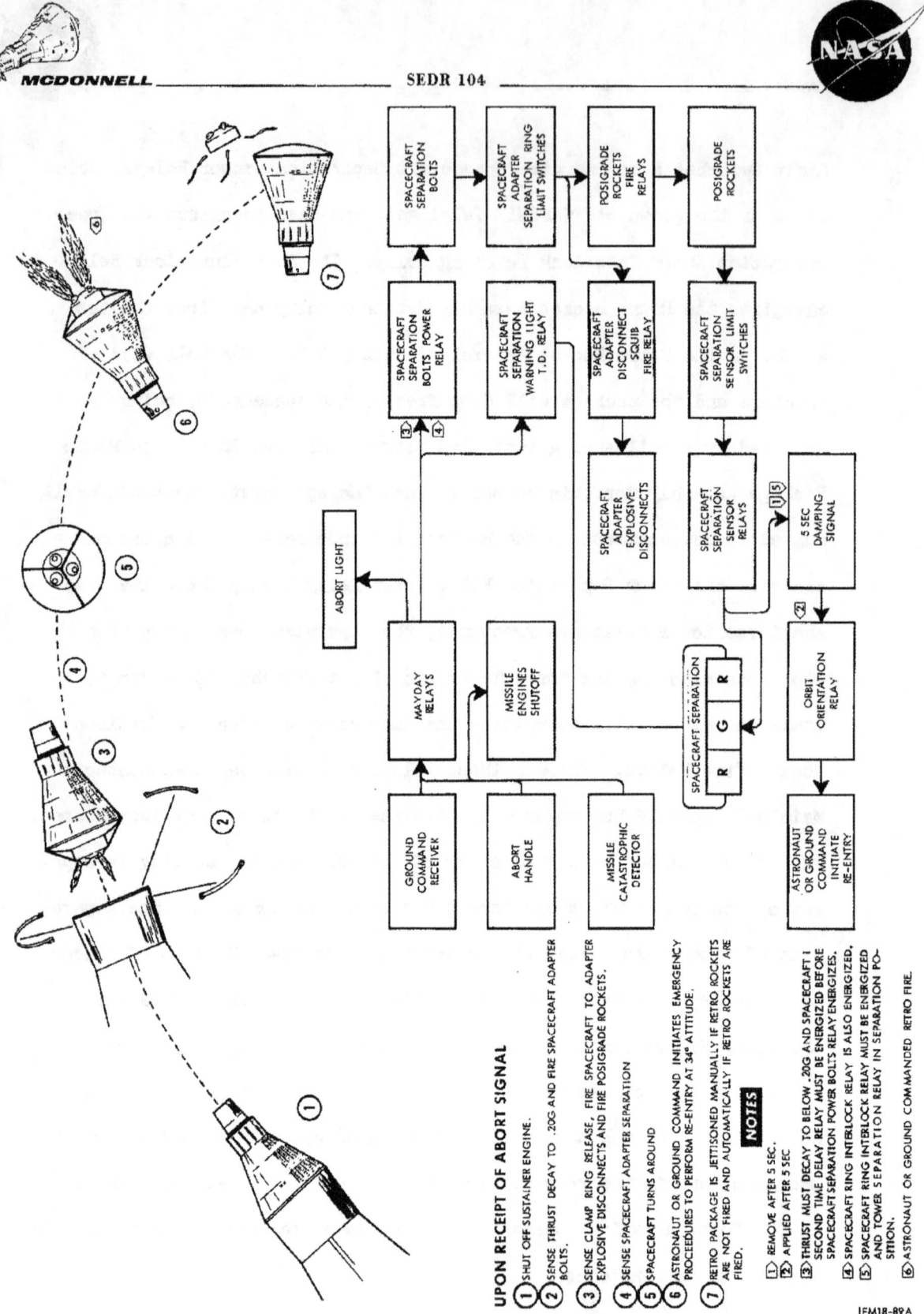

Figure 6-7 Escape System After Tower Separation

SEDR 104

6-17. ESCAPE AFTER TOWER SEPARATION

The methods of initiating an abort after staging are identical to the methods named for the escape after liftoff and are: (1) Ground command receiver abort signal; (2) Astronaut Abort Handle; (3) Booster Catastrophic Failure Detection System. Any of the three methods will energize the Mayday Relays. The sequence which occurs by the energizing of these relays is described in the following paragraph.

6-18. OPERATION

The signal which energizes the Mayday Relays also is transmitted to the missile to shut down the sustainer engine (See Figure 6-7). Through contacts of the energized Mayday Relays, a power circuit is completed to the ABORT Light on the main instrument panel and the .20g contacts of the Thrust Cutoff Sensor are armed. As thrust decays to .20g, the contacts close and energize the Spacecraft Separation Bolts Power Relay firing five Spacecraft Separation Bolts squibs and separating the Spacecraft Adapter Clamp Ring. The sequence following clamp ring separation is the same as the normal sequence (Refer to Paragraph 6-7). Re-entry may be accomplished by any of the emergency procedures (i.e., astronaut or ground initiated). Refer also to Paragraph 6-9. If the abort is initiated before the spacecraft has obtained the correct velocity for orbital flight and it is not desired to fire the Retro Rockets, the Retro Package must be jettisoned manually. It should be noted that even if the spacecraft does not attain orbital velocity, the quickest way for re-entry is by emergency firing of Retro Rockets.

SECTION VII

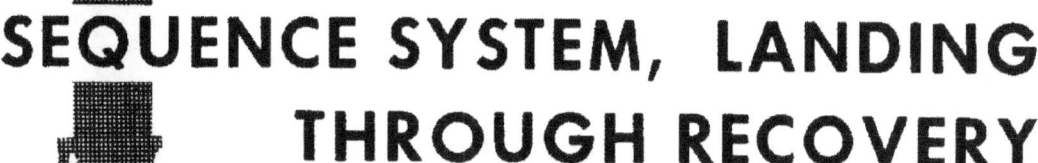

SEQUENCE SYSTEM, LANDING THROUGH RECOVERY

TABLE OF CONTENTS

TITLE	PAGE
Automatic Sequence Description	7-3
Automatic Sequence Operation	7-4
Emergency Sequence Description	7-7
Emergency Sequence Operation	7-10
System Components	7-12

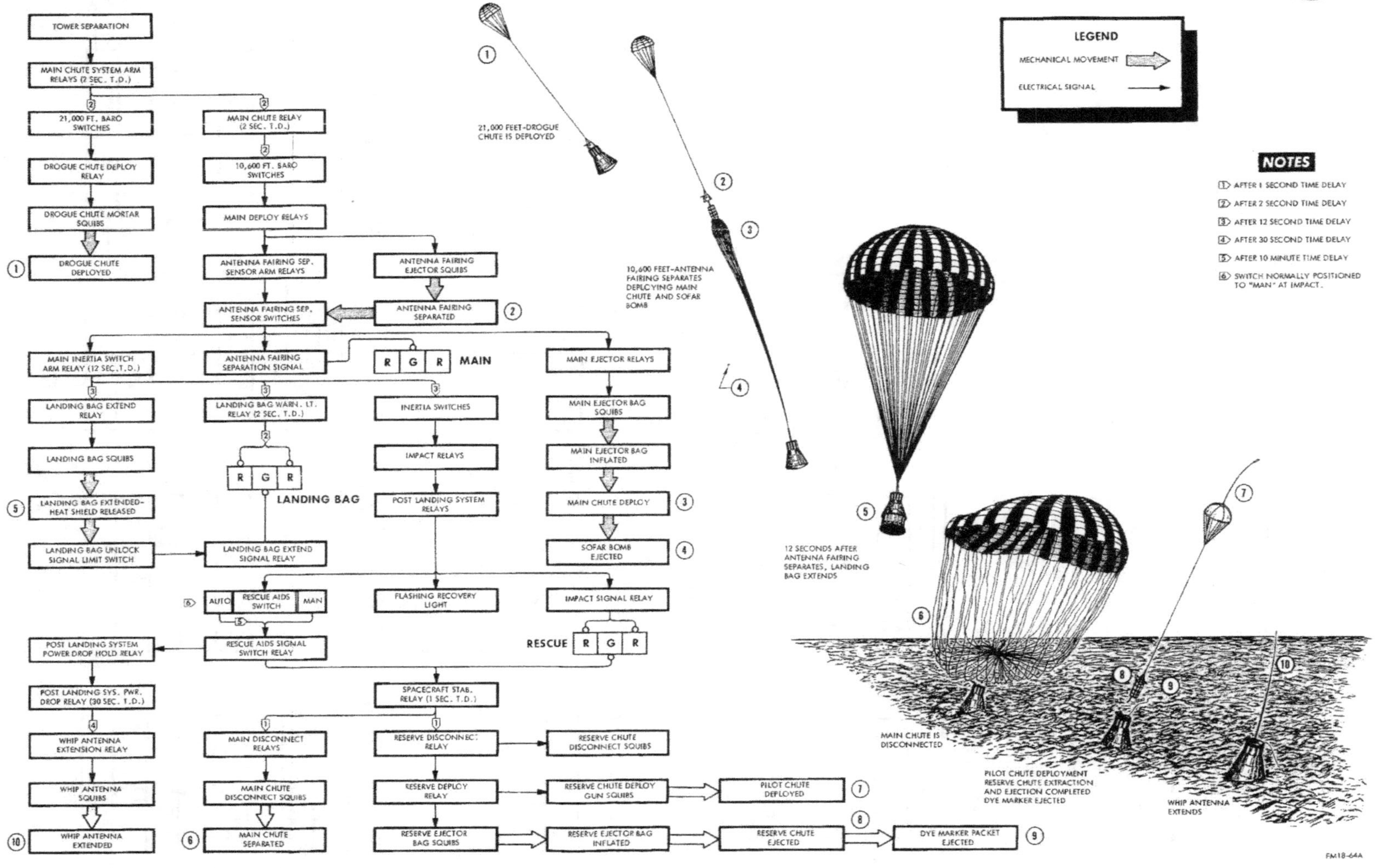

Figure 7-1 Automatic Landing and Recovery System

 SEDR 104

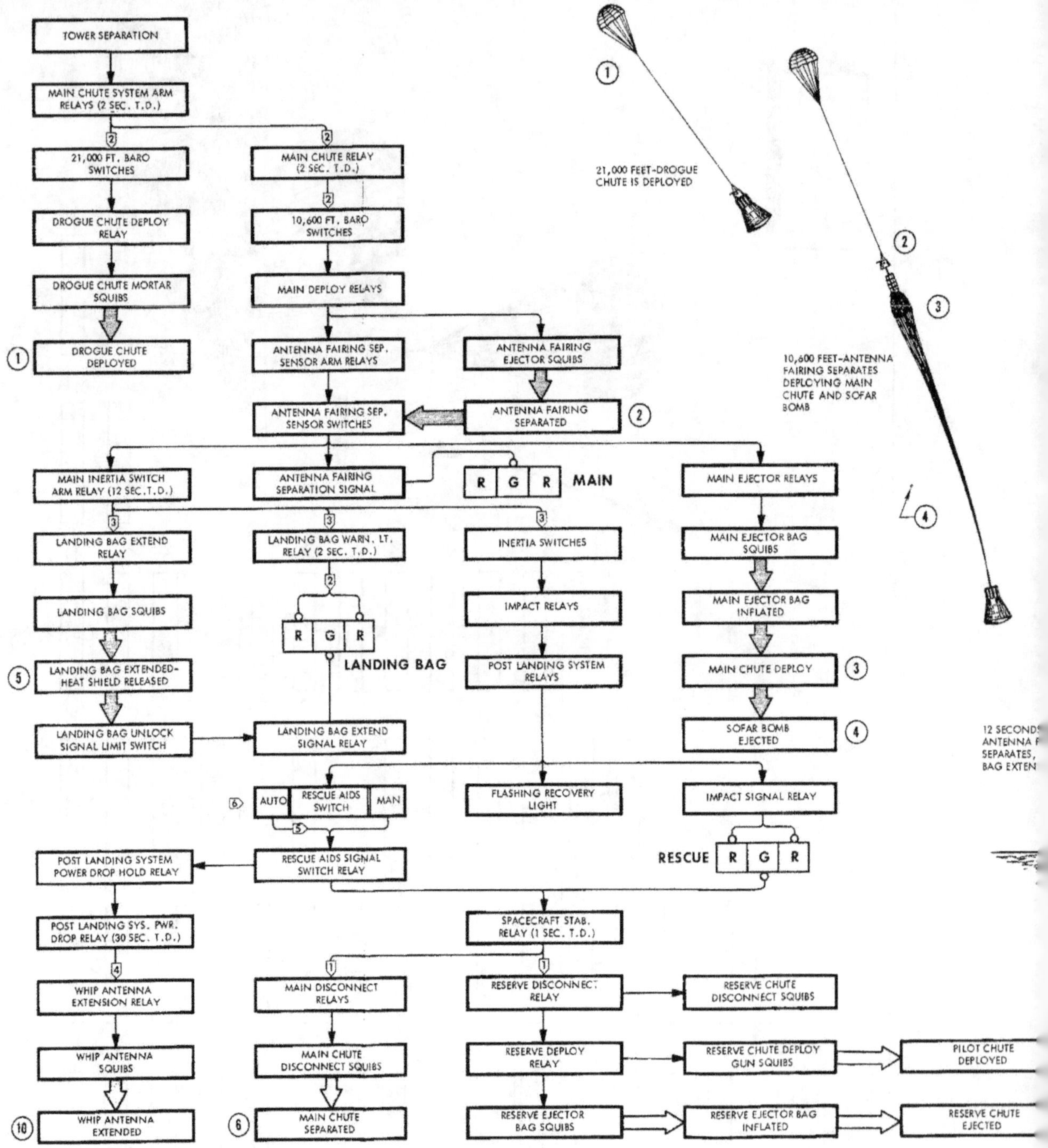

Figure 7-1 Automatic Landing and Recovery System

SEDR 104

LEGEND
MECHANICAL MOVEMENT ⇨
ELECTRICAL SIGNAL →

NOTES
1. AFTER 1 SECOND TIME DELAY
2. AFTER 2 SECOND TIME DELAY
3. AFTER 12 SECOND TIME DELAY
4. AFTER 30 SECOND TIME DELAY
5. AFTER 10 MINUTE TIME DELAY
6. SWITCH NORMALLY POSITIONED TO "MAN" AT IMPACT.

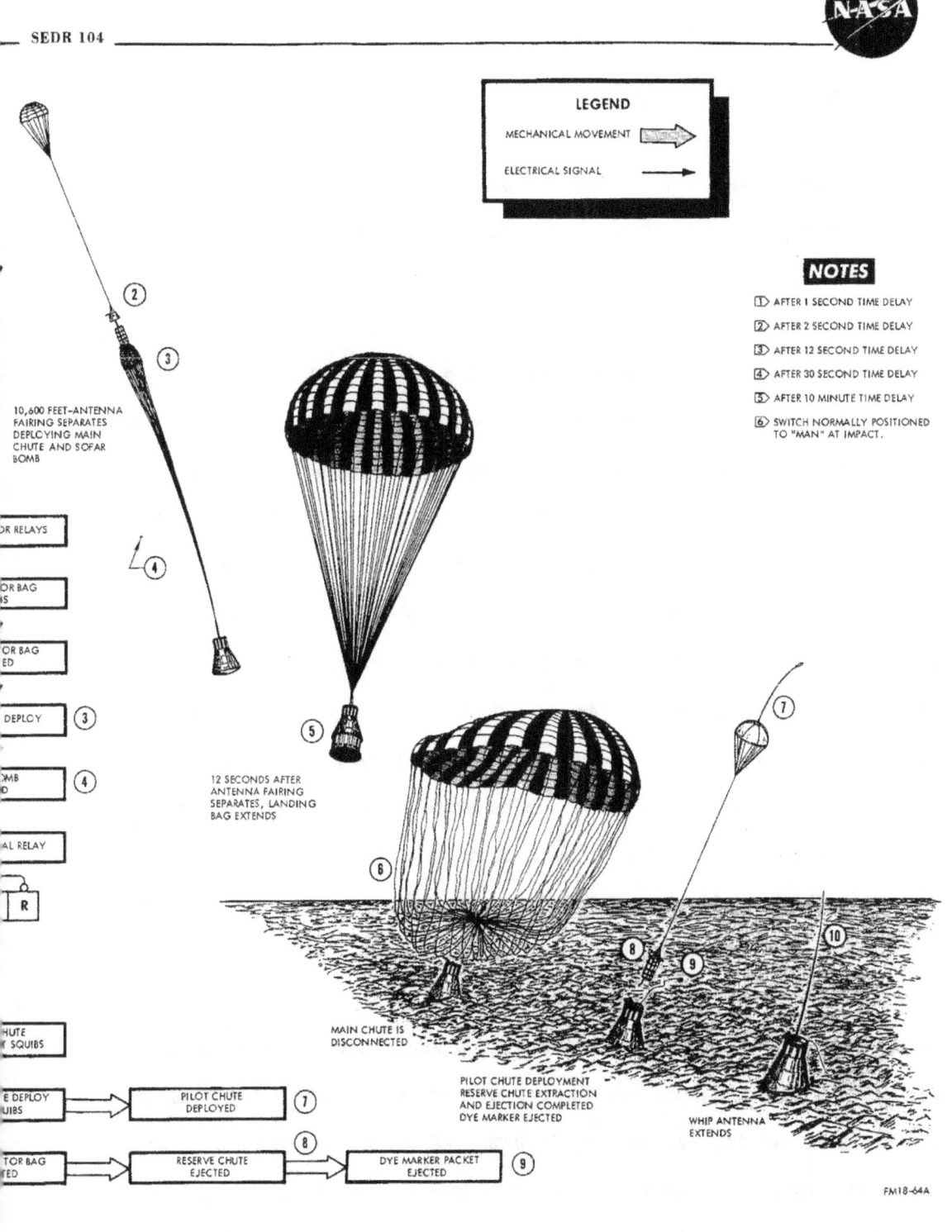

10,600 FEET—ANTENNA FAIRING SEPARATES DEPLOYING MAIN CHUTE AND SOFAR BOMB

12 SECONDS AFTER ANTENNA FAIRING SEPARATES, LANDING BAG EXTENDS

MAIN CHUTE IS DISCONNECTED

PILOT CHUTE DEPLOYMENT RESERVE CHUTE EXTRACTION AND EJECTION COMPLETED DYE MARKER EJECTED

WHIP ANTENNA EXTENDS

PILOT CHUTE DEPLOYED

RESERVE CHUTE EJECTED

DYE MARKER PACKET EJECTED

FM18-64A

SEDR 104

VII. SEQUENCE SYSTEM LANDING THROUGH RECOVERY

7-1. AUTOMATIC SEQUENCE DESCRIPTION

The landing and recovery sequence system provides automatic electrical and mechanical sequencing to land the spacecraft safely after an abort or normal re-entry, and to initiate locating aids for assistance in the subsequent recovery. The primary (completely automatic) system incorporates a drogue parachute, used initially to decelerate and stabilize the spacecraft in the initial phase of recovery and a main parachute for further deceleration. Redundant circuits have been incorporated to eliminate the possibility of single-point failures. The landing is accomplished utilizing a 63-foot-diameter parachute which is deployed at 10,600 feet. In the event of a main chute failure, a 63-foot-diameter reserve chute may be deployed by the astronaut's manual selection. Both main and reserve chutes are reefed to limit shock loads at initial opening. The reefing line is severed automatically after a predetermined time delay and the chute will open fully, lowering the spacecraft at the prescribed landing speed. A sound fixing and ranging (SOFAR) bomb is attached to the main chute risers and is released as a function of main chute deployment. When the risers pull taut, the bomb is ejected and falls to the ocean ahead of the spacecraft. When the bomb reaches a depth of 2,500 feet below sea level, it is detonated thus providing a means for determining the approximate location of the spacecraft landing area by sound fixing and ranging stations. After main chute deployment, the landing impact bag is extended, providing a cushioning effect for the landing impact. After impact, the main chute is automatically disconnected followed by reserve chute ejection; however, chute disconnect is delayed as a function of the Rescue Aids Switch: in automatic (AUTO), a 10-minute delay is accomplished; in manual (MAN), a one-second delay is accomplished. Therefore, if the astronaut desires immediate chute separation at impact, he must position the Rescue Aids Switch to the "MAN" position at impact to by-pass the 10-minute delay (or any portion of the 10-minute delay remaining after impact). The astronaut will then egress normally taking with him the survival kit which contains a life raft and other survival aids.

7-2. AUTOMATIC SEQUENCE OPERATION

On tower separation, power is removed from the Tower Separation Relays allowing the Main Chute System Arm 2-Second Time Delay Relays to be energized (See Figure 7-1). After the 2-second time delay has run out power flows through the closed contacts energizing the Main Chute 2-Second Time Delay Relays, and after a 2-second delay arms the No. 1 10,600 ft. Baroswitch through the 10,600 ft. Arm Relay contacts. The No. 1 and No. 2 Baroswitches are wired in series thereby requiring both switches to be closed before normal main chute deployment can be accomplished. The No. 2 21,000 ft. Baroswitch is armed through the closed contacts of the Main Chute System Arm 2-Second Time Delay Relays. Upon descent to 21,000 ft., both the No. 1 and No. 2 21,000 foot Baroswitches, which are wired in series, close thereby energizing the Drogue Deploy Relay, firing the Drogue Chute Mortar, deploying the Drogue Parachute. The Drogue Chute stabilizes and decelerates the spacecraft. At approximately 10,600 ft., the 10,600 ft. Baroswitches actuate the Main Deploy Relays resulting in the removal of the squib ground circuits and firing of all four squibs of the Antenna Fairing Ejector. Also, the Main Deploy Warning Light 2-Second Time Delay Relay is actuated, and at the end of 2-second delay the red Main Deploy Telelight is illuminated. The Antenna Fairing Separation Sensor Arm Relays are energized through the

closed contacts of the Main Deploy Relays arming the two Antenna Fairing Separation Sensor Switches.

The firing of the four Antenna Fairing Ejector Squibs causes the Antenna Fairing to separate from the spacecraft. A lanyard, connected from the Antenna Fairing to the Main Chute, extracts the Main Chute from the chute compartment. The Main Chute opens initially in the reefed condition to limit shock loads. Four seconds after the chute is deployed, the reefing line is severed by a pyrotechnic charge in the reefing line cutters allowing the parachute to open fully.

The separation of the fairing from the spacecraft allows the Antenna Fairing Separation Sensor Switches to function. Through the switches, power is routed to energize the Main Ejector Relay firing the Main Ejector Bag Squibs. When the squibs fire, they generate a gas, filling the Ejector Bag at the bottom of the Main Chute compartment aiding the ejection of the chute. At the same time the Antenna Fairing Separation Signal Relays are energized, illuminating the green Main Deploy Telelight and removing power from the red telelight. Power is also directed through the Antenna Fairing Separation Sensors to energize the Main Inertia Switch Arm 12-Second Time Delay Relays. After the 12-second time delay has run out, the energized contacts allow power to be supplied to energize the Landing Bag Extend and Landing Bag Warning Light 2-Second Time Delay Relays as well as to the Inertia Switches. The closed contacts of Landing Bag Extend Relay fire the squibs of the Landing Bag Valve releasing the heat shield and extending the impact landing bag. As the 2-second time delay runs out,

SEDR 104

the Landing Bag Warning Light Relay illuminates the red Landing Bag Telelight. Upon heat shield separation, the Landing Bag Unlock Signal Limit Switches are actuated and through its closed contacts, power from the Auto Landing Bag Fuse is directed to energize the Landing Bag Extend Signal Relay illuminating the green Landing Bag Telelight and extinguishing the red light. The force of impact on landing operates the Inertia Switches which provide power to the coils of the Impact Relays. Through the closed contacts of the Impact Relays, power is supplied to energize the Post Landing System Relays. Through the activated contacts of the Post Landing System Relays, power is transmitted to energize the Impact Signal Relay. Also through the closed contacts of the Post Landing System Relays, the Flashing Recovery Light circuit is completed setting the light in operation. When the Impact Signal Relay is energized, the green Main Deploy and Landing Bag Telelights are extinguished and the red Rescue Aids Telelight is illuminated. With the Rescue Aids Switch in "MAN", at the end of the 1-second time delay, the Spacecraft Stabilization Relay is activated allowing the Main Disconnect and Reserve Disconnect Relays to be energized; with the switch in "AUTO", this action is delayed 10 minutes. Through the energized contacts of Main Disconnect Relays, the Main Chute Disconnect Squibs are fired, releasing the main chute from the spacecraft. The Reserve Disconnect Relays fire the Reserve Chute Disconnect Squibs releasing the Reserve Chute and energizing the Reserve Deploy Relays. The Reserve Deploy Relays fire the Reserve Chute Ejector Bag Squibs. The Reserve

Chute Ejector Bag Squibs activate the gas generator which has a one-second delay in ignition time before inflating the ejector bag expelling the Reserve Chute and the dye marker. When the Rescue Aids switch is placed in the "MAN" position or after the 10-minute delay with the switch in "AUTO", the Rescue Aids Switch Signal Relay and the Post Landing System Power Drop Hold Relay are energized. The energized Rescue Aids Switch Signal Relay removes power from the red Rescue Aids Telelight and illuminates the green light. The Post-Landing System Power Drop Hold Relay energizes the Post-Landing System Power Drop 30-Second Time Delay Relay. At the end of the 30-second time delay the Whip Antenna Extend Relay is energized firing the Whip Antenna Extend Squibs activating the gas cartridge extending the active element of antenna to its full length. When the Post-Landing System Relays are energized on impact and, depending on the position of the Rescue Aids Switch, power is applied immediately or after a 10-minute time delay to the coil of the Post Landing System Power Drop Hold Relay and to the Post Landing System Drop 30-Second Time Relay. After the 30-second delay, the Post Landing System Power Drop Relay is energized which removes power from the remaining components except the Whip Antenna Extend Relay.

7-3. **EMERGENCY SEQUENCE DESCRIPTION**

The emergency provisions of the landing system basically consist of manually-operated back-up systems initiated by the astronaut. The appropriate button, pull-ring and switches are located on the Left Hand Con-

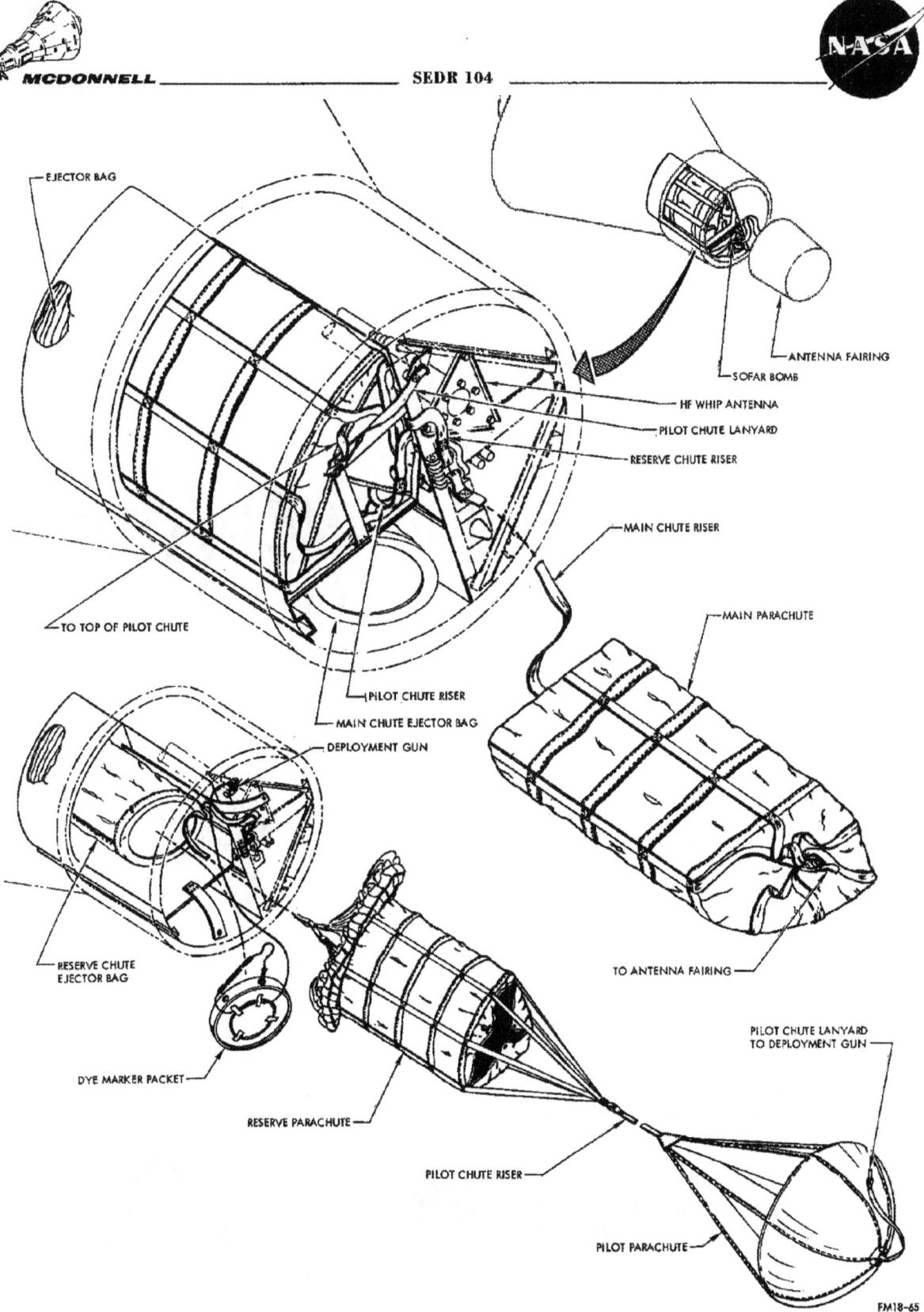

Figure 7-2 Main and Reserve Parachute System (Sheet 1 of 2)

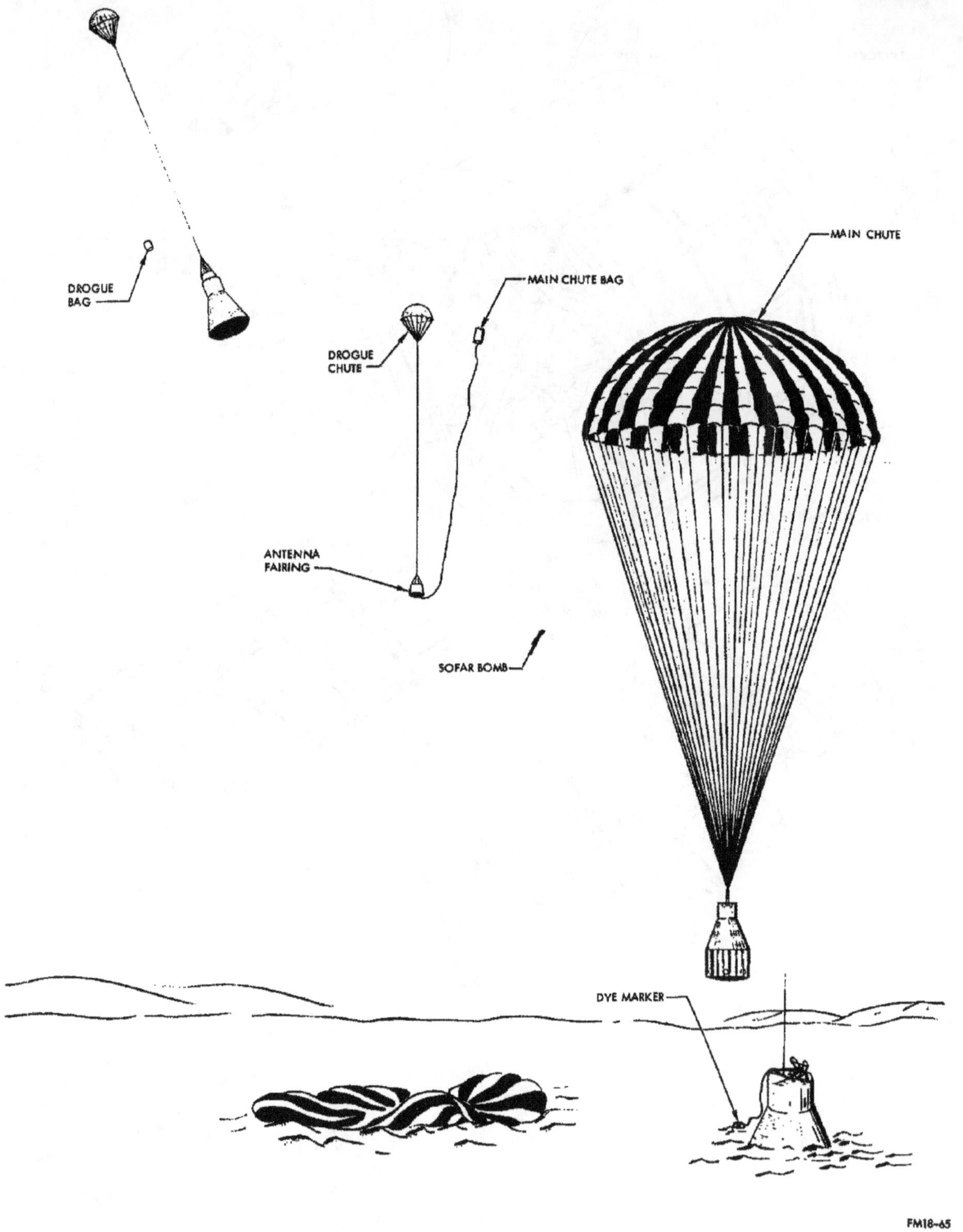

Figure 7-2 Main and Reserve Parachute System (Sheet 2 of 2)

SEDR 104

sole. The emergency system controls manually-initiated deployment of the Drogue, Main and Reserve Chutes, extension of the Landing Bag, and initiation of rescue aids.

7-4. EMERGENCY SEQUENCE OPERATION

On descending to 21,000 ft., if Drogue Chute failure is detected by lack of opening shock and by a visual check through the window, the astronaut will depress the Drogue button (See Figure 7-3). Depressing the button allows the Emergency Drogue Deploy Relay to be energized and the Drogue Chute Mortar Squibs to be fired deploying the Drogue Chute. If the green Main Deploy Telelight fails to illuminate, failure of the Maint Chute to deploy may be detected by a lack of opening shock, a visual check and no decrease in rate of descent. Upon determining that the Main Chute has not deployed, the astronaut will place the Recovery Arm Switch on the Control Panel to the manual position. If Main Chute Deploy is still not sensed, operating the Main Deploy Pull Ring energizes the Emergency Main Deploy Relay, firing the Antenna Fairing Ejector Squibs, ejecting the Antenna Fairing and deploying the Main Chute through the normal automatic sequence. When the green Main Deploy Telelight is illuminated and the rate of descent is greater than 32-feet-per-second, the chute is visually checked for damage. If the chute is damaged or did not deploy, actuating the Reserve Deploy Pull Ring will energize the Reserve Deploy Relays. Through the energized contacts of the Reserve Deploy Relays, power is applied to the Main Chute Disconnect firing the squibs disconnecting the chute from the spacecraft. At the same time, the Reserve Chute Ejector Bag Squibs are fired generating a gas after a 1-second delay and inflating the ejector bag which aids in deploying the Reserve Chute.

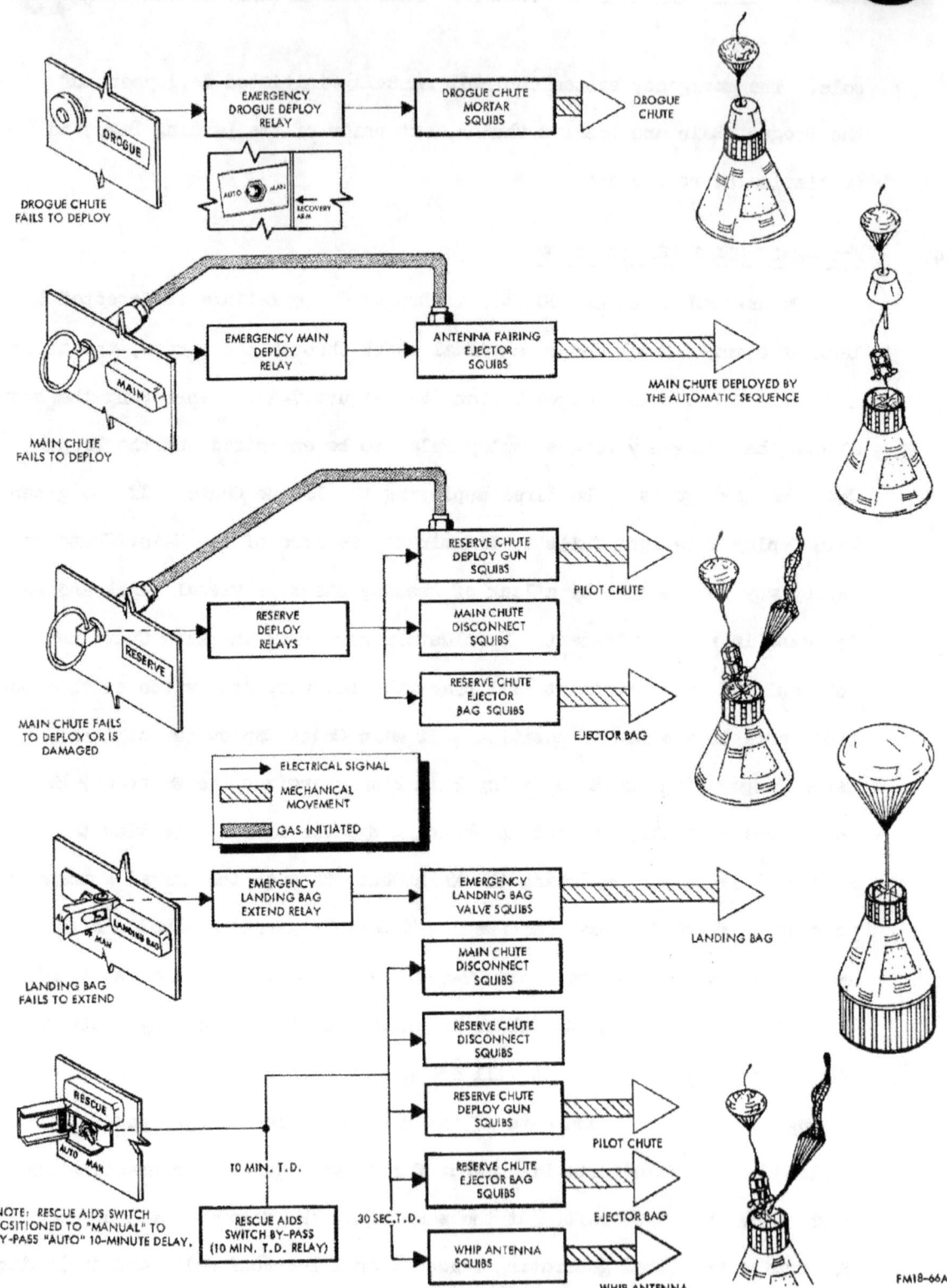

Figure 7-3 Landing and Recovery Emergency System

Twelve seconds after the Main Chute is deployed, the green Landing Bag Telelight should be illuminated. If the light does not come on, place the Landing Bag Switch in the "MAN" position, energizing the Emergency Landing Bag Extend Relay, firing the Emergency Landing Bag Valve Squibs, releasing the heat shield and extending the impact landing bag. Ten minutes after impact the Rescue Aids Switch By-Pass Relay is energized. When activated, the relay by-passes the Rescue Aids Switch and energizes the relays which supply power to fire the Main Chute Disconnect Squibs as well as the squibs for the Reserve, Disconnect, Deploy Gun, and the Ejector Bag in the same manner as if the switch were placed in the "MAN" position.

7-5. SYSTEM COMPONENTS

7-6. DROGUE PARACHUTE

The drogue parachute assembly (See Figure 7-5) consists of a conical ribbon-type drogue canopy with integral riser, drogue deployment bag, drogue mortar, sabot, and drogue mortar cover. The drogue parachute canopy is a conical ribbon parachute having 8 gores of 2-inch wide, 460-lb. tensile strength ribbons and 8 tubular nylon suspension lines of 1,000-lb. tensile strength each. The parachute is constructed to a diameter of 6.85 feet and permanently reefed (restricted) to an effective diameter of 6.0 feet by means of pocket bands. The constructed total porosity is 27.9% and the effective porosity (through reefing) is 36.3%. The 30-ft. long integral riser is made from three layers of 3,000-lb. tensile strength low-elongation hot-stretched Dacron webbing. The drogue parachute stabili-

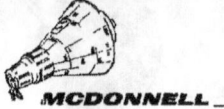

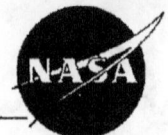

1 ANTENNA FAIRING EJECTOR GUN
2 PARACHUTE DISCONNECTS (2)
3 RECOVERY LIGHT
4 ANTENNA LANYARD
5 BARO SWITCHES (4)
6 MAIN CHUTE AND BAG
7 EJECTION BAGS (2)
8 GAS GENERATORS (2)
9 SHARK REPELLANT
10 DROGUE MORTAR
11 WHIP ANTENNA
12 PILOT CHUTE DEPLOYMENT GUN
13 PILOT CHUTE
14 RESERVE CHUTE AND BAG
15 SEA MARKER
16 INERTIA SWITCH
17 SOFAR BOMB.

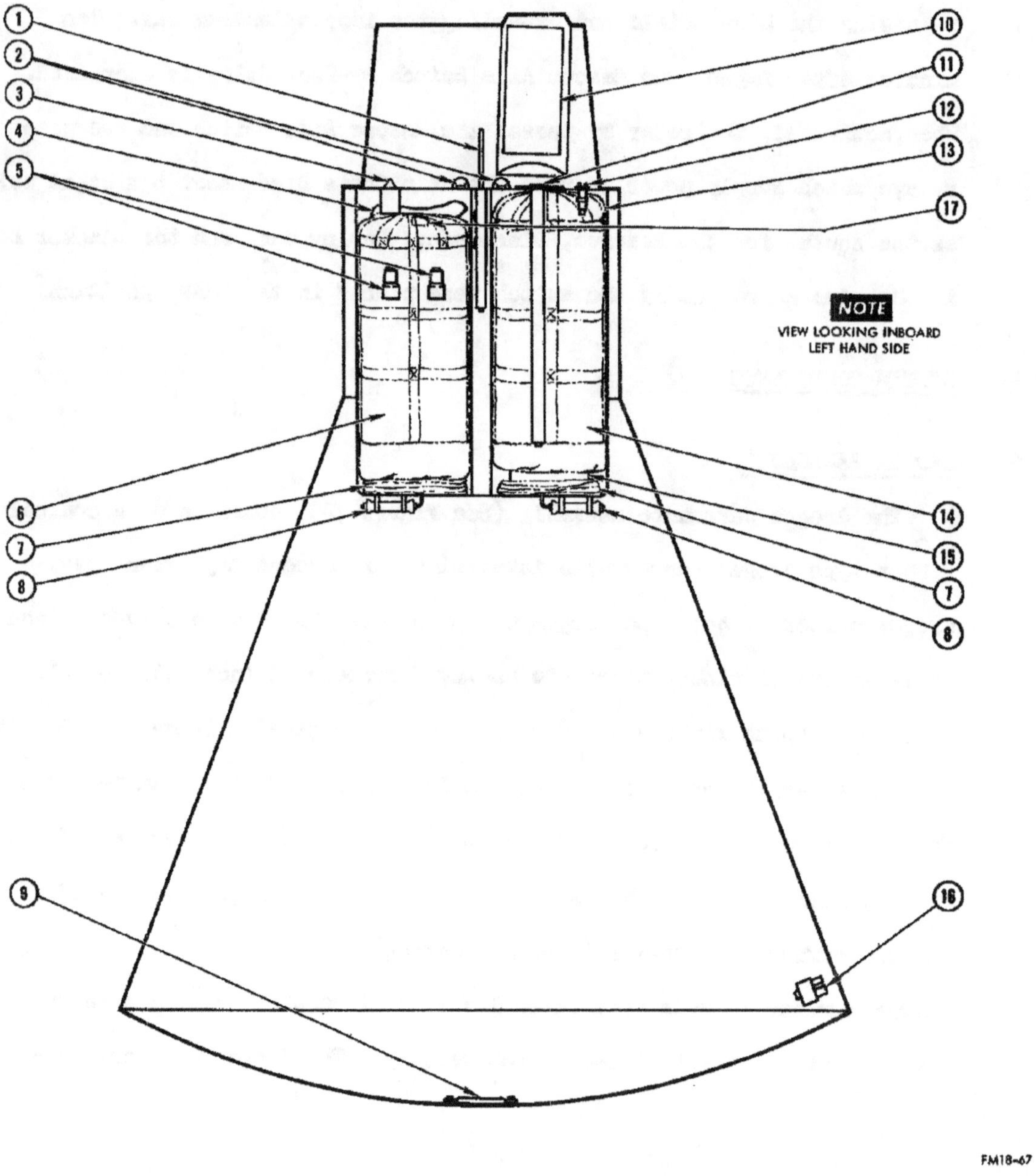
1
2
3
4
5
6
7
8
9
10
11
12
13
17
14
15
7
8
16
NOTE
VIEW LOOKING INBOARD
LEFT HAND SIDE
FM18-67

Figure 7-4 Landing and Recovery System Installation.

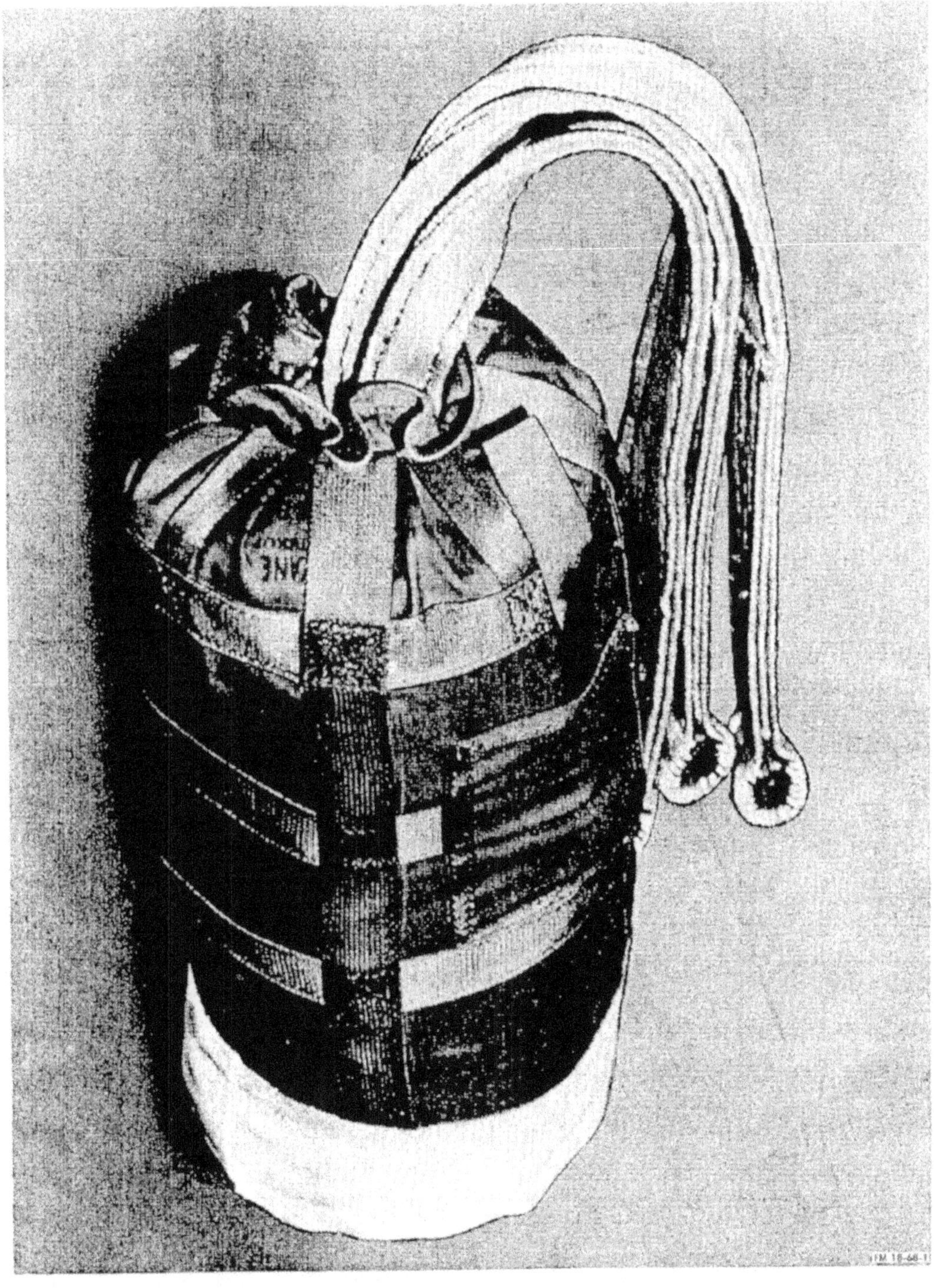

SEDR 104

Figure 7-5 Drogue Parachute

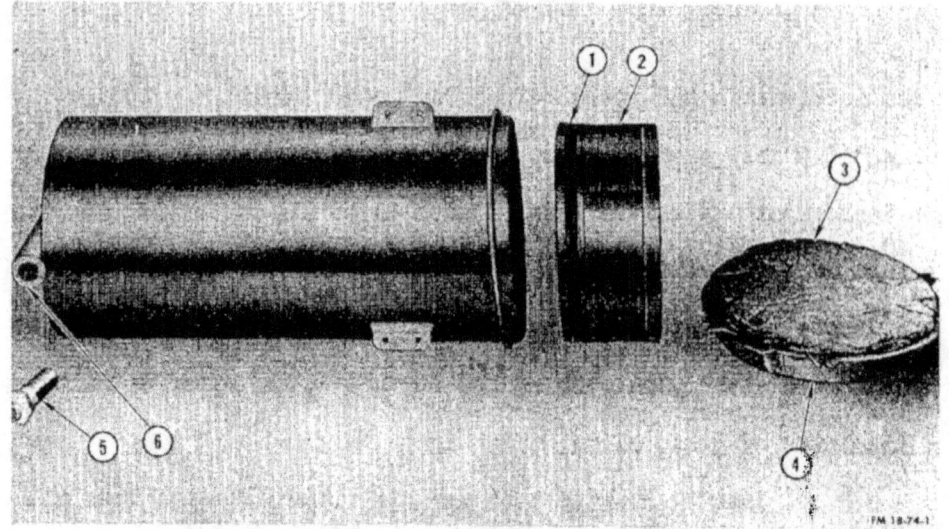

1. "O" RING
2. SABOT
3. INSULATION
4. COVER
5. CARTRIDGE
6. CHAMBER

Figure 7-6 Drogue Chute Mortar Assembly

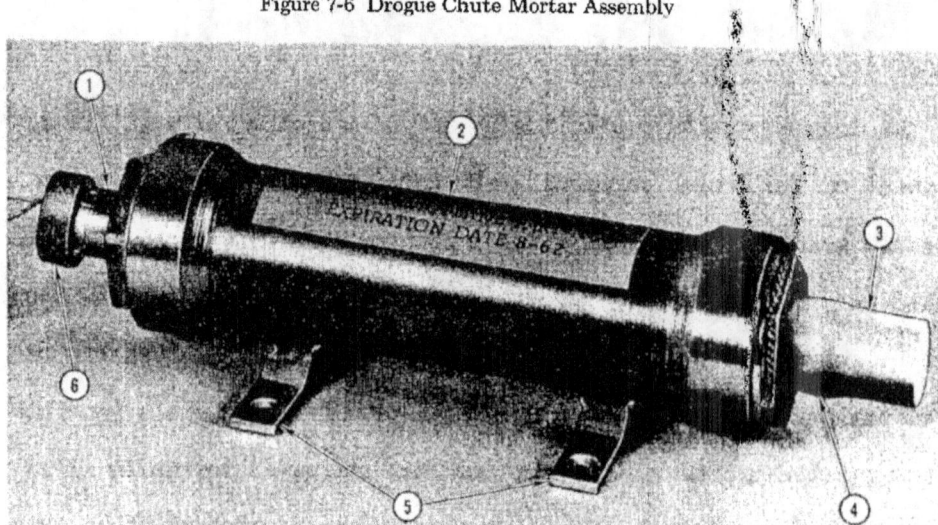

1. ELECTRICAL CONNECTOR
2. MAIN CHAMBER
3. PROTECTIVE CAP
4. TUBE FITTING
5. ATTACHMENT LUGS
6. SHORTING PLUG

Figure 7-7 Main and Reserve Chute Gas Generator

SEDR 104

zes and decelerates the spacecraft. The canopy weighs 2.9 lbs. without riser and 5.9 lbs. including the 30-ft. Dacron riser. The drogue parachute deployment bag serves a dual function of (1) protecting the drogue parachute during ejection and (2) providing means for orderly deployment of the drogue parachute. The bag is manufactured of cotton sateen fabric reinforced with nylon webbing and covered at the upper end with a heat insulator of glass cloth. The bag is weighted at the upper end with a 0.5 lb. lead disc which assists in stripping the bag from the canopy at the completion of line and riser stretchout. Inside the bag are 4 cotton tapes to which the riser is secured during packing in order to provide orderly riser deployment. The mouth of the bag is closed with a light cotton cord.

7-7.   DROGUE CHUTE MORTAR AND SABOT

The drogue parachute ejection mortar is a device for positive deployment of the drogue parachute with sufficient energy to overcome local pressure gradients and gravitational forces. The drogue parachute is packed in a protective bag and stowed in the mortar tube on top of a lightweight sabot (See Figure 7-6). The sabot functions as a piston to eject the parachute pack, when pressured from below by gases generated from a pyrotechnic charge. The propellant charge is initially fired into a breech chamber of small volume, to produce high pressure which is subsequently vented through a small orifice and into the main chamber at relatively lower pressures. In this manner, reaction loads are kept to a minimum, since the pressure energy is not expended instantaneously. The pressure sealing quality of the sabot is derived from an "O" ring, installed in a groove near the base. Two small holes are located in the "O" ring groove to vent air trapped in the mortar tube underneath the sabot on installation. For proper operation, the "O" ring and the inner wall of the mortar tube, which is always in contact with the "O" ring, are lubricated before installation. The drogue parachute pack is retained in its stowed position within the mortar tube by a thin metal "Rene-41" cover which is attached to the upper surface of the antenna housing. Three cutout sections, provided in the sides of the cover, permit routing of the steel cable risers into the drogue chute can. The cover is designed to constrain the chute in its compartment against negative decelerations and also to require minimal forces to break loose from its attachments at the time of deployment. Pressure of the chute pack causes the cover to deflect in such a manner that attachment tabs pull out from under attaching screw heads through a slotted hole designed for this purpose. The energy required to expel the drogue chute from its compartment is provided from high pressure gases, generated by ignition of a pyrotechnic charge. The cartridge is loaded with 66 grains of powder, contained in a propellant can attached to a steel body which houses the ignition wiring, and terminates in an electrical connector. The ignition circuitry consists of two separate and individual bridges, either of which is capable of igniting the power charge upon application of the proper current.

7-8. MAIN PARACHUTE

The main parachute assembly consists of: main parachute canopy, riser, deployment bag, and parachute disconnect. The main parachute

Figure 7-8 Main Parachute and Packing Box

1. STRUCTURE
2. SHORTING WIRE
3. SQUIB CARTRIDGE
4. BUSHING
5. SHEAR PIN
6. PISTON
7. LEAD BUFFER
8. ARM

Figure 7-9 Main and Reserve Parachute Disconnect

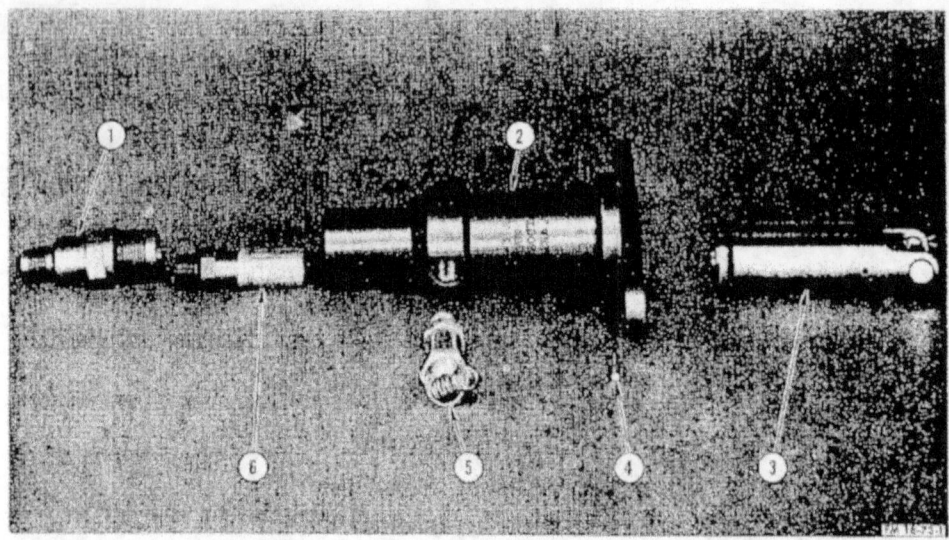

1. FIRING MECHANISM 4. SHEAR PIN
2. BODY   5. ELECTRICAL CARTRIDGE
3. PROJECTILE   6. MAIN CARTRIDGE

Figure 7-10 Pilot Chute Deployment Gun

Figure 7-11 Parachute Ejector Bag

SEDR 104

canopy is a 63-foot nominal diameter ringsail type. The ringsail parachute is fabricated from 2.25- and 1.1-ounce per square yard nylon parachute cloth into 48 gores with 48 suspension lines of 550-pound tensile strength. The main parachute is packed in a deployment bag which provides a low snatch force and orderly deployment (See Figure 7-8). The bag is manufactured from cotton sateen fabric, reinforced with nylon webbing and covered at the upper end with Thermoflex and glass cloth insulation. Inside the bag, midway along its length, is a pair of transverse locking flaps. Their function is to separate the canopy fabric from possible entanglement with the lines and to cause full line stretch-out before canopy deployment.

7-9. PARACHUTE DISCONNECT

Both main and reserve parachutes are attached to the spacecraft by a device designed to sustain the parachute loads during descent and to disconnect the parachute. The chute is disconnected 10 minutes (+ 1 second) after impact with the Rescue Aids Switch in "AUTO" or 1 second after positioning the Rescue Aids Switch to "MAN" (after impact). The assembly consists of 5 separate details installed in a mounting structure which is an integral part of the spacecraft. The parachute riser is looped around the arm which transmits the load to the structure through the piston. The shear pin restrains the piston from any motion tending to displace it. When the chute disconnect signal is completed, a squib cartridge is fired and the resultant expanding gas forces the piston forward into the arm recess, cutting the shear pin in the process. Full displacement of the piston removes parachute load transmission to structure, allowing the arm to rotate around the pivot pin. The loop of the parachute riser slips off the arm and the disconnect function is complete. The lead buffer serves to absorb energy of moving piston and prevents rebound of the piston back into the locked position.

7-10. **RESERVE PARACHUTE**

The reserve parachute assembly consists of: the pilot chute deployment gun and lanyard, pilot parachute, reserve parachute canopy, reserve parachute deployment bag, and reserve parachute disconnect. The reserve parachute deployment bag is similar to the main parachute deployment bag with the addition of flaps at the upper end of the bag to contain the packed pilot chute. The reserve parachute disconnect is identical with that used to disconnect the main parachute. The reserve parachute canopy is identical with the main parachute canopy.

7-11. **PILOT PARACHUTE**

The pilot parachute is a flat, circular type, 72 inches in diameter with a 30 ft. bridle. It is manufactured at 3.5-ounce per square yard fabric in the canopy and 2.25-ounce fabric in the vanes.

7-12. **PILOT CHUTE DEPLOYMENT GUN**

The pilot chute deployment gun (See Figure 7-10) initiates the first step in the sequence of reserve parachute deployment. Either gas pressure or an electrical impulse will cause the gun to fire, thus expelling a 12-ounce projectile which is attached to the reserve parachute pilot chute.

SEDR 104

The pilot chute inflates and in turn pulls out the reserve landing chute, completing the sequence. Whether fired electrically or pneumatically, a one-second time delay is provided between receipt of the impulse and detonation of the main charge. This delay permits the main parachute (if deployed and damaged) to separate from the spacecraft, to avoid entanglement with the reserve parachute to be deployed. The gun is basically a tubular body which contains the main firing cartridge and the projectile assembly. The projectile assembly is held in place by a pin which is sheared when the projectile is expelled. The main cartridge, which generates the gas pressure to eject the projectile, is fired as follows: (1) Gas pressure, through the gas firing mechanism (supplied when RESERVE PULL-RING is operated), drives a firing pin into the primer cap at the base of the main cartridge, initiating a time delay train, causing a subsequent detonation of the charge. A minimum of 750 psi gas pressure is required for pneumatic operation. (2) An electric impulse is received at the time delay igniter installed through the side of the gun. After a one-second delay, the igniter fires through the wall of the main cartridge and detonates it instantaneously. Firing characteristics of the igniter cartridge are as follows: All Fire Current 2.5 amps per bridge, All Fail Current 0.5 amps per bridge. The ignition circuit consists of two individual bridges terminating in a 4-pin receptacle. Muzzle velocity of the projectile is 250-300 ft/sec.

7-13. PARACHUTE EJECTOR BAGS

The ejector bags are inflatable air cells made of lightweight rubberized nylon fabric (See Figure 7-11). The design inflated shape is that

of a cylinder, 11 inches in diameter and approximately 35 inches in height. The upper end of the bag is slanted at full inflation to promote jettison of the parachute pack overboard after impact.

7-14. PARACHUTE EJECTOR GAS GENERATOR

This is a device to provide a rapid and sufficient volume of gas to inflate the main and reserve parachute ejector bags (See Figure 7-7). The reserve parachute gas generator is similar to that used for the main parachute except the additional feature of 1.25 second delay in ignition time. The generator functions to produce gas by the relatively slow burning of a solid-powder propellant in the main chamber. The gas is directed from the main chamber into the ejector bags through a 3/8 inch-diameter stainless steel tube. The tube serves also as a heat exchanger to reduce temperatures to within tolerable values prior to entry into the ejector bag. The generator body is equipped with lugs for mounting to the parachute container with four bolts. Ignition circuit characteristics are as follows: All Fire Current 2.5 amps, All Fail Current 0.5 amps.

7-15. DYE MARKER PACKET

The dye marker packet is a post landing recovery aid which performs its function by dissolving in water, thus producing a highly visible yellow-green patch. Approximately 1 pound of fluorescein dye is packed into a soluble plastic bag, which in turn is packed into an outer aluminum container. The entire packet assembly is ejected overboard, at the time of reserve chute ejection. The fluorescein dye forms a spot on the ocean surface which is visible from an airplane 10,600 feet high at a distance of 10 miles on a clear day.

SEDR 104

CAUTION

The dye marker package should be stored
in a dry place and not be exposed to water.

7-16. **RECOVERY LIGHT**

To aid in the visual location of the spacecraft after landing, a flashing light is installed in the recovery compartment. The intensity of the light is such that it will be visible in normal darkness for 40 nautical miles and up to an altitude of 12,000 feet. The flashing rate is approximately 15 flashes per minute. Powered by a battery pack located within the spacecraft, the light's circuit will be closed through energized contacts of the Post Landing System Relays which are activated by the closing of the Inertia Switches on impact followed by the energizing the Impact Relays. The light will operate for approximately 28 hours.

7-17. **WHIP ANTENNA**

To provide operation of the HF voice receiver-transmitter and HF recovery beacon after impact, a Whip Antenna is used. The active element is stowed in a collapsed condition in the recovery compartment and when extended is approximately 16 feet long. The antenna is extended by a gas cartridge which is activated when the Post-Landing System Power Drop 30-Second Time Delay Relay energizes the Whip Antenna Extend Relay. The Post Landing System Power Drop 30-Second Time Delay is started immediately upon impact with the Rescue Aids Switch in "MAN" and after a 10-minute delay with the switch in "AUTO". While it is extending a galling action takes place between the segments of the active element holding it rigid in the extended position.

MCDONNELL — SEDR 104

Figure 7-12 Baroswitch

Figure 7-13 Inertia Switch

MCDONNELL — SEDR 104

7-18. BAROSWITCHES

There are two pairs of Baroswitches used in the recovery system (See Figure 7-12). In these switches, an over-center spring is included in the design to minimize chatter during vibration and shock and to prevent contact oscillation. The switches are located in the recovery compartment, one pair is set to close at 21,000 feet; the other at 10,600 feet.

7-19. INERTIA SWITCH

The inertia switch is essentially a spring device actuated by mass (See Figure 7-13). A landing shock of 7.5 plus or minus 1.13g's minimum will produce momentary closing of two electrical contacts, thus completing an electrical circuit. This switch is used in conjunction with a latching relay which receives an electrical pulse and by latching into the closed position, provides continuous electrical continuity. The inertia switches used consist of four separate snap-action switches and two separate masses, all housed in a common case.

7-20. SOFAR BOMB

A post-landing recovery aid. SOFAR is an abbreviated form for "sound fixing and ranging". This component performs its function when it detonates by hydro-static pressure at a predetermined water depth. Shock waves from the explosion are received by sound detection devices aboard picket ships or shore bases and a position fix on the capsule is thus made. The maximum range of the Mercury SOFAR Bomb is 3000 miles. The SOFAR Bomb which is tossed overboard by action of the main chute ejection system is preset to detonate at 3500 ft. (See Figure 7-14).

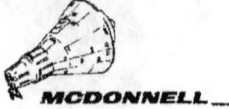

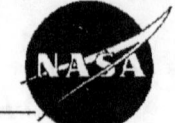

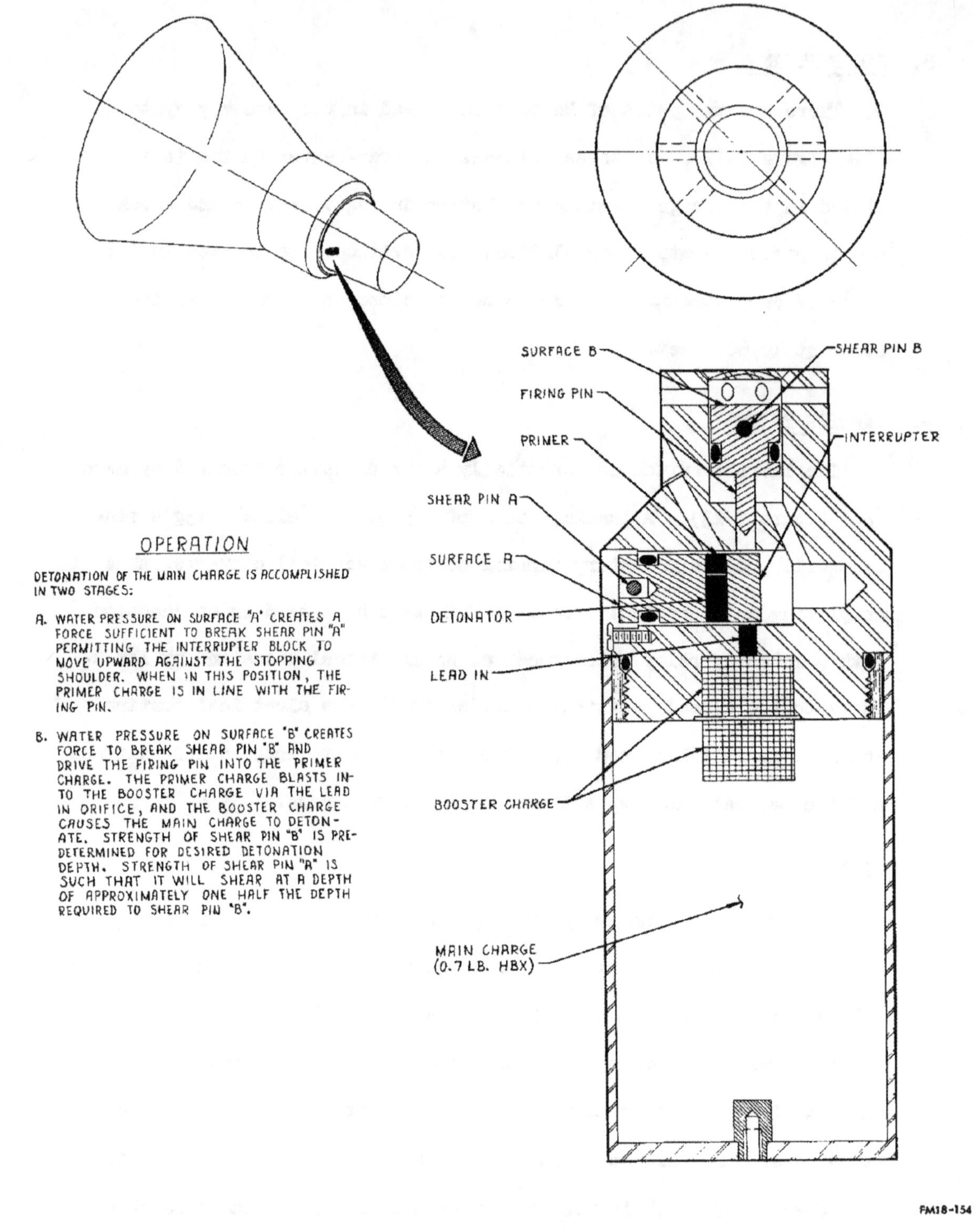
SURFACE B
SHEAR PIN B
FIRING PIN
PRIMER
INTERRUPTER
SHEAR PIN A
SURFACE A
DETONATOR
LEAD IN
BOOSTER CHARGE
MAIN CHARGE (0.7 LB. HBX)
OPERATION
DETONATION OF THE MAIN CHARGE IS ACCOMPLISHED IN TWO STAGES:
A. WATER PRESSURE ON SURFACE "A" CREATES A FORCE SUFFICIENT TO BREAK SHEAR PIN "A" PERMITTING THE INTERRUPTER BLOCK TO MOVE UPWARD AGAINST THE STOPPING SHOULDER. WHEN IN THIS POSITION, THE PRIMER CHARGE IS IN LINE WITH THE FIRING PIN.
B. WATER PRESSURE ON SURFACE "B" CREATES FORCE TO BREAK SHEAR PIN "B" AND DRIVE THE FIRING PIN INTO THE PRIMER CHARGE. THE PRIMER CHARGE BLASTS INTO THE BOOSTER CHARGE VIA THE LEAD IN ORIFICE, AND THE BOOSTER CHARGE CAUSES THE MAIN CHARGE TO DETONATE. STRENGTH OF SHEAR PIN "B" IS PREDETERMINED FOR DESIRED DETONATION DEPTH. STRENGTH OF SHEAR PIN "A" IS SUCH THAT IT WILL SHEAR AT A DEPTH OF APPROXIMATELY ONE HALF THE DEPTH REQUIRED TO SHEAR PIN "B".
FM18-154

Figure 7-14 Sofar Bomb Schematic

SECTION VIII

ESCAPE AND JETTISON ROCKET SYSTEMS

TABLE OF CONTENTS

TITLE	PAGE
System Description	8-4
Escape Tower	8-4
Escape Rocket	8-4
Jettison Rocket	8-7
Astronaut Controls	8-9
System Operation	8-12

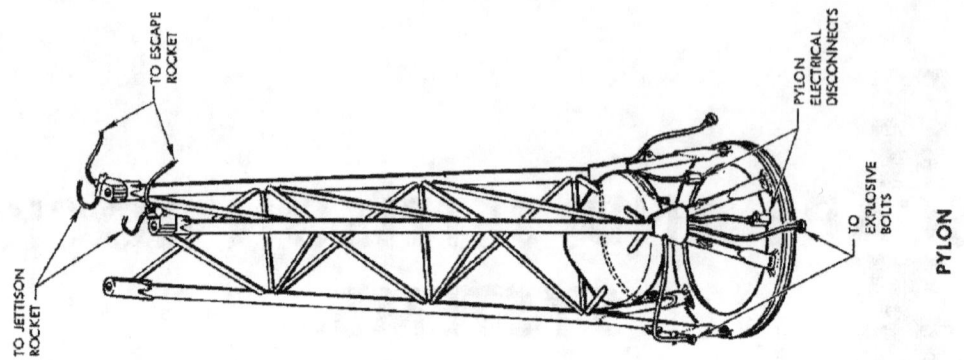
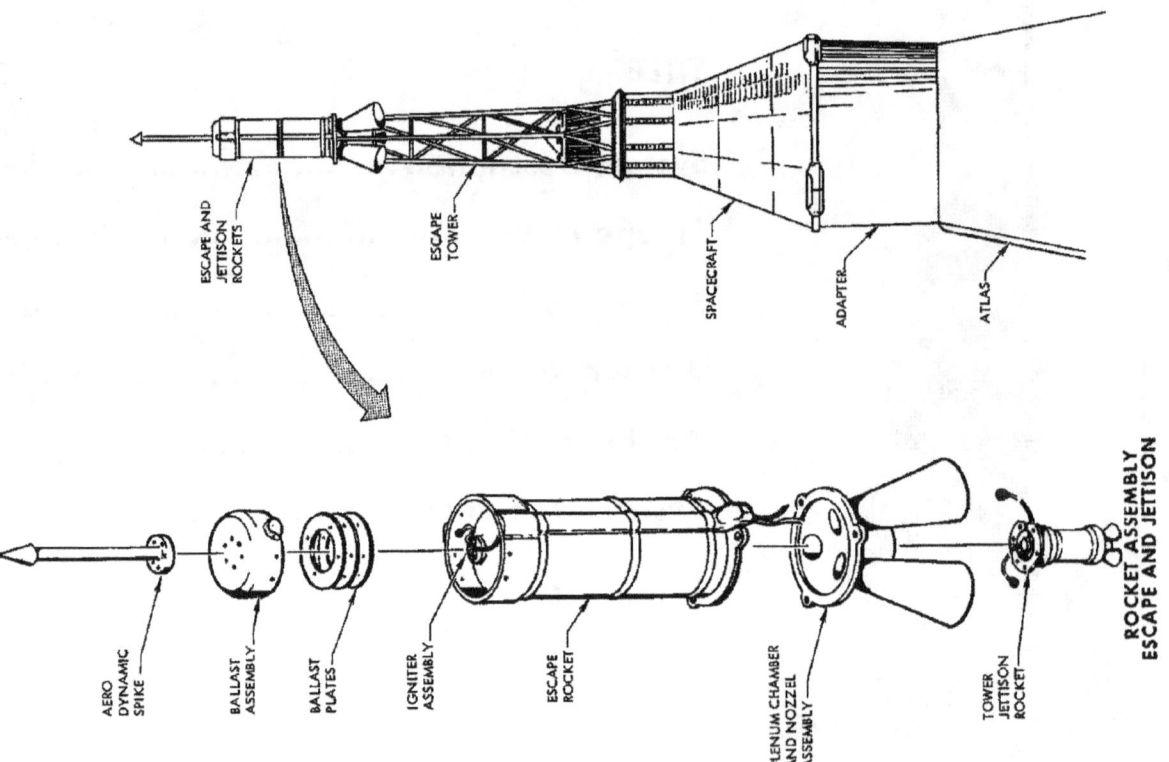

Figure 8-1 Escape System (Sheet 1 of 2)

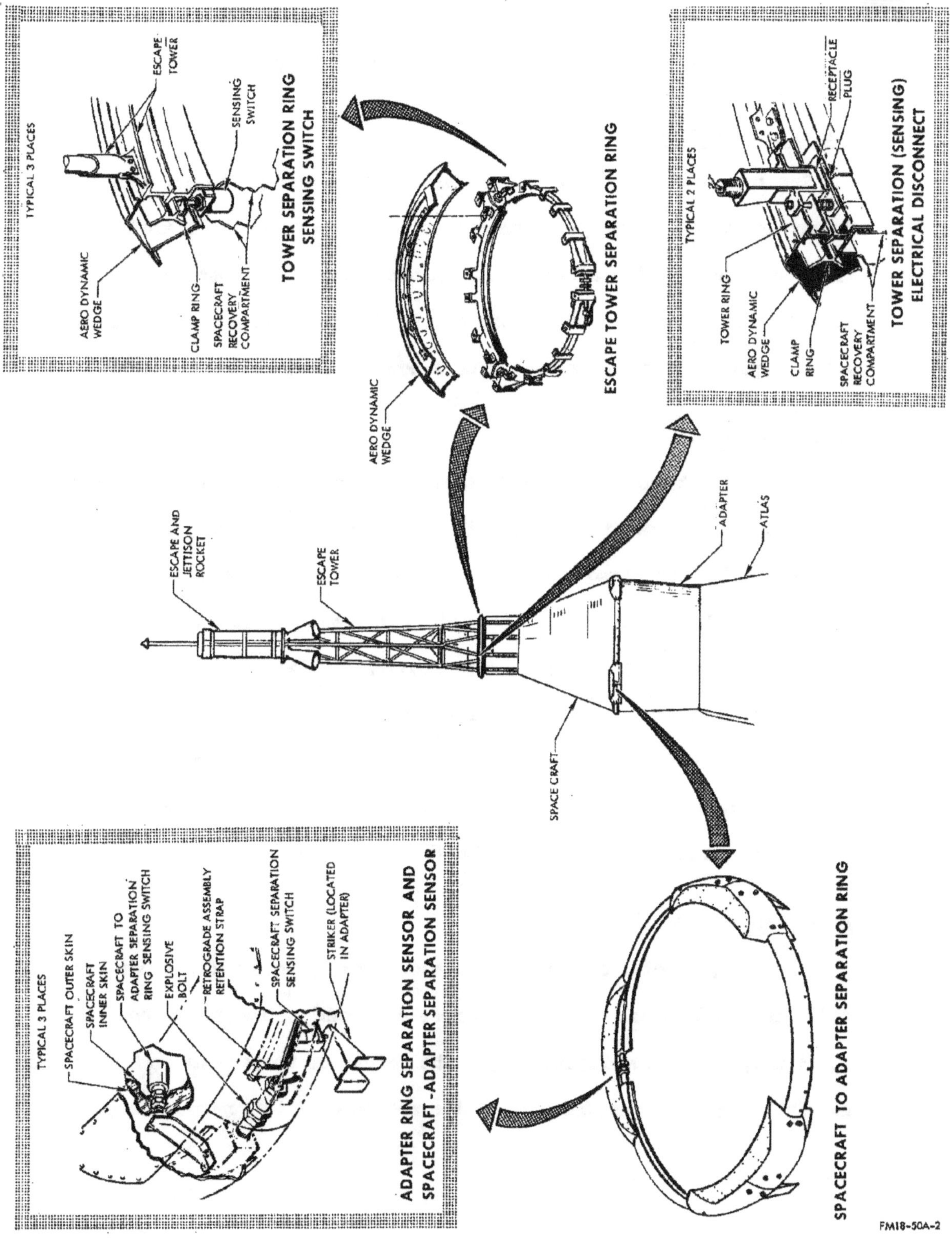

MCDONNELL — SEDR 104

Figure 8-1 Escape System (Sheet 2 of 2)

SEDR 104

VIII. ESCAPE AND JETTISON ROCKET SYSTEM

8-1. **SYSTEM DESCRIPTION**

The spacecraft escape system consists of the escape tower, escape rocket and escape tower jettison rocket. In addition, there are controls available to the astronaut to initiate an escape sequence and to provide a manual backup of certain events which normally occur by automatic sequencing during an escape or a normal mission.

8-2. **ESCAPE TOWER**

The escape tower is a welded steel structure approximately ten feet in length. The tower is attached to the spacecraft recovery compartment by a three-segment clamp ring. The segments of the clamp ring are held together by three explosive bolts. The upper end of the escape tower provides a mounting base for the escape rocket. The escape tower also provides for the routing of the electrical wiring necessary to provide ignition of the escape rocket, escape tower jettison rocket and the clamp ring explosive bolts. For a more detailed discussion of the escape tower structure, refer to Section 3, Major Structural Assemblies.

8-3. **ESCAPE ROCKET**

The escape rocket consists of an electrically actuated igniter assembly, a $\frac{1}{4}$ - inch 4130 steel case, rocket nozzle assembly, plenum chamber and a solid propellant (see Figure 8-2). The length of the escape rocket is approximately 70 inches. The diameter of the rocket case is approximately 15

 MCDONNELL SEDR 104

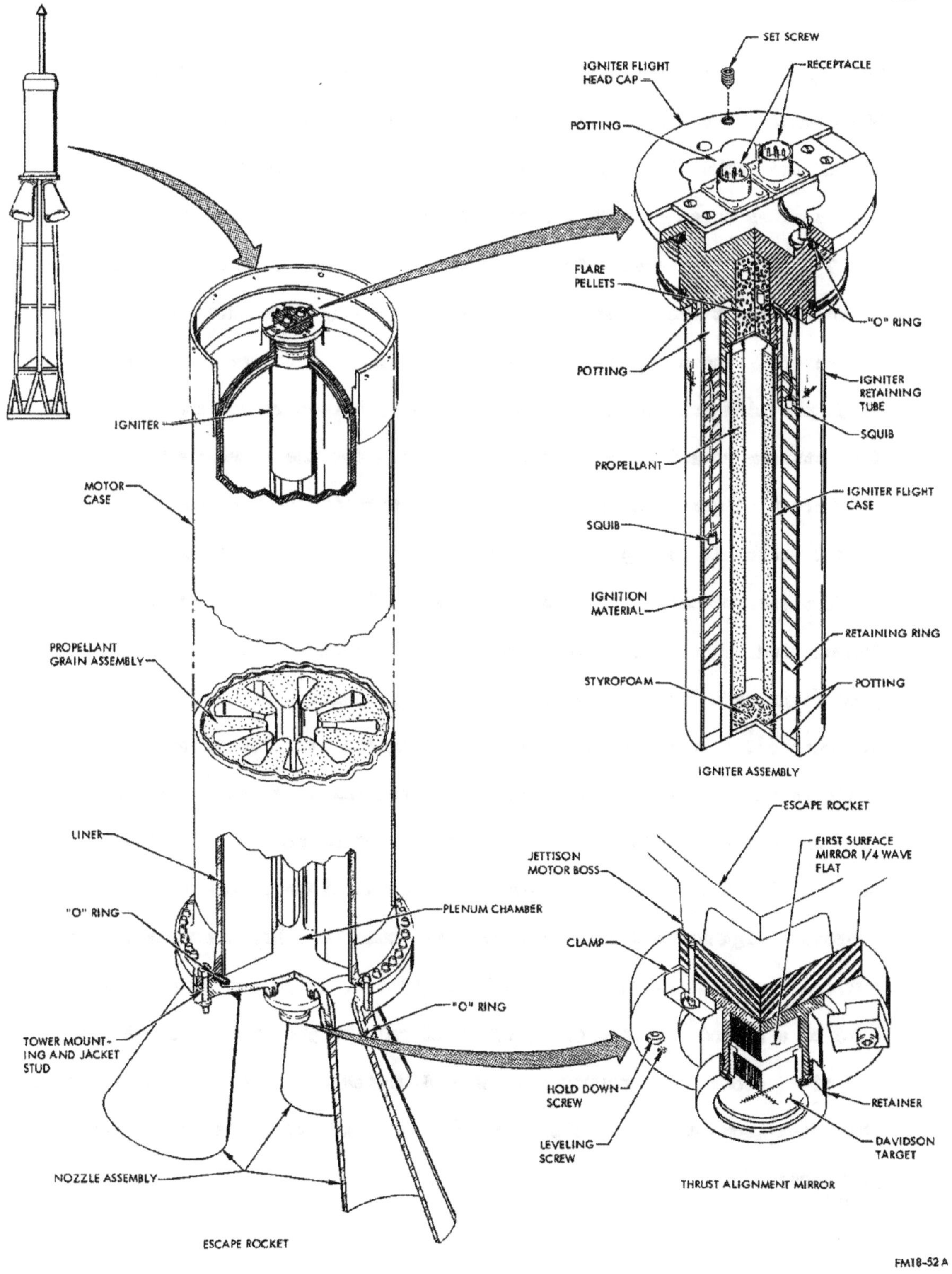

Figure 8-2 Escape Rocket

8-5

inches. The weight of the motor prior to firing is approximately 650 pounds. For aerodynamical stability, ballast is added to the top of the rocket (see Figure 8-1, Sheet 1 of 2). The rocket motor incorporates three equally spaced nozzles. The nozzles are canted at 19 degrees from centerline of rocket case to centerline of nozzle so as to direct the rocket blast outward and away from the tower and spacecraft. The aft closure of the rocket motor incorporates a boss which provides for the installation of the jettison rocket motor. The jettison rocket motor boss also provides for the attachment of the thrust alignment mirror. The optical sighting of the resultant thrust vector is accomplished by the thrust alignment mirror.

The escape rocket propellant is a polysulfide ammonium perchlorate formulation. The United States Bureau of Explosives classifies the propellant as a "Class B Explosive". The propellant is sensitive to pressure and a spark or flame may easily ignite it. The propellant grain is an internal burning nine-point star which is cast directly into and bonded to the case. With the nine-point port design, the possibility of thrust misalignment is reduced. This is due to the improved alignment between the star ports and the exhaust nozzles. The nominal resultant axial thrust at 70 degrees F. is 52,000 pounds for 0.78 of a second; it then drops off uniformly to 5000 pounds in the next 0.6 of a second. The thrust will then diminish at a reduced rate to zero. The total impulse of the motor, at sea level, is approximately 56,500 pound-seconds.

The escape rocket igniter is a head mounted dual unit with two completely independent initiation systems to each unit. The dual initiation system to each unit has independent circuitry from different batteries. One unit is cylindrical in shape, and is a central dynaflow type of long duration. This unit is essentially a miniature rocket motor. It incorporates a small propellant grain which can be initiated by either of two squibs surrounded by boronpotassium nitrate pellets. Surrounding the first unit is the second unit, composed of an annular plastic tube filled with a metal-oxidant mixture in which are located two sets of five squibs. Any one squib is capable of initiating the unit.

8-4. **JETTISON ROCKET**

The jettison rocket is a qualified Thor retro unit. The rocket consists of an electrically actuated igniter, motor case and a tri-nozzle assembly. The nozzles are canted at 30 degrees from centerline of rocket case to centerline of nozzle. The rocket weighs 22.0 pounds, has a length of 18 inches, a diameter of 5.5 inches, and produces 600 pounds of thrust for 1.55 seconds at 70 degrees F. at sea level. The rocket has been successfully fired from -75 degrees F to 175 degrees F., and from sea level to vacuum.

The jettison rocket igniter is a head mounted unit with dual ignition capabilities. This unit is cylindrical in shape with a hexagonal head and threads into the top of the jettison rocket. The igniter contains approximately 7 grams of USF-2D ignition pellets which are ignited by any one of

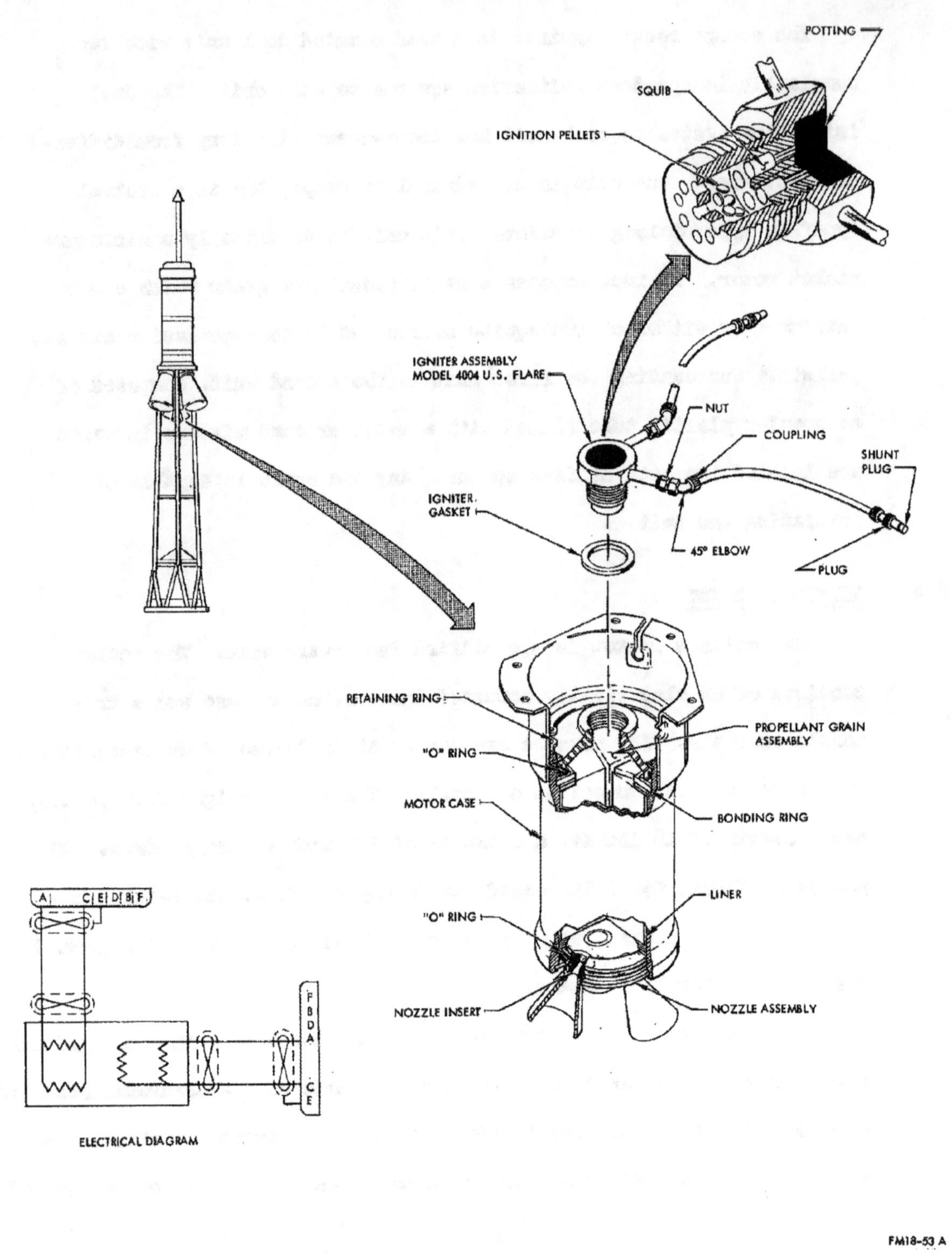

Figure 8-3. Jettison Rocket

four squibs. The four squibs are arranged in two pairs of two squibs each. Each pair has independent circuitry from a different power source.

8-5. ASTRONAUT CONTROLS

The astronaut's controls for the escape system consist primarily of the abort handle; in addition, there are two pull-rings to provide manual backup of automatic functions.

The abort handle's primary function is to initiate the abort sequence. The handle is also used as a restraint handle during launch. Location of the abort handle is forward of the astronaut's support couch left arm rest. For an astronaut initiated abort, the release button located in the top of the handle must be depressed, allowing the handle to be rotated outboard. When moved to the abort (outboard) position, an electrical switch is actuated, which acts to detonate the spacecraft-to-adapter clamp ring bolts. The escape sequence is then initiated, providing that the main umbilical has been disconnected. Before umbilical release, the abort handle is inoperative.

Two pull-rings are located on the left hand console of the main instrument panel and provide the astronaut with the capability to manually detonate the spacecraft-to-adapter clamp ring and the escape tower clamp ring. Located immediately adjacent to the pull-rings are two tele-light indicators designated "JETT TOWER" and "SEP CAPSULE". These indicators illuminate green when the escape tower has been jettisoned and when the spacecraft has separated from the booster-adapter, respectively.

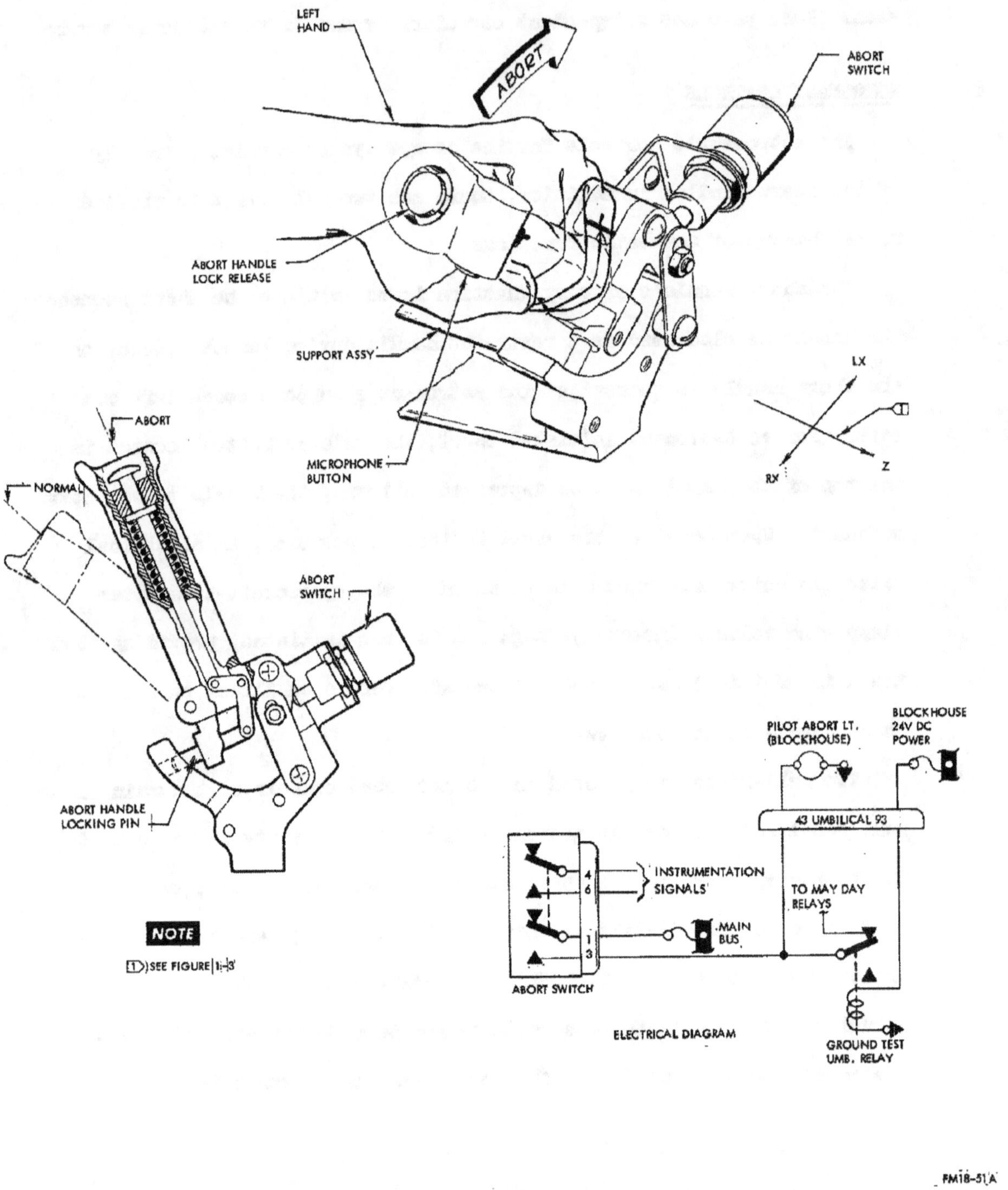

Figure 8-4. Abort Handle

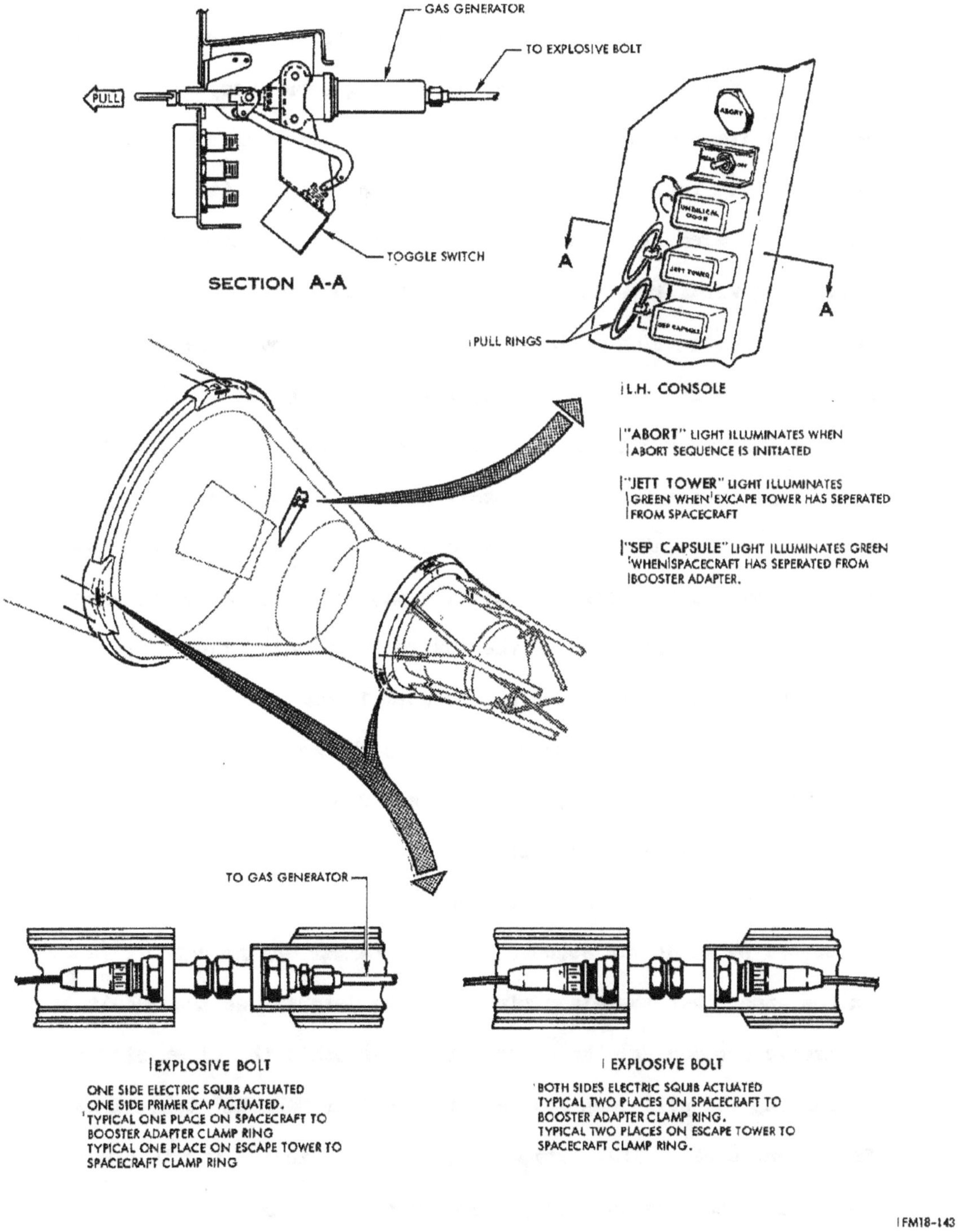

Figure 8-5 Emergency Clamp Ring Controls

8-6. SYSTEM OPERATION

Under normal mission conditions, when the escape system is not used, the escape tower clamp ring is detonated at approximately T + 154 secs. When the escape tower clamp ring separates, the escape rocket ignites and carries the escape tower clear of the spacecraft. Under normal mission conditions, the escape tower jettison rocket is not fired. Normally, the clamp ring bolts are detonated by automatic sequencing which applies electrical power to the squibs in the clamp bolts. If detonation of the escape tower clamp bolts fails to occur as the result of automatic sequencing, the astronaut may jettison the tower manually by pulling the pull-ring adjacent to the "JETT TOWER" tele-light on the instrument panel. Activating the pull-ring closes a toggle switch and applies an alternate source of electrical power to the clamp bolt squibs. Activating the pull-ring also initiates a gas generator located behind the instrument panel. The gas generator is connected to a percussion cap located in one of the three explosive bolts. The rapid expansion of gas fires the explosive bolt, thus allowing the clamp ring to be detonated even with a loss of electrical power in all the squib circuits.

Under normal conditions, the spacecraft-to-adapter clamp ring is detonated by automatic sequencing at sustainer engine cut-off (SECO). In the event that the clamp ring fails to detonate as the result of automatic sequencing, the astronaut may detonate the clamp ring manually by pulling the pull-ring adjacent to the "SEP CAPSULE" tele-light on the instrument panel. This action initiates a series of events similar to those described for the escape tower clamp ring above.

The spacecraft electrical system provides for an abort any time after the gantry is removed. When an abort is initiated (See Section 6 for various methods of initiating an abort), the spacecraft-to-adapter ring is detonated, the escape rocket ignited, and the spacecraft is propelled away from the booster. If the abort is made off the pad, the escape rocket will carry the spacecraft to an altitude of approximately 2500 feet. At the peak of the escape trajectory, as determined by a maximum altitude sensor, the escape tower clamp ring is detonated, the escape tower jettison rocket ignited, and the escape tower propelled away from the spacecraft. Two seconds after tower jettisoning, the drogue chute is deployed. Two seconds after drogue chute deployment, the antenna fairing is released. Twelve seconds after antenna fairing release, the heat shield is released, extending the landing impact skirt.

SECTION IX

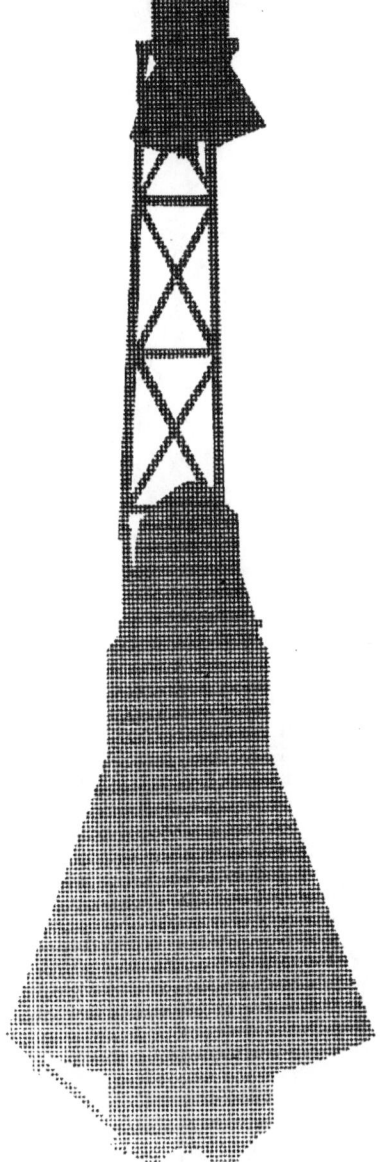

POSIGRADE ROCKET SYSTEM

TABLE OF CONTENTS

TITLE	PAGE
System Description	9-3
Posigrade Rocket	9-3
Rocket Igniter	9-3
System Operation	9-5

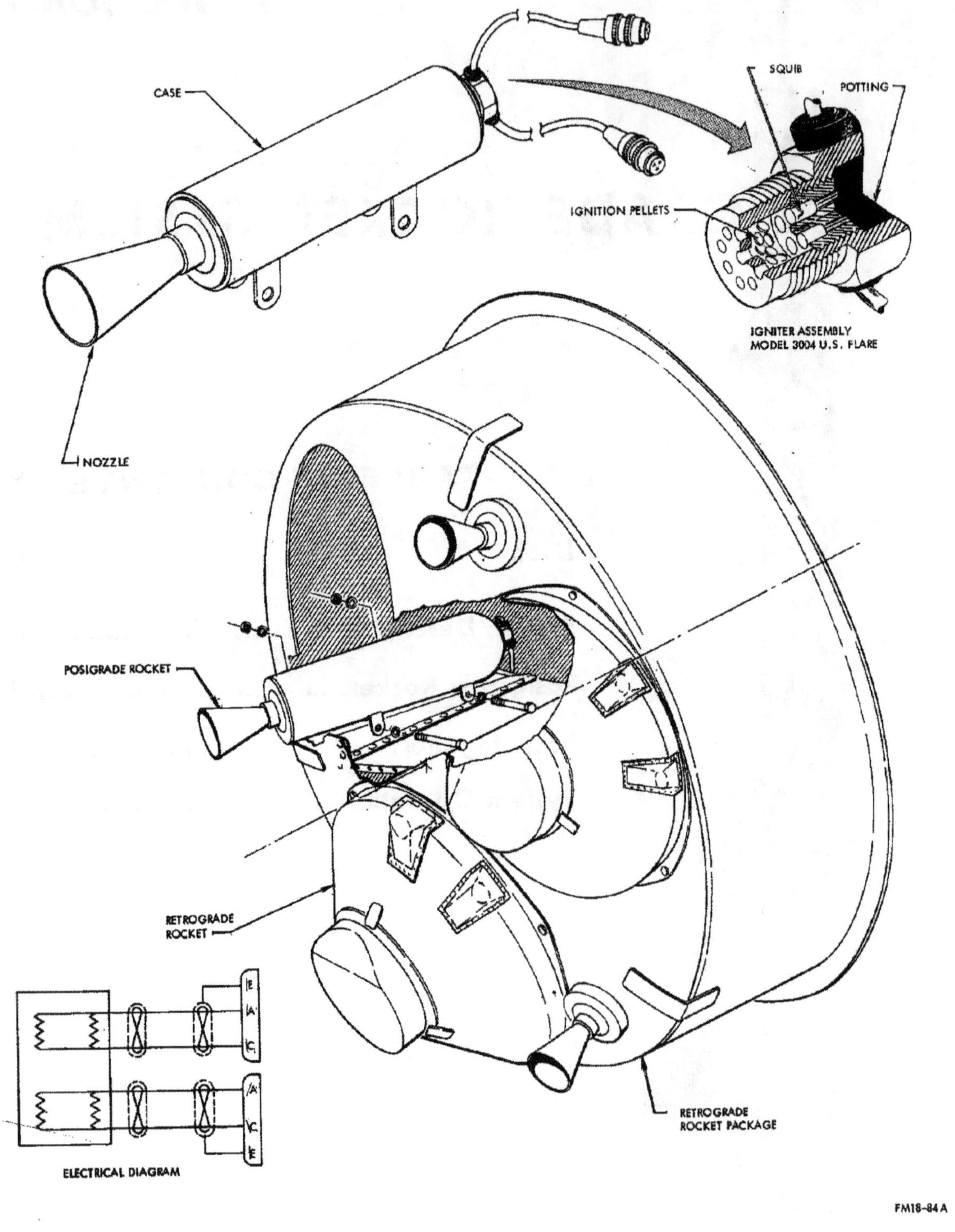

Figure 9-1. Posigrade Rocket System

IX. POSIGRADE ROCKET SYSTEM

9-1. SYSTEM DESCRIPTION

The posigrade rocket system consists primarily of the three posigrade rockets and igniters mounted in the retrograde package and the associated wiring necessary to ignite the rockets at the proper time.

9-2. POSIGRADE ROCKET

The posigrade rocket primarily consists of a nozzle assembly and case, a solid propellant and an electrically actuated igniter. The posigrade rocket is a cylindrical device measuring approximately 14.7 inches in length, 2.8 inches in diameter and weighing approximately 5.24 pounds. This rocket is basically an Atlas retro-rocket with minor changes for increased reliability. Reliability has been gained by two methods: first, dual ignition of the igniter squibs from two different buses; second, only one of the three rockets is necessary to accomplish successful separation. Due to the wide temperature range of the rockets, a temperature control system is not required. Posigrade rocket thrust is 416 pounds $\pm$ 5% at sea level. Firing time is 1.01 sec.

9-3. ROCKET IGNITER

The posigrade rocket igniter is a head mounted unit with dual ignition capabilities. The igniter is cylindrical in shape with a hexagonal head for threading it into the top of the posigrade rocket. This unit contains approximately three grams of ignition pellets which are ignited by either of two pairs of squibs. Each pair has independent circuitry from a different power source and any one squib is capable of igniting the pellets.

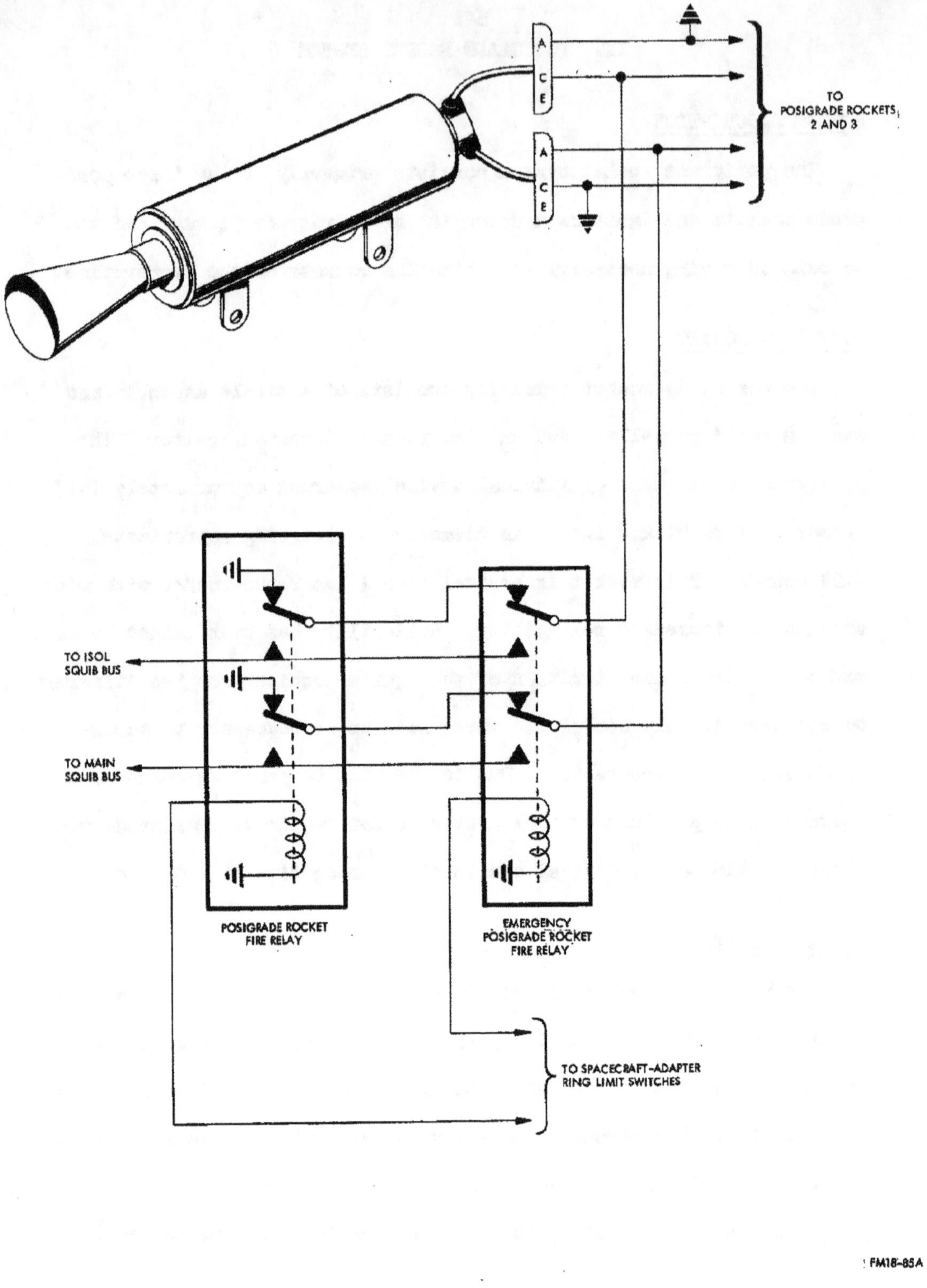

Figure 9-2. Posigrade Rocket Ignition System

9-4. SYSTEM OPERATION

The purpose of the posigrade rockets is to accomplish separation between the spacecraft and booster. Under normal mission conditions, the posigrade rockets are activated by the spacecraft-to-booster clamp ring separation, which occurs at sustainer engine cut-off. The posigrade rockets propel the spacecraft away from the booster at a rate of 15 feet per second. The three rockets are fired simultaneously; however, should two of them fail, the remaining unit would successfully cause separation.

The posigrade rockets are also used to separate the spacecraft from the booster in the event an abort is initiated after tower separation but prior to sustainer engine cut-off.

SECTION X

RETROGRADE ROCKET SYSTEM

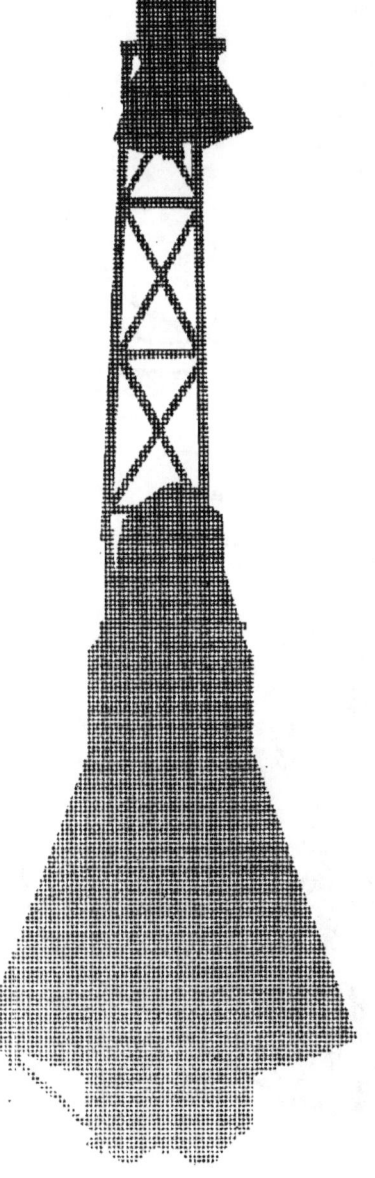

TABLE OF CONTENTS

TITLE	PAGE
System Description	10-3
System Operation	10-5

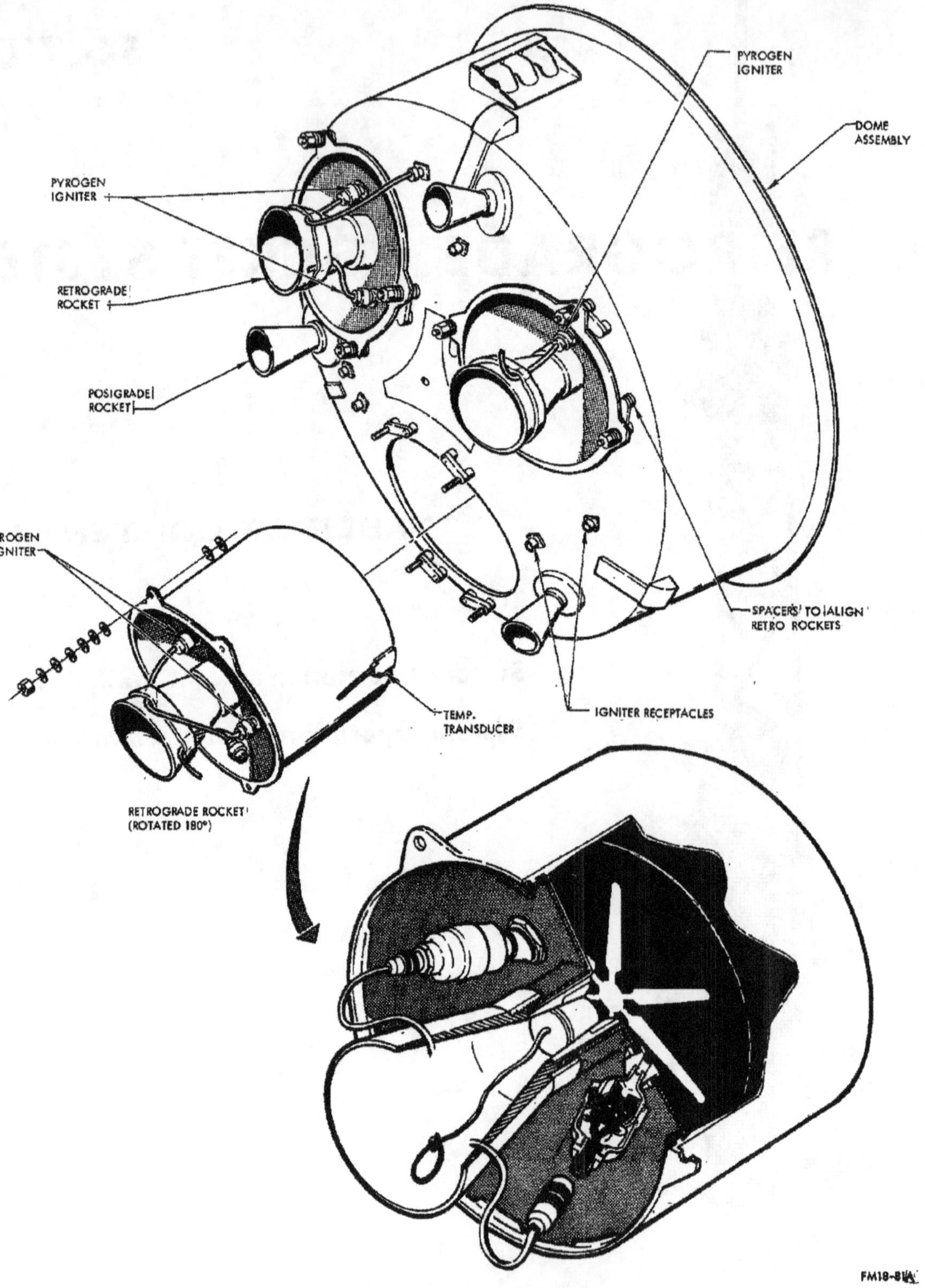

Figure 10-1 Retrograde Rocket System

SEDR 104

X. RETROGRADE ROCKET SYSTEM

10-1. SYSTEM DESCRIPTION

The retrograde rocket system consists primarily of the three retro-rockets, their pyrogen igniters, and the associated wiring necessary for rocket ignition. The retro-rockets are housed in the jettisonable retrograde package along with the posigrade rockets (See Figure 10-1).

10-2. ROCKET MOUNTING

The retro-rockets are mounted in the retro package. The retro package is held against the center of the heat shield by three retention straps. The retention straps are attached at three equally spaced points on the base of the spacecraft and converge to a common point in the center of the retro package where they are joined together by an explosive bolt. Sixty seconds following retrograde firing signal, the bolt detonates, the straps are released and a coil spring ejects the retro package away from the spacecraft (See Figure 10-2). To protect the rockets, particularly from micro-meteorites, each rocket has a metal cover over its exposed nozzle end. The cover is blown off by the rocket blast at time of light-off. Mounting of the rockets is so designed as to direct the resultant thrust vector towards the spacecraft's predetermined center of gravity at time of firing.

10-3. RETRO PACKAGE ELECTRICAL WIRING

The retrograde package is supplied electrical power through three electrical umbilicals which are equally spaced around the base of the

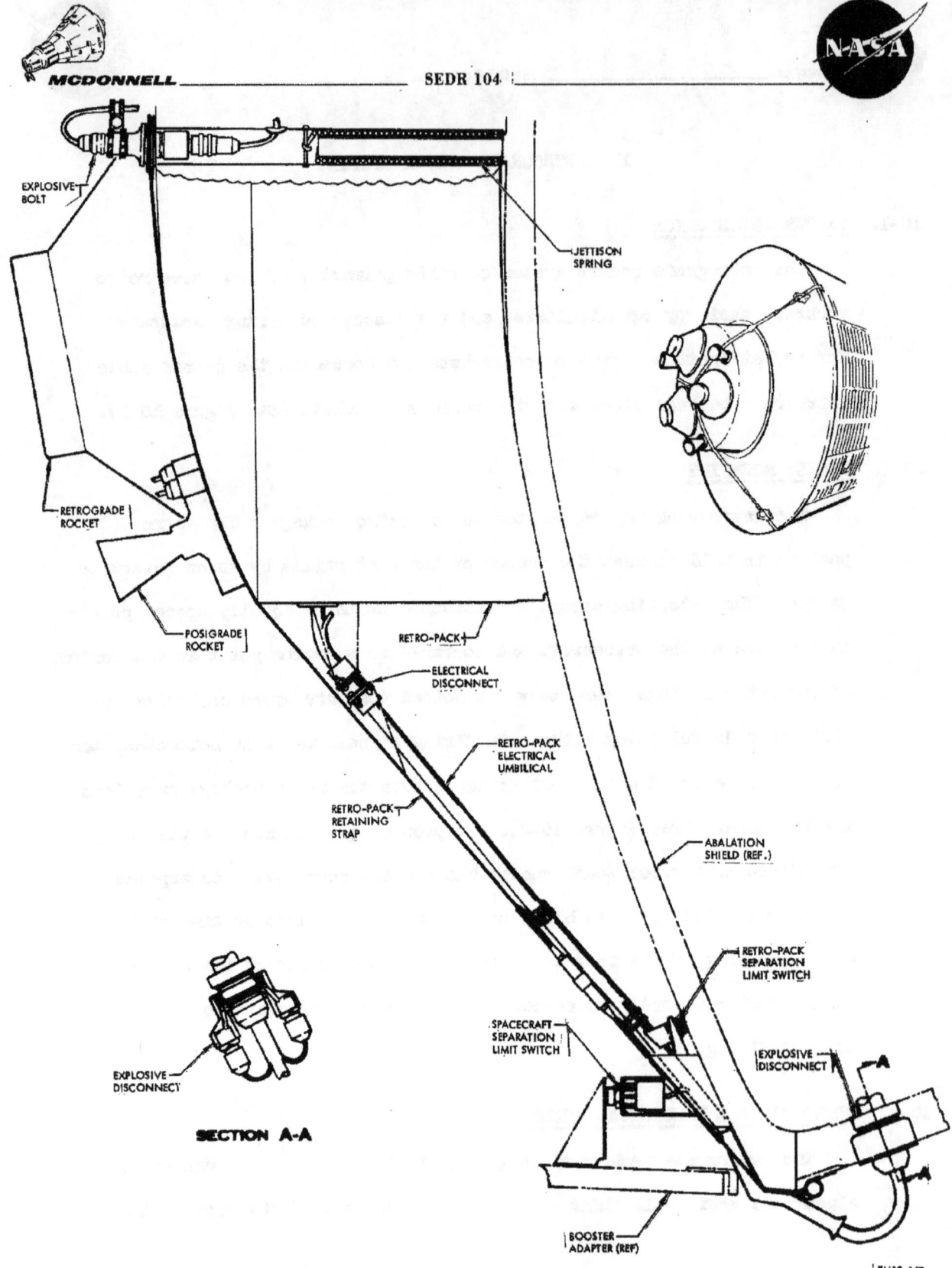

Figure 10-2 Retro-Pack Retention System

spacecraft. The electrical umbilicals leave the spacecraft through explosive disconnects, follow the three retro package retention straps to the retro package and enter through rubber grommets. A three connector test point is located on the side of the retro package and allows all electrical leads to be broken between the point at which they enter the retro package and the unit to which they supply power. (See Figure 10-3).

10-4. **RETROGRADE ROCKET**

Leading particulars are: total weight approximately 68 pounds, length 15.4 inches, diameter 12 inches. Due to the importance of the retrograde system to the over-all mission, a redundant rocket firing system has been employed. An attached temperature transducer monitors retro-rocket temperature. Each retro-rocket has a total impulse of 12,960 lb. - sec. $\pm$ 5%, in a vacuum. Average thrust is 1070 pounds for a 11.26 second firing time.

10-5. **SYSTEM OPERATION**

The purpose of the retrograde rocket system is to slow the spacecraft prior to re-entry. Actual firing of the rockets is preceded by a 30-second period during which the spacecraft is positioned to the retrograde attitude. Retrograde attitude is defined as follows: $34° \pm 1°$ pitch, $0° \pm 1°$ roll or yaw. The firing sequence will not begin, normally, until the retrograde attitude limits have been attained, and will be temporarily interrupted should the spacecraft exceed the attitude limits after the sequence has begun. Should the need arise, however, the above limits may be manually overridden. Firing of the rockets, which occurs at five-second intervals can be initiated by any of the following: (1) satellite clock runout; (2) Astronaut selection; (3) ground command.

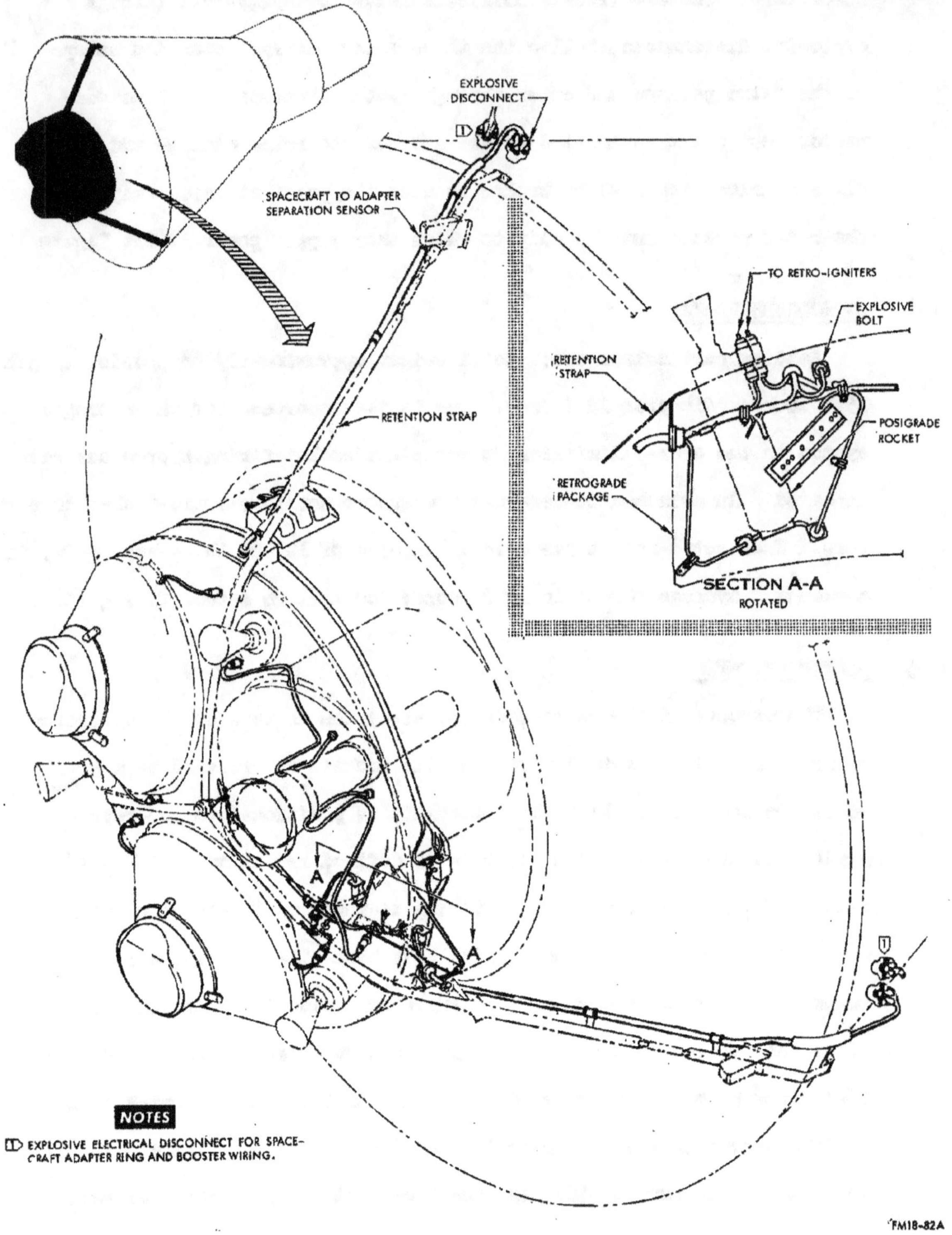

Figure 10-3. Retro-Package Electrical Installation

10-6. ROCKET FIRING

All three rocket fire relays receive 24 V d-c simultaneously; however, the No. 2 and No. 3 rocket fire relays have a five- and ten-second time delay, respectively. The left rocket fires first, five seconds after the left rocket ignites the bottom rocket fires and five seconds after the bottom rocket ignites the right rocket fires. Each retro-rocket fires for a total of ten seconds.

The retro-rockets are fired sequentially to avoid the ineffective results from a failure of either of the first two rockets. Consequently, if the No. 1 rocket failed to a degree which would disrupt the Retrograde Attitude Permission Relay, the No. 2 Retro Fire Relay would become de-energized. The spacecraft could then be repositioned automatically or manually by the Reaction Control System and upon regaining the retrograde attitude, the No. 2 Retro Fire Relay will be re-energized through the Retrograde Attitude Permission Relay. With the No. 2 Retro Fire Relay energized, the No. 2 rocket will be fired and the retro sequence continued. The same sequence of events would occur if the No. 2 rocket were to fail.

10-7. RETRO-ROCKET EMERGENCY OVERRIDE

There are five telelights on the astronaut's left console which concern the retrograde system. The first one is RETRO SEQ. and is a green function light. This light will illuminate when the retro sequence is started, either by the satellite clock or by the button adjacent to the light. The purpose of the button is to initiate re-entry prior to satellite clock runout or failure of same.

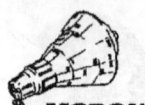

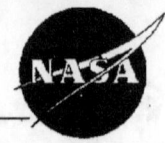

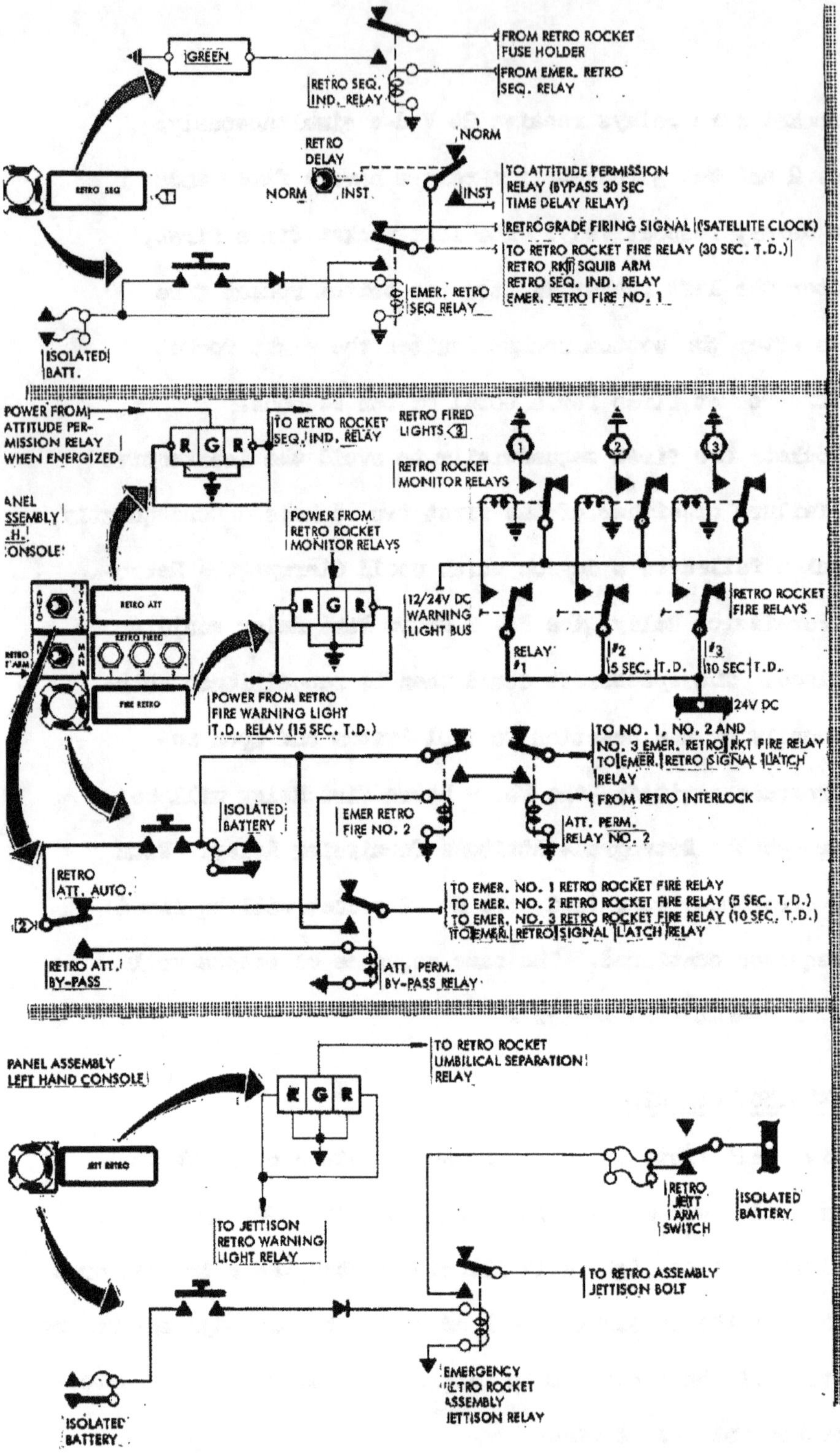

REASON FOR SELECTION

1 FAILURE OF SATELLITE CLOCK TO FUNCTION.

2 EARLY RE-ENTRY BEFORE CLOCK RUN-OUT.

1 THE "RETRO ATTITUDE" LIGHT DOES NOT ILLUMINATE OR ILLUMINATES RED, BUT VISUAL AIDS INDICATE SPACECRAFT IS IN RETRO ATTITUDE.

1 "JETTISON RETRO" LIGHT ILLUMINATES RED.

NOTES

1 GREEN LIGHT ONLY.

2 IF THE "RETRO ATT." LIGHT IS GREEN AND THE "FIRE RETRO" LIGHT ILLUMINATES RED THE "RETRO ATT" SWITCH SHOULD BE LEFT IN THE "AUTO" POSITION AND THE "FIRE RETRO" BUTTON DEPRESSED. CIRCUITRY WILL PASS THROUGH ENERGIZED ATTITUDE PERMISSION RELAY BUT WILL BY-PASS THE 30 SEC. T.D. SEQ. FIRE RELAY.

3 LIGHTS DISABLED

Figure 10-4 Retrorocket Over-ride Firing System Schematic

The next three telelights in the retrograde sequence are RETRO ATT., RETRO FIRED and FIRE RETRO. For a normal flight, the Retro Attitude Switch adjacent to the RETRO ATT. Light should be in "AUTO", this requires the spacecraft to be in the retrograde attitude before the retro-rockets will fire. With the Retro Attitude Switch in the "BYPASS" position, the retro-rockets can be fired without the spacecraft being in the retrograde attitude.

The RETRO ATT Telelight illuminates red when the spacecraft is not in retrograde attitude and illuminates green when retrograde attitude is attained, after Tr (time-retrograde) signal.

Originally the three RETRO FIRED Telelights illuminated individually as each retro-rocket was sequentially fired. Experience has shown that visual indication of retro-rocket firing is unnecessary. Therefore, the three RETRO FIRED Telelights have been disabled.

The FIRE RETRO Telelight will illuminate green when all three retro-rockets have fired. The FIRE RETRO Telelight will illuminate red if the three retro-rockets have not ignited within 15 seconds after the fire signal is sent to the No. 1 retro-rocket.

If the RETRO ATT. Telelight illuminates red, the astronaut must check the spacecraft attitude in order to determine if the spacecraft is in the correct retrograde attitude. If the spacecraft is found to be in the correct attitude, then the astronaut should position the Retro Attitude Switch to the "BY-PASS" position and also push the FIRE RETRO button.

Ten seconds later the FIRE RETRO Telelight should illuminate green. However, if the astronaut determined that the spacecraft was not in the correct attitude, the Reaction Control System should be employed in order to correctly position the spacecraft in the retrograde attitude (See Section 5). When the correct retrograde attitude is attained, the RETRO ATT. Telelight should illuminate green.

If the RETRO ATT. Telelight illuminates green, but the FIRE RETRO Telelight illuminates red, the button adjacent to the FIRE RETRO Telelight should be pushed leaving the Retro Attitude Switch in the "AUTO" position.

The fifth telelight is the JETT. RETRO Telelight. This telelight will illuminate green 60 seconds after the No. 1 retro-rocket is ignited. In the event this light illuminates red, the adjacent button should be depressed to supply an alternate source of power to the retro-package jettison bolt. If the retro-package cannot be jettisoned by the automatic or override method, it will be ejected sometime during re-entry when the extreme heat encountered will detonate the explosive bolt or burn the retention straps to allow the coil spring to eject the package.

10-8. RETRO ROCKET ARM SWITCHES

Located on the left hand console are two switches which control power to the retro-rocket ignition squibs and to the retrograde package jettison squib.

The Retro Rocket Squib Arm Switch has an "AUTO" and a "MAN" position. With the switch in the "AUTO" position power is supplied to the retro-

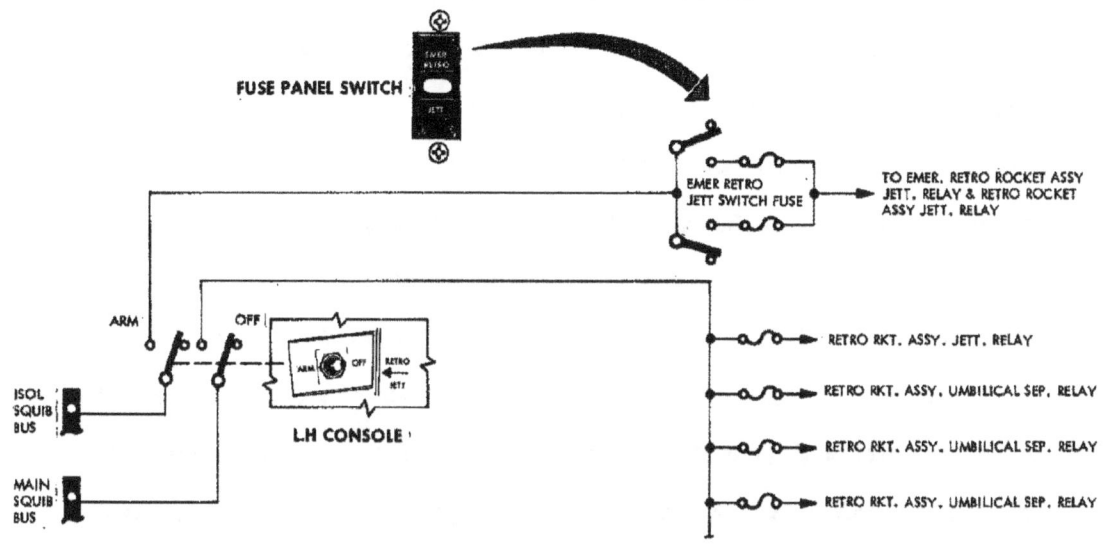

Figure 10-5 Retrorocket Arm Switches

Rocket Fire Relay contacts when the Retro Rocket Squib Arm Relay is energized. By placing the switch in the "MAN" position power is supplied directly to the Retro Rocket Fire Relay contacts through the switch, bypassing the Retro Rocket Squib Arm Relay. (See Figure 10-5).

The Retro Rocket Jettison Arm Switch has an "ARM" position and an "OFF" position. With the switch in the "OFF" position, no power is available to fire the explosive bolt which holds the retro-package to the spacecraft and no power is available to fire the disconnects in the three umbilicals leading to the retro-package. By placing the switch in the "ARM" position, power is supplied to the contacts of the Retro Rocket Assembly Jettison Relay and the Retro Rocket Assembly Umbilical Separation Relay (See Figure 10-5).

10-9. RETROGRADE SYSTEM WARNING LIGHTS

Two amber warning lights, located on the main instrument panel, concern the Retrograde System. The RETRO WARN Warning Light illuminates 30 seconds prior to the beginning of the retrograde sequence. The RETRO RESET Warning Light is illuminated by ground command and pertains to the reset of retrograde time on the satellite clock.

SECTION XI

ELECTRICAL POWER AND INTERIOR LIGHTING SYSTEMS

TABLE OF CONTENTS

TITLE	PAGE
Electrical Power System	
System Description	11-3
System Operation	11-4
System Units	11-11
Interior Lighting	
System Description	11-16

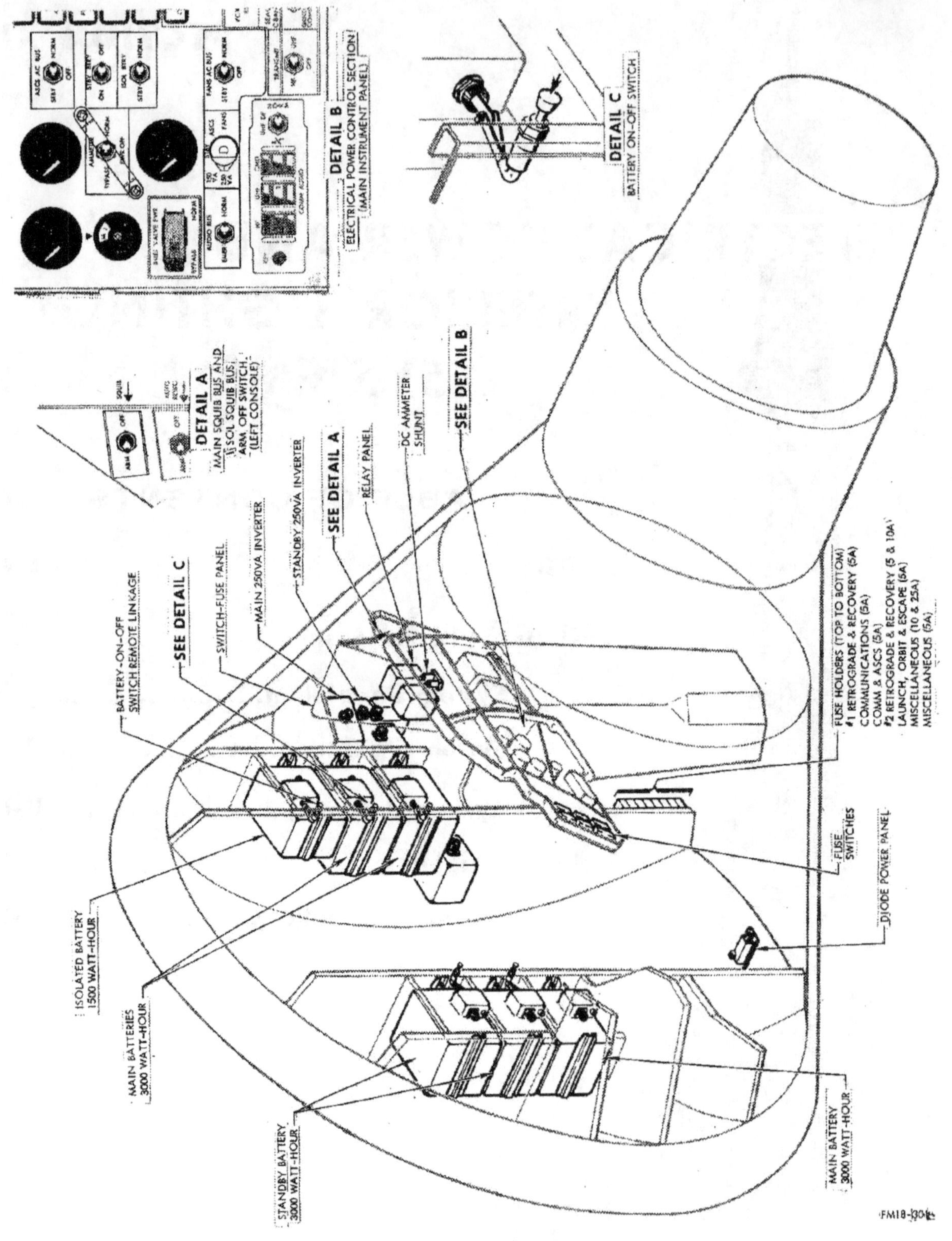

MCDONNELL — SEDR 104

Figure 11-1. Electrical Power Supply System

SEDR 104

XI. ELECTRICAL POWER AND LIGHTING SYSTEMS

11-1. **ELECTRICAL POWER SYSTEM**

11-2. **SYSTEM DESCRIPTION**

The spacecraft power supply consists of three, 3000 WH main batteries, two, 3000 WH standby batteries and one, 1500 WH isolated battery. The standby batteries have paralleled taps brought out at 6, 12 and 18 volts to power the spacecraft 6 V standby, 12 V standby and 18 V standby busses. The isolated battery also has taps brought out at 6 V and 18 V to power the spacecraft 6 V isolated and 18 V isolated busses. External d-c power is supplied through fuses or diodes in the spacecraft prior to launch (See Figure 11-1 and 11-2). An ammeter is used to indicate battery current when the batteries are in use. A d-c voltmeter is utilized to read individual battery voltages as well as the main and isolated 24 V bus voltage. Main and isolated 24 V bus voltage may be read on the d-c voltmeter when the main bus is powered from external power or the battery power.

The d-c electrical loads are supplied through fuses with the exception of vital control circuits which incorporate a solid conductor in place of a fuse. Some of the d-c circuits utilize two fuses in a switch-fuse arrangement. A three position, center OFF switch, provides normal operation in the No. 1 position or a rapid switchover to the No. 2 position for emergency use. Some of the more important d-c circuits are provided with a solid conductor in the emergency or No. 2 position of the following circuits:

(a) Emerg. Capsule Separation Control.

(b) Tower Separation Control.

11-3

    (c) Emerg. Main Chute Deploy.

    (d) Retro. Jett. Control.

    (e) Retro Manual Control.

    (f) Reserve Chute Deploy.

    (g) Emerg. Reserve Chute Deploy.

    (h) Retro Seq.

    (i) Envir. Control.

A-c power is produced by two main inverters and one standby inverter. The two main inverters are rated at 115 volts, 400 cycles. One main inverter has a capacity of 250 VA while the other main inverter is rated at 150 VA. The standby inverter has a capacity of 250 VA at 115 volts, 400 cycles.

The a-c loads are not fused because of inherent overload protection in the inverters.

11-3. **SYSTEM OPERATION**

11-4. **D-C POWER CONTROL**

The spacecraft main 24 volt d-c power supply consists of three, 3000 watt-hour batteries. The main batteries are connected in parallel to each other by an ON-OFF switch on each battery. Each battery contains an internal diode in the positive leads as protection against discharge through a faulty or discharged battery. The main batteries feed the main bus directly when the individual BTRY, ON-OFF switches are in the ON position.

The main 24 volt d-c bus is connected directly to the inputs of the two filters for the main 250 VA inverter and the main 150 VA inverter.

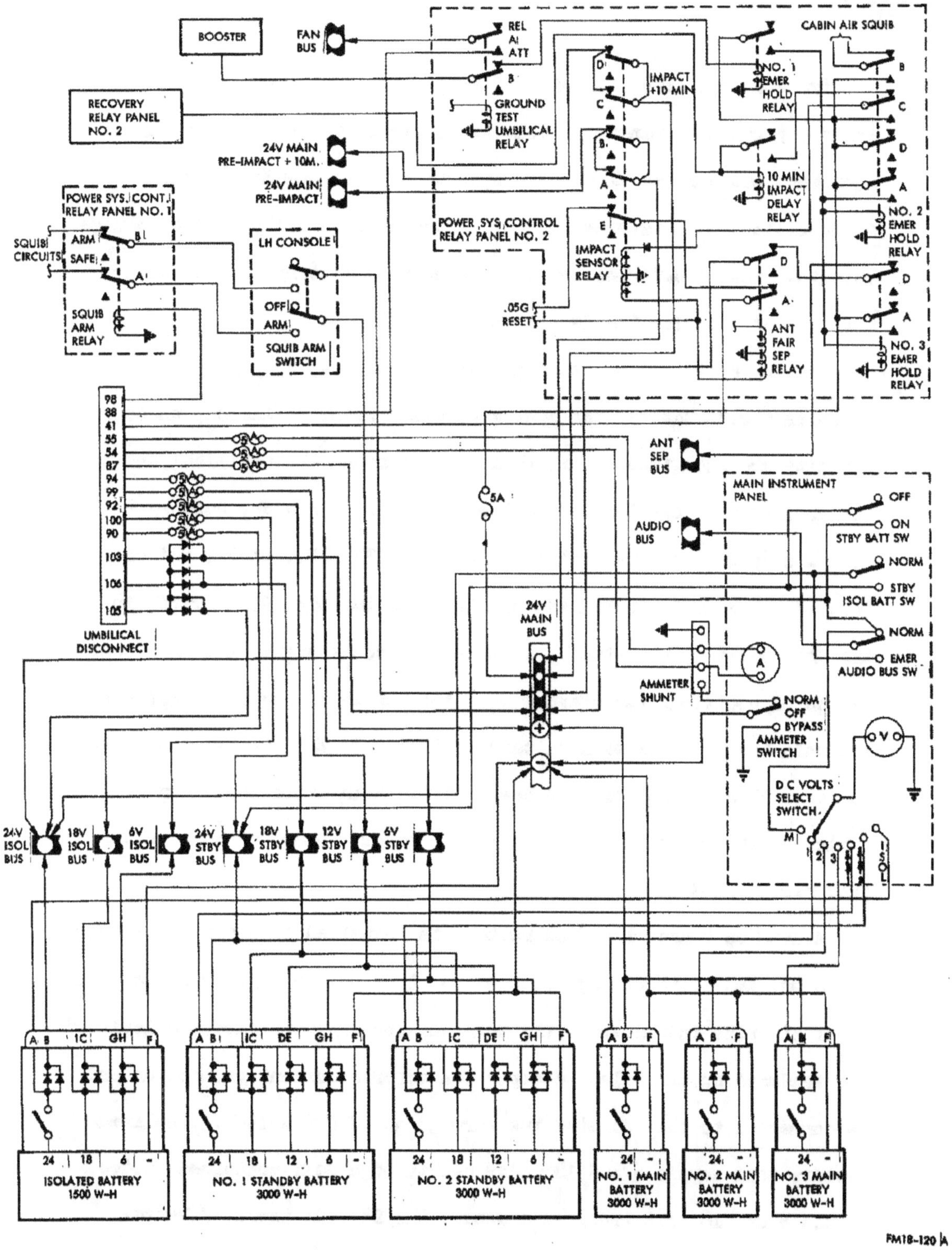

Figure 11-2. D-C Power Control Schematic

SEDR 104

Two standby batteries, of 3000 watt-hours capacity each, provide standby power and lower tap voltage requirements for communications equipment and warning lights dim operation. The standby batteries incorporate diode, reverse current protection, on all positive voltage outputs. ON-OFF switches are located on each battery to provide a means to de-energize the 24 volt output.

Standby battery taps, through internal diodes, supply 6, 12 and 18 volts d-c to the various system busses. Prior to launch, these circuits are energized by external power through the umbilical and the external power fuses located in the spacecraft. Standby battery 24 volt d-c application is controlled by the STBY BTRY, ON-OFF switch located on the main instrument panel. With the STBY BTRY switch in the OFF position, the standby batteries do not energize the main bus. In the ON position, the standby 24 V d-c bus is connected directly to the main bus. If the main battery is depleted, then the standby batteries will energize the main bus. If the main batteries are not depleted, then both battery groups will supply power to the main bus. (See Figure 11-2).

A 1500 watt-hour isolated battery is installed to provide emergency audio bus and squib firing voltages and to supply emergency voltages to other circuits in the event the main and standby batteries are depleted. The isolated battery also incorporates reverse current protection and is connected to the isolated bus through an ON-OFF switch. Isolated battery taps supply 6 and 18 volts d-c through self-contained diodes, to the 6 and 18 volt isolated busses. The 24 volt isolated battery

output is available through the ARM position of the SQUIB switch and through the EMER position of the AUDIO BUS switch to the associated busses. The isolated 24 volt d-c output may also be connected in parallel with the 24 volt output of the standby batteries through the STDBY position of the ISOL-BTRY switch.

External d-c power is supplied through the umbilical cable to spacecraft circuitry. This power is used for pre-launch operations in order to converve the spacecraft battery supply. Normally 6, 12, 18 and 24 volts d-c are supplied through seven inputs. External power voltages of 6, 12 and 18 volts are fed through fuses in the spacecraft to the 6, 12 and 18 volt busses. The 24 V external power is routed through diodes in the spacecraft, to the 24 V busses. Only four external power supplies are used to supply the 6, 12, 18 and 24 volt requirements. The d-c ammeter is used, with the ammeter switch in the NORM position to indicate d-c current from the batteries in the circuit.

11-5. A-C POWER AND CONTROL

11-6. MAIN INVERTERS

Main 115 volt, 400 cycle a-c power is supplied by two inverters of 150 volt-amperes and 250 volt-amperes each. The a-c load is divided into two groups namely the ASCS a-c bus and the FANS a-c bus. The 250 VA inverter supplies the ASCS a-c bus and the 150 VA inverter supplies the FANS a-c bus. The cabin fluorescent lights are energized from the FANS bus. The main d-c bus powers the 150 VA (fans bus) inverter through a line

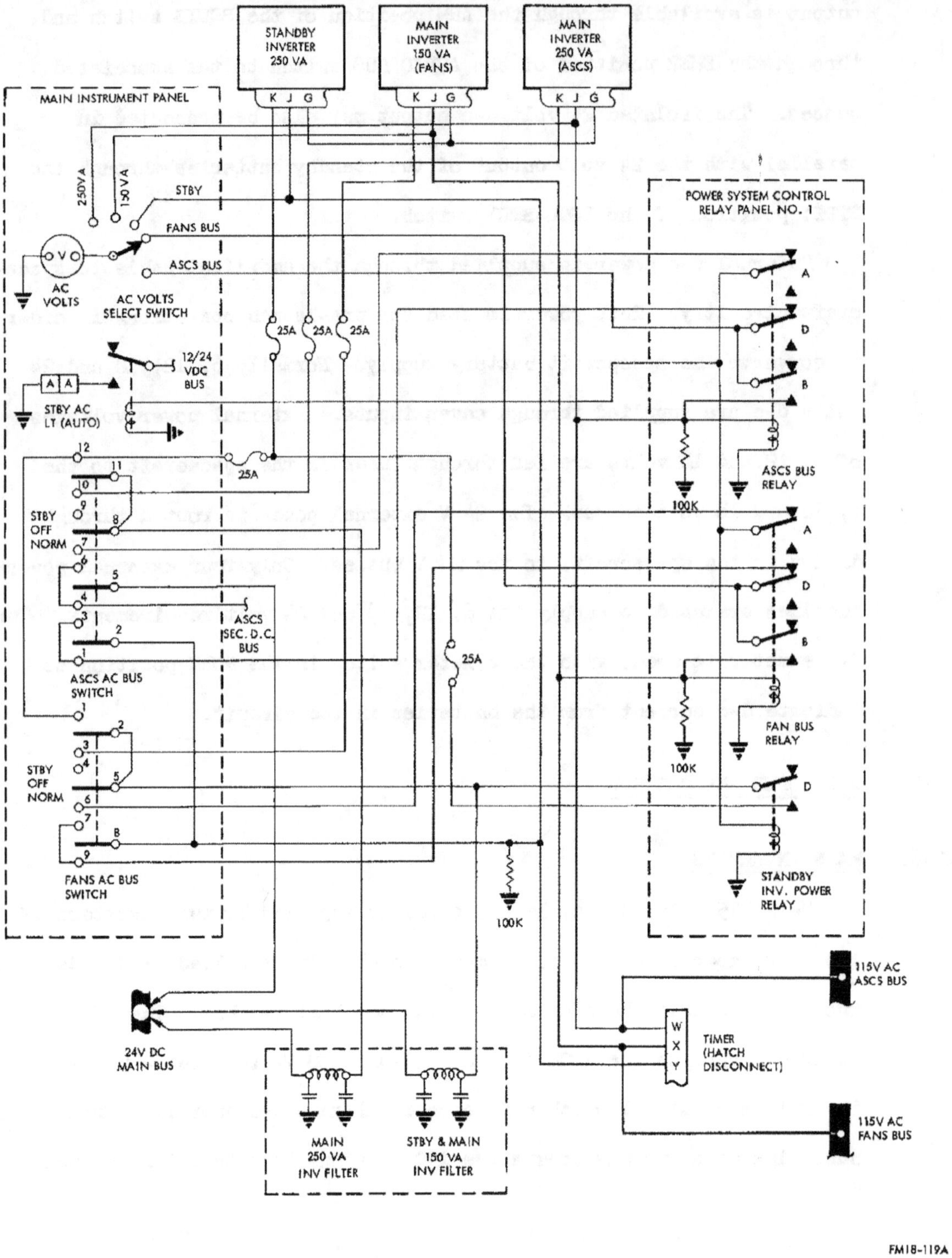

MCDONNELL SEDR 104

Figure 11-3 A-C Power Control Schematic

11-8

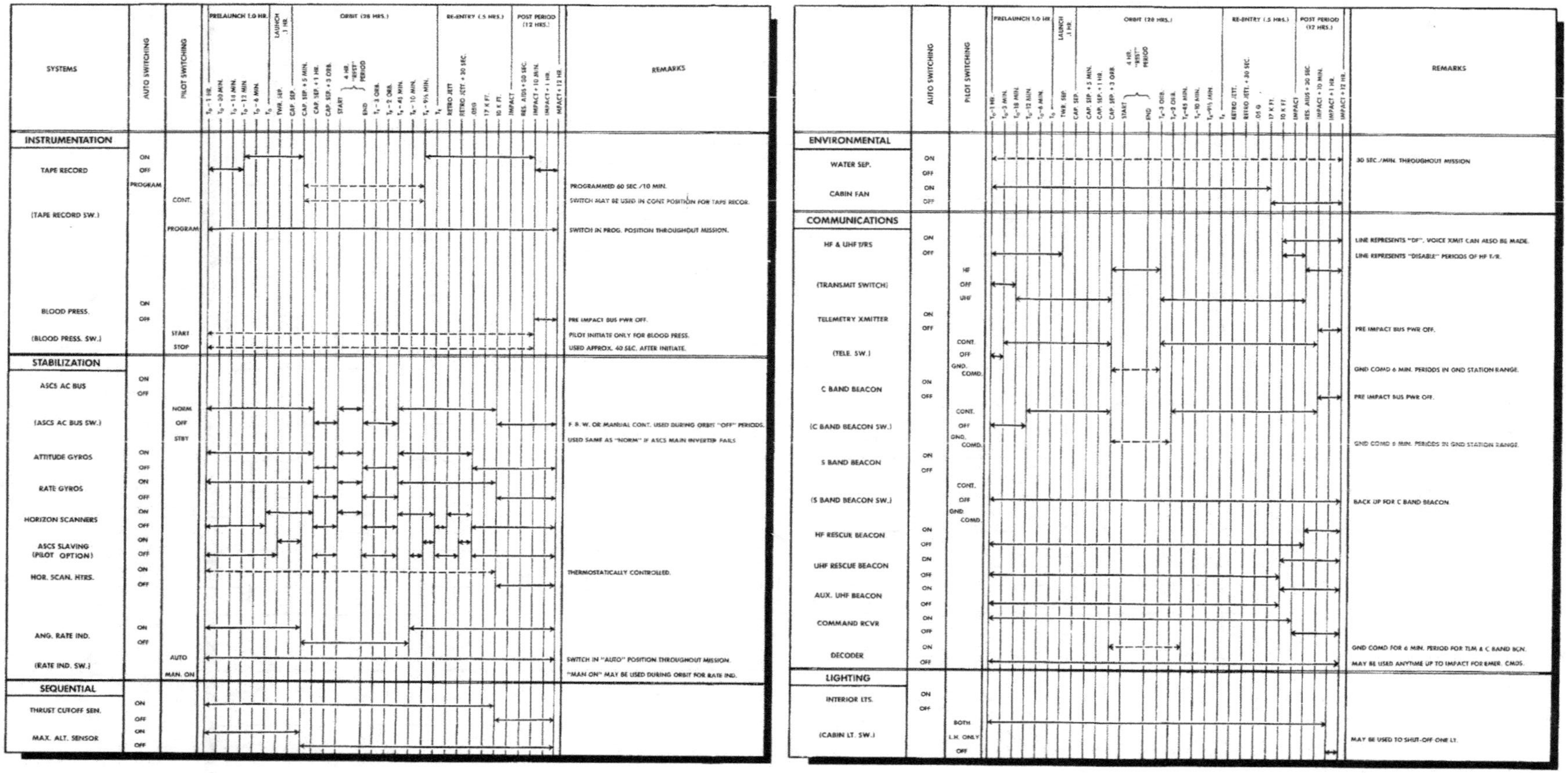

Figure 11-4 Typical System Equipment Operating Sequence

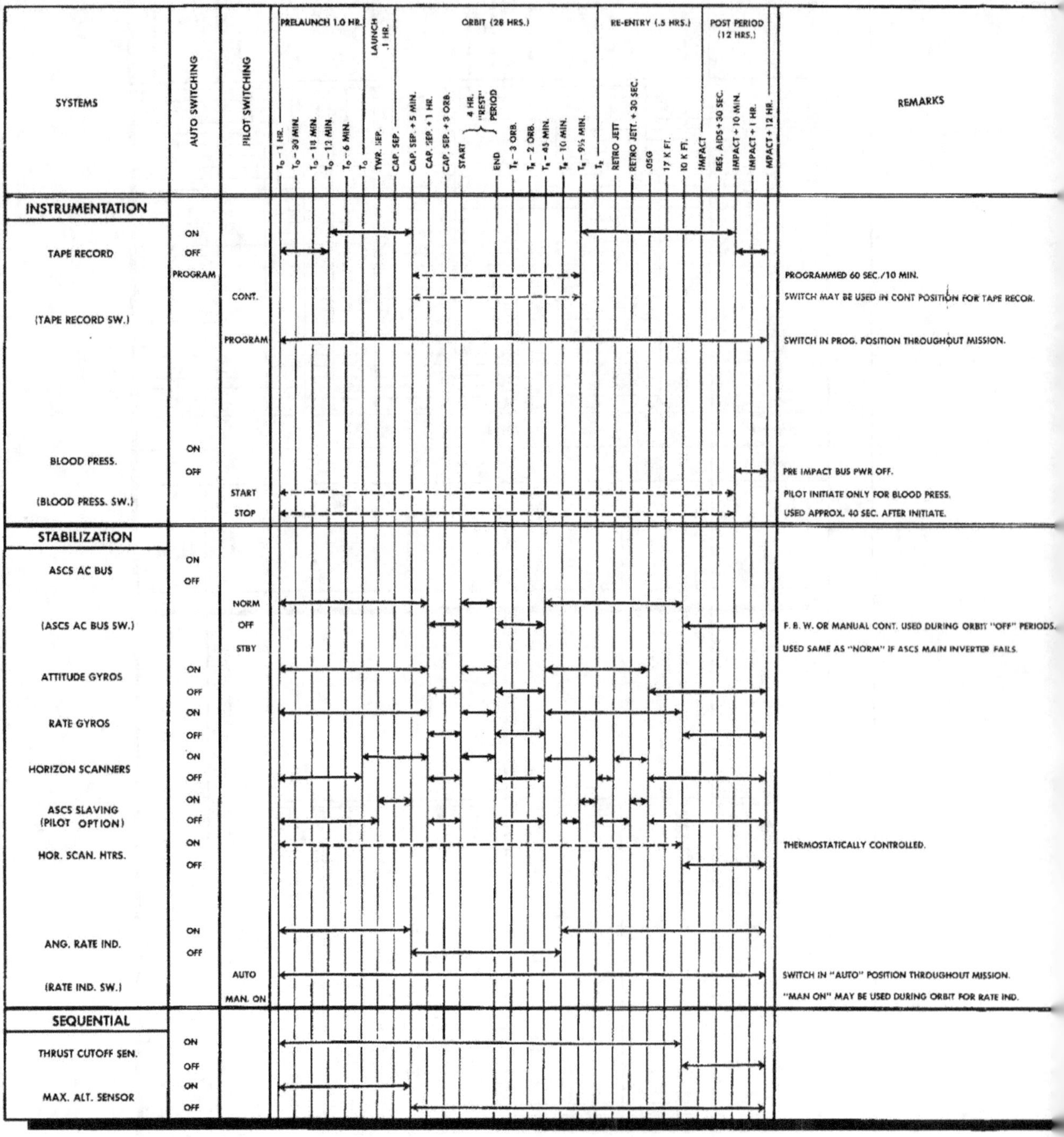

Figure 11-4 Typical System Equipment Operating Sequence

11-9

SYSTEM
EDR 104

	AUTO SWITCHING	PILOT SWITCHING	PRELAUNCH 1.0 HR.	LAUNCH .1 HR.		ORBIT (28 HRS.)		RE-ENTRY (.5 HRS.)	POST PERIOD (12 HRS.)	REMARKS
ENVIRONMENTAL										
WATER SEP.	ON / OFF		←---→							30 SEC./MIN. THROUGHOUT MISSION
CABIN FAN	ON / OFF		←---→							
COMMUNICATIONS										
HF & UHF T/RS	ON / OFF		←------→					←→		LINE REPRESENTS "DF", VOICE XMIT CAN ALSO BE MADE. LINE REPRESENTS "DISABLE" PERIODS OF HF T/R.
(TRANSMIT SWITCH)		HF / OFF / UHF	←→ ←-----------------→							
TELEMETRY XMITTER	ON / OFF		←---→					←→		PRE IMPACT BUS PWR OFF.
(TELE. SW.)		CONT. / OFF / GND. COMD.	←---→ ←------------------→ ←----→							GND COMD 6 MIN. PERIODS IN GND STATION RANGE.
C BAND BEACON	ON / OFF		←---→					←→		PRE IMPACT BUS PWR OFF.
(C BAND BEACON SW.)		CONT. / OFF / GND. COMD.	←---→ ←------------------→ ←----→							GND COMD 6 MIN. PERIODS IN GND STATION RANGE.
S BAND BEACON	ON / OFF		←---→							
(S BAND BEACON SW.)		CONT. / OFF / GND. COMD.								BACK UP FOR C BAND BEACON.
HF RESCUE BEACON	ON / OFF							←-------→		
UHF RESCUE BEACON	ON / OFF		←---→							
AUX. UHF BEACON	ON / OFF		←---→							
COMMAND RCVR	ON / OFF		←---→							
DECODER	ON / OFF					←----→				GND COMD FOR 6 MIN. PERIOD FOR TLM & C BAND BCN. MAY BE USED ANYTIME UP TO IMPACT FOR EMER. CMDS.
LIGHTING										
INTERIOR LTS.	ON / OFF									
(CABIN LT. SW.)		BOTH / L.H. ONLY / OFF							←→	MAY BE USED TO SHUT-OFF ONE LT.

FM18-118A

filter circuit and a 25 ampere fuse. The 250 VA inverter (ASCS bus) is also powered from the main d-c bus through a line filter and 25 ampere fuse. The d-c power is controlled through the NORM position of the ASCS AC BUS, and FANS AC BUS switches on the main instrument panel. The outputs of the FANS and ASCS inverters feed the solenoids of the fans bus relay and the ASCS bus relay. These energized relays feed the inverter output through the closed contacts of the relays to power the FANS and ASCS busses. (See Figure 11-3).

11-7. STANDBY INVERTER

Standby 115 V, 400 cycle a-c power is supplied by one 250 volt ampere standby inverter. The standby inverter will supply a-c power to either, or both, ASCS and FANS a-c busses by selecting the STANDBY position on the respective ASCS AC BUS or FANS AC BUS switch located on the main instrument panel.

In the event of failure of either the 250 VA or 150 VA main inverters the respective fan bus relay or ASCS bus relay will be de-energized. This action will automatically energize the standby inverter relay which in turn will apply d-c power from the filter to the standby inverter. The a-c output from the standby inverter is then directed through contacts in the de-energized ASCS or fans bus relays to their respective busses. A warning light on the main instrument panel indicates when the standby inverter is switched into operation by reason of failure of either of the main inverters. (See Figure 11-3).

11-8. POWER DISTRIBUTION

11-9. D-C POWER DISTRIBUTION

D-c power is taken from three separate battery groups, namely the main battery, standby battery and isolated battery. Various sub-busses which operate from these sources and the bus separation method are as follows:

(a) Main d-c bus directly to the main batteries.

(b) Main 24 V squib bus through SQUIB switch from main bus.

(c) Main retro squib bus from main bus through main retro squib power drop relay.

(d) ASCS main 24 V bus from main bus through separation relay.

(e) Pre-impact plus 10 minute from main bus through impact relay.

(f) Standby d-c bus directly to standby battery.

(g) Isolated d-c bus directly to isolated battery.

(h) Isolated 24 V squib bus through SQUIB switch from isolated bus.

(i) Isol retro squib bus from main squib bus through isolated retro squib power drop relay.

(j) Audio bus from main bus or isolated bus through AUDIO BUS switch.

(k) Standby 6, 12 and 18 volt busses directly to taps on standby battery.

(l) Isolated 6 and 18 volt busses directly to taps on isolated battery.

(m) Warning lights 12/24 V d-c bus.

11-10. SYSTEM UNITS

11-11. BATTERIES

Each cabin battery consists of series connected silver-zinc rechargeable cells having a nominal potential rating of 24 volts and a

MCDONNELL — SEDR 104

NOTES
1. TOTAL ACTUAL AND INDICATED POWER AND HEAT DISSIPATION LOAD 12,589.8 WATT-HOURS. (MAIN BATTERIES SUPPLY 7,553.8 WATT-HRS. AND STANDBY BATTERIES SUPPLY 5,035.9 WATT-HRS.).
2. STANDBY BATTERIES ARE PARALLELED WITH MAIN BATTERIES THROUGHOUT ENTIRE MISSION.
3. PEAKS TO SHOW SQUIB FIRING SEQUENCE ONLY.

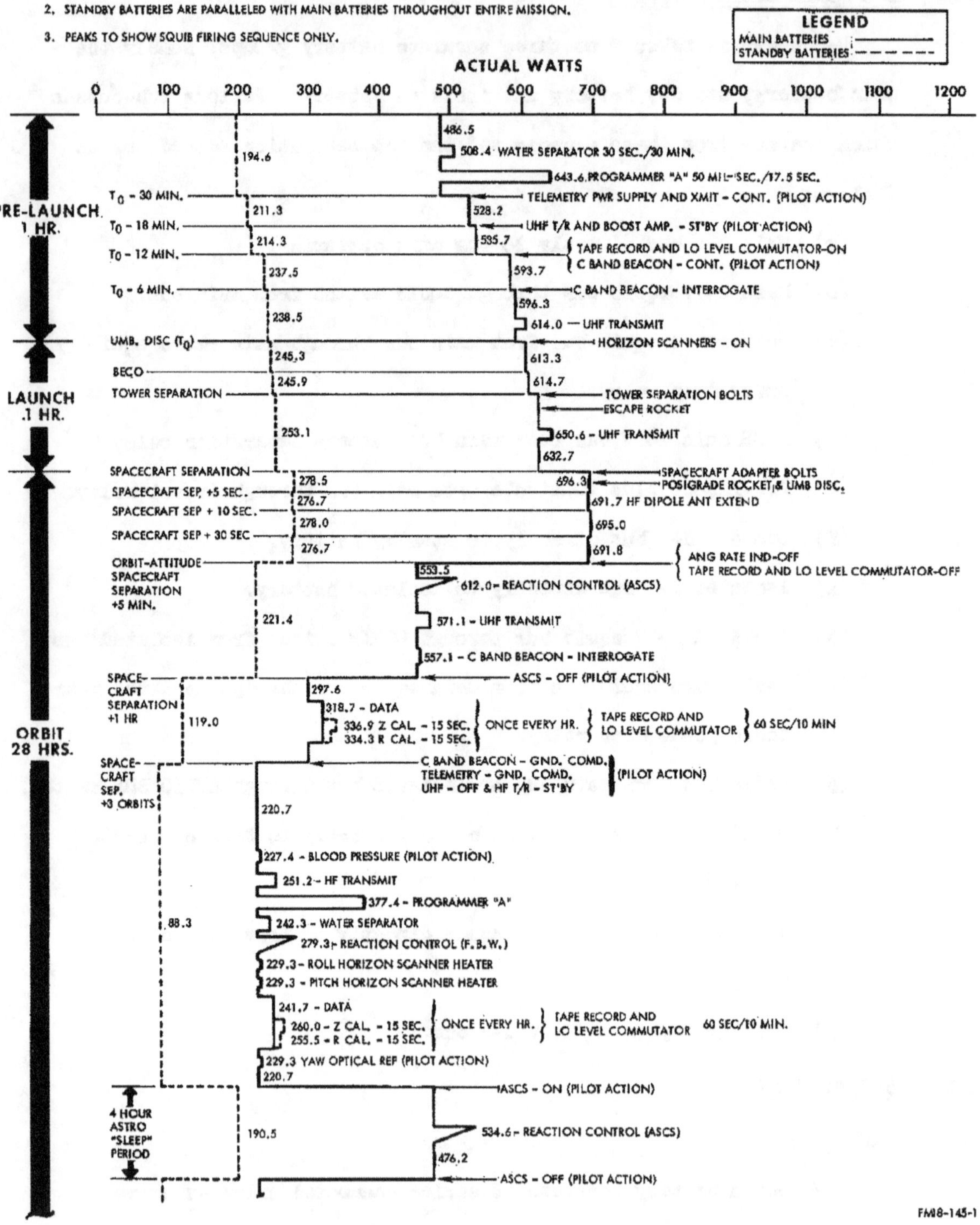

Figure 11-5 D-C Watt-Hour Loading (Sheet 1 of 2)

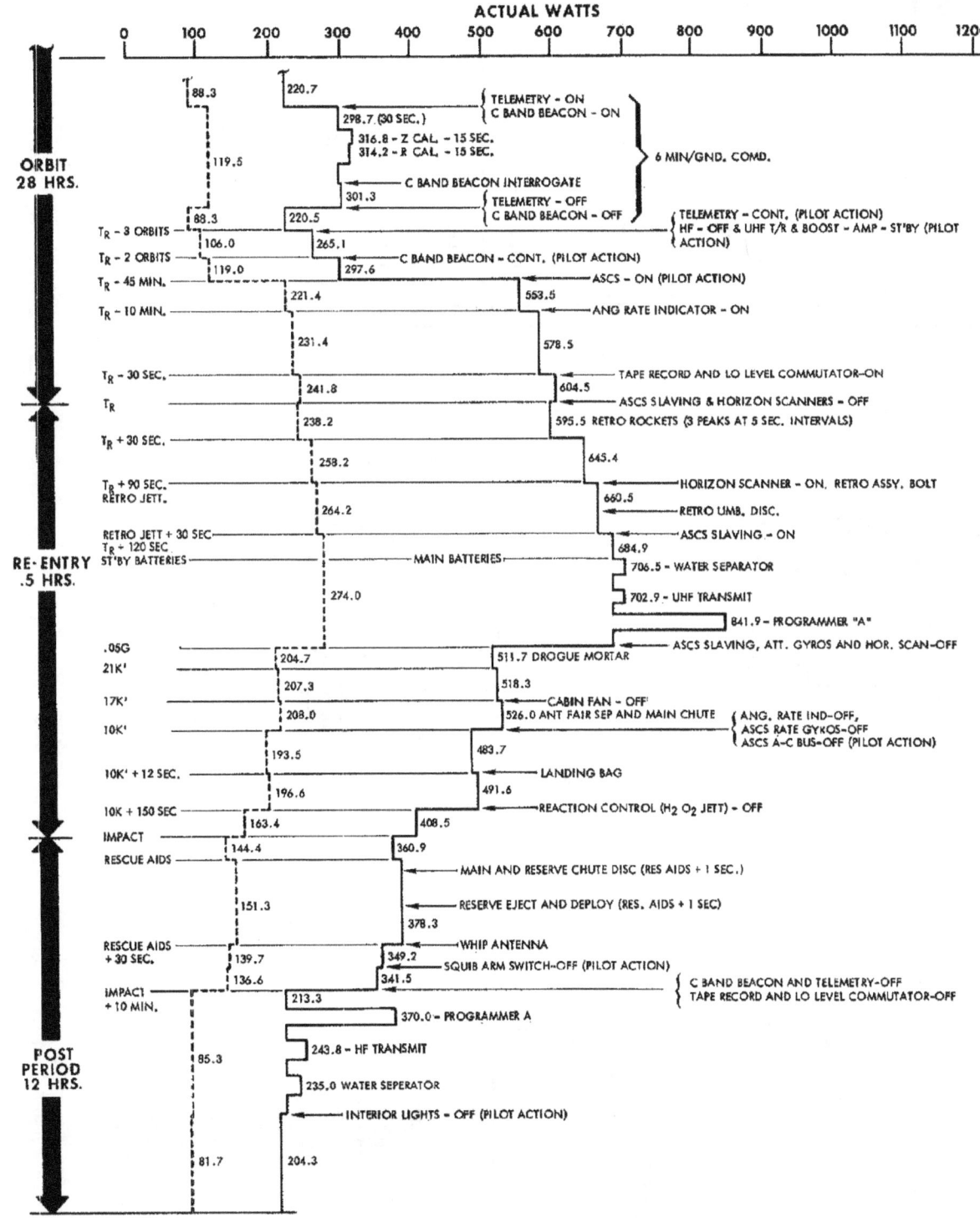

Figure 11-5 D-C Watt-Hour Loading (Sheet 2 of 2)

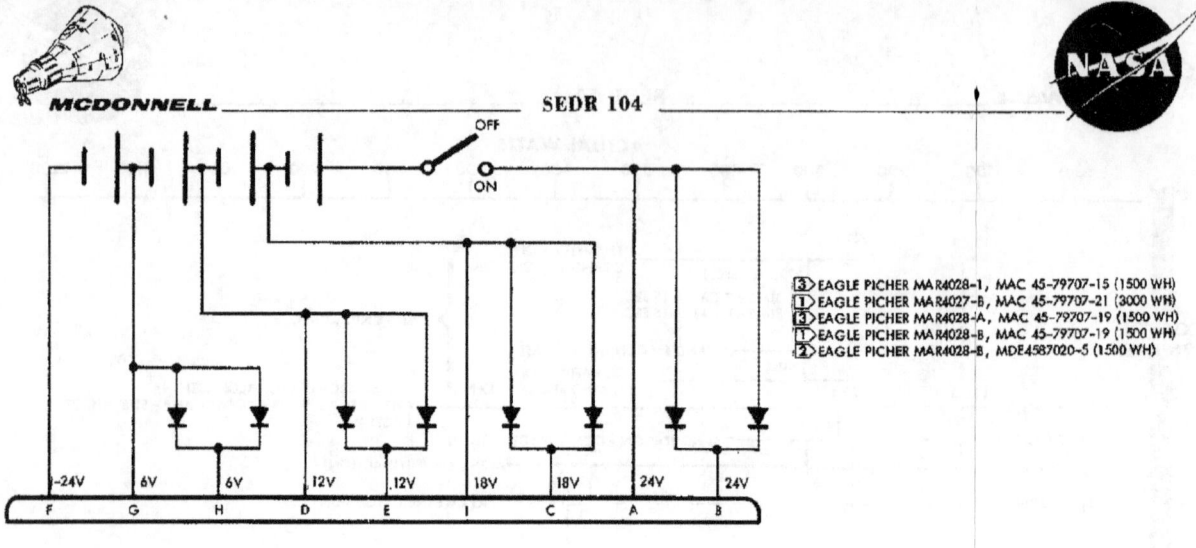

NOTES

1. MAY BE INSTALLED IN SPACECRAFT FOR GROUND TESTS OR FLIGHT.
2. MAY BE INSTALLED IN SPACECRAFT FOR GROUND TESTS.
3. DO NOT INSTALL IN SPACECRAFT. MAY BE ACTIVATED AND USED FOR POWER SYSTEMS TESTS, GROUND CART FOR "THRU THE HATCH POWER" AND BATTERY CHARGER TESTS.

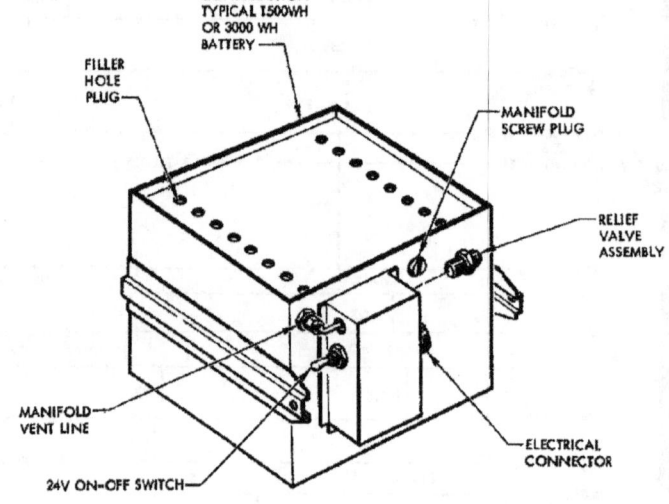

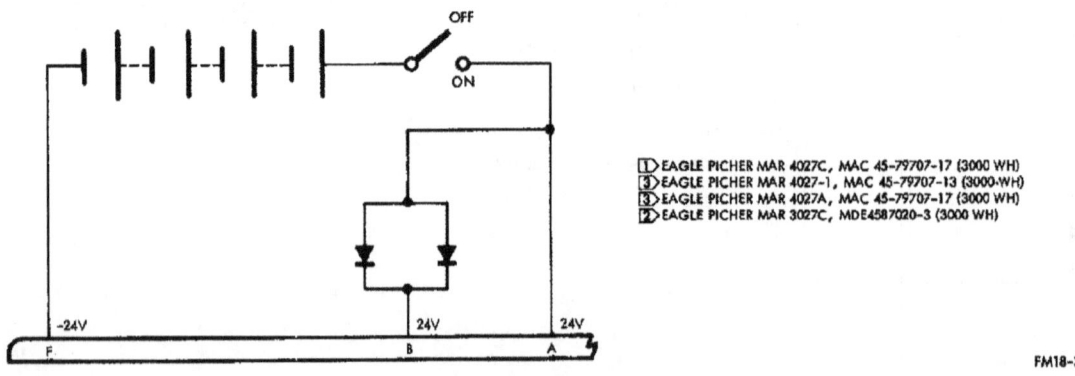

Figure 11-6 Typical Battery Assembly

minimum capacity rating of 3000 watt-hours for the three main batteries, 3000 watt-hours for the two standby batteries and 1500 watt-hours for the isolated battery. Each battery is equipped with a pressure relief valve designed to maintain internal pressure from 5.5 to 14.9 psi. The pressure relief valve is mounted external to the battery case. The battery switch is in the positive 24 volt output.

The battery electrolyte consists of a 40 percent solution of reagent grade potassium hydroxide and distilled water and is used to activate the dry charged battery initially. After the first discharge cycle, the battery may be recharged by a constant current battery charger. The batteries furnish power for all electrical equipment in the spacecraft; therefore, proper servicing and maintenance is of extreme importance. The batteries are designed for five complete cycles of discharge and charge; however, for highest reliability, units should not exceed four cycles or an activated life of 60 days prior to flight. For the internal wiring of the batteries see Figure 11-6.

11-12. INVERTERS 250 VOLT-AMPERE AND 150 VOLT-AMPERE

The d-c to a-c inverters installed in the spacecraft are of a solid state design capable of operating continuously at full rated power output in a ambient atmosphere of 160°F. or at 80°F. at 5 psia 100% oxygen. Inverters are cooled by the use of heat sinks and baffles. The output is 115 volts a-c $\pm$ 5%, single phase to ground, with a frequency of 400 cycles $\pm$ 2.5% and essentially sinusoidal in waveform.

SEDR 104

11-13. **D-C AMMETER 0-50 AMPERE**

    The d-c ammeter is located on the main instrument panel and provides the astronaut with an indication of total current drain from all batteries (see Figure 11-1). The basic ammeter movement has a 50 millivolt sensitivity. A shunt of suitable resistance is connected across the input of the meter providing a low resistance path to ground with the proper voltage drop at 50 amperes for a meter movement to full scale deflection.

11-14. **D-C VOLTMETER 0-30 VOLTS**

    A d-c voltmeter, and selector switch, are located on the main instrument panel (see Figure 11-1). Approximate battery condition can be determined by placing the D-C VOLTS switch to the appropriate positions and reading the individual battery voltages. Main and isolated voltages may also be determined by placing the D-C VOLTS switch to the appropriate M or I position.

11-15. **A-C VOLTMETER**

    An a-c voltmeter and a five position selector switch are mounted on the main instrument panel. The five positions of the a-c voltmeter switch are 250 VA, 150 VA, STBY, ASCS and FANS. (See Figure 11-1).

11-16. **INTERIOR LIGHTING**

11-17. **SYSTEM DESCRIPTION**

    Interior lighting for the spacecraft consists of two fluorescent cabin lights, and a series of warning telelights. See Figure 11-7 for

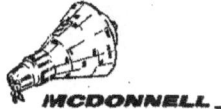

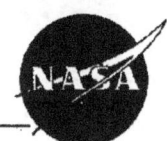

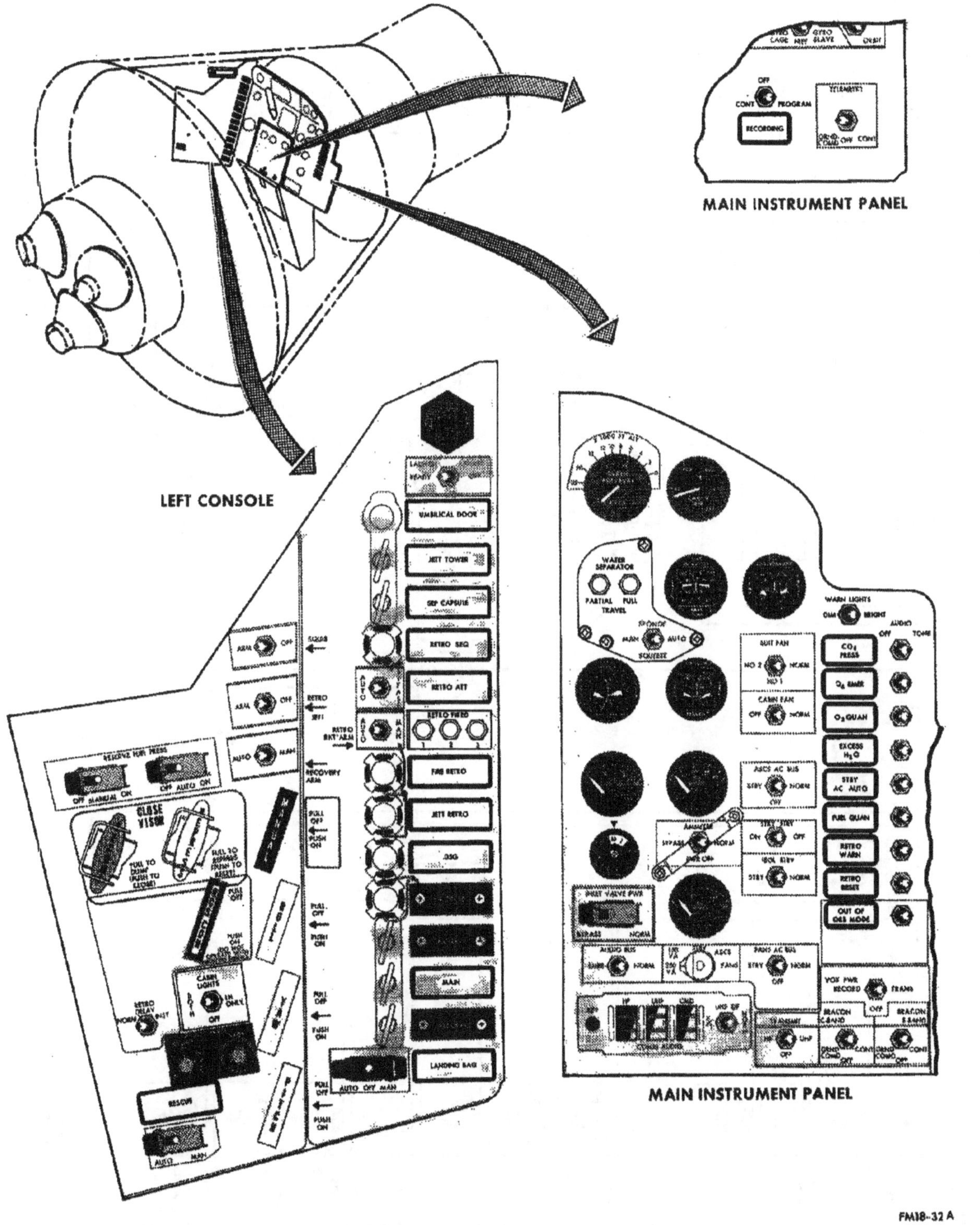

FM18-32 A

Figure 11-7 Interior Lights and Warning Lights System

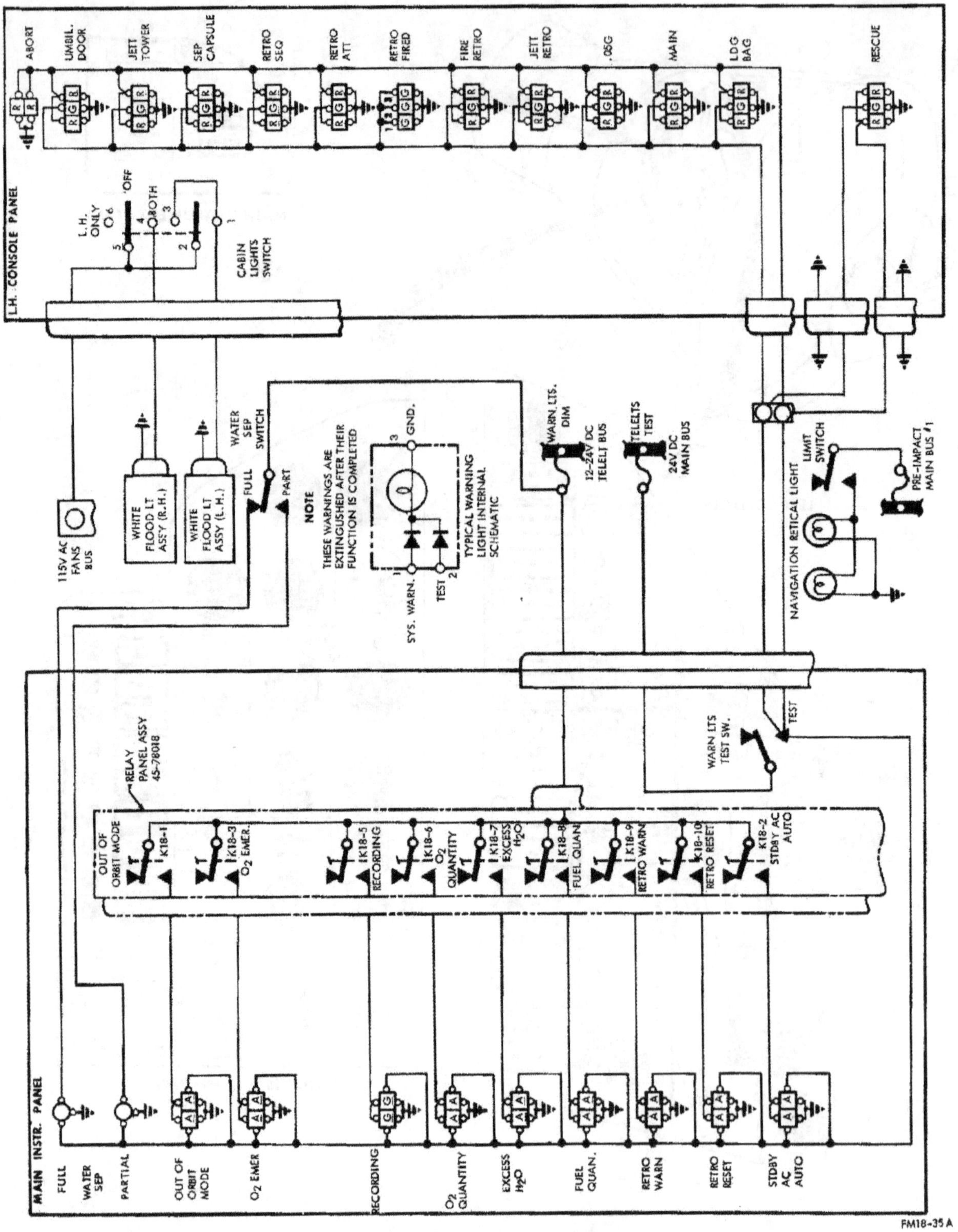

Figure 11-8 Interior Lights and Warning Lights Schematic

location and arrangement of cabin lights and telelights.

11-18. **CABIN FLOOD LIGHTS**

Two fluorescent cabin flood lights are mounted on brackets to the right and left and above the astronaut. Power for the cabin lights is supplied from the 115 V a-c inverter fans bus and controlled by a three position switch located on the left console. The switch positions are marked BOTH, LF ONLY and OFF. The cabin flood lights are of high actinic value, especially suitable for camera usage. The lights produce little heat and have a low wattage consumption of 7 watts each (see Figure 11-7).

SECTION XII

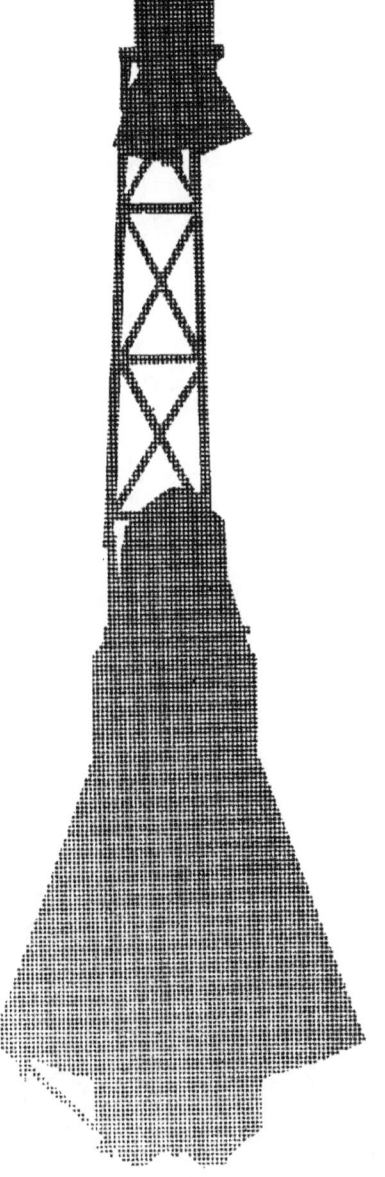

COMMUNICATION SYSTEM

TABLE OF CONTENTS

TITLE	PAGE
System Description	12-5
System Operation	12-10
System Units	12-30

 **MCDONNELL** SEDR 104

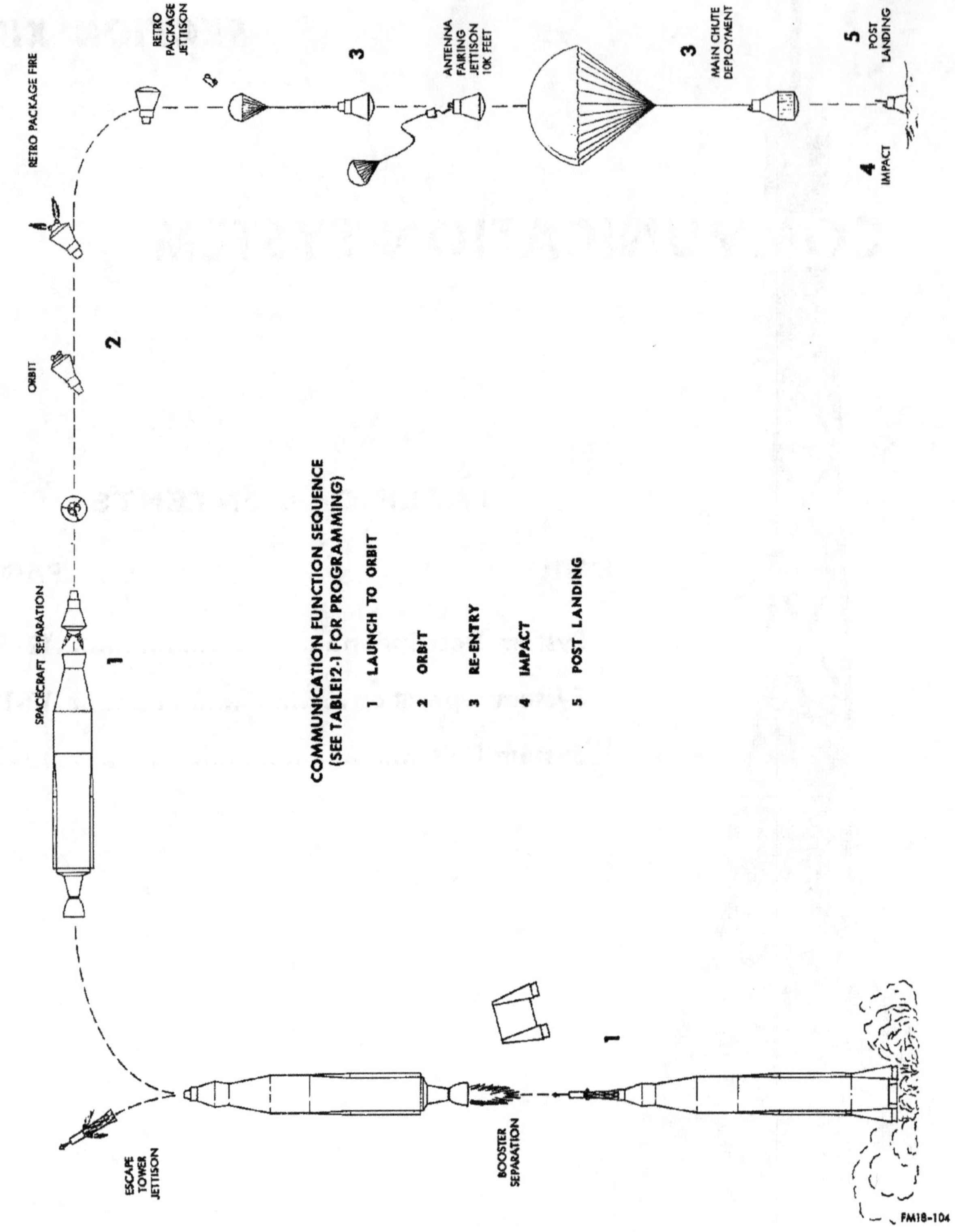

Figure 12-1. Communications Sequence

12-2

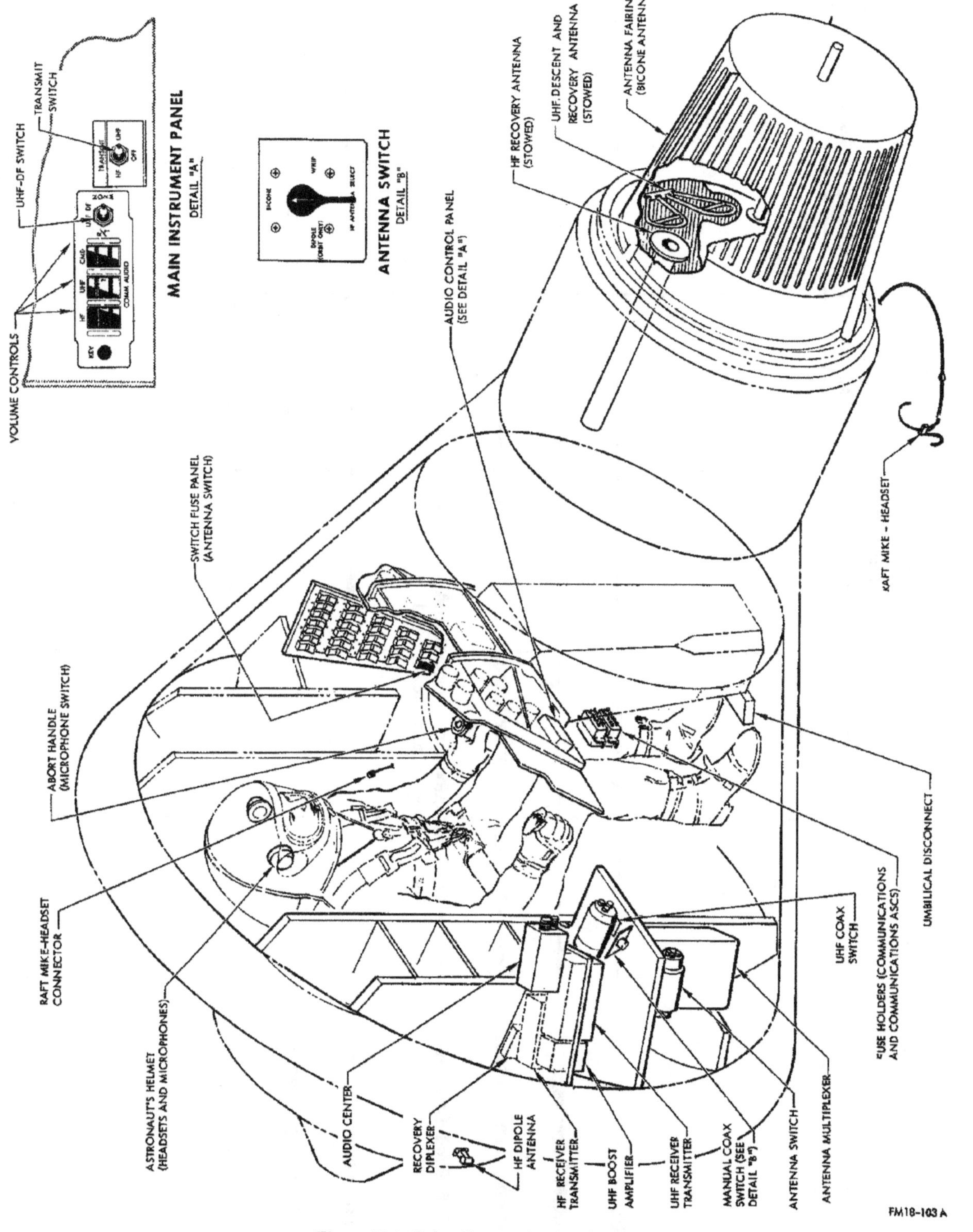

MCDONNELL — SEDR 104

Figure 12-2 Voice Communication System

12-3

TABLE 12-1
COMMUNICATION FUNCTIONS

FUNCTION	PHASE				
	LAUNCH	ORBIT	RE-ENTRY	IMPACT	POST LANDING
HF R/T		←——Astronaut Selection——→			
UHF COMM Receiver	←——————————————————————————————→				
Telemetry Transmitter	←——————10 Minutes After Impact——————→				
C and S Band Beacons	←——On Continuous or Controlled by Astronaut and/or Ground Command——→				
UHF Recovery Beacon			←——————————Bicone Ant. Separation——→		
H.F. Recovery Beacon			←————————Bicone Ant. Separation——→		
Aux. UHF Rescue Beacon			←——————————————→		

12-4

SEDR 104

XII. COMMUNICATIONS SYSTEM

12-1. SYSTEM DESCRIPTION

12-2. VOICE COMMUNICATION

The astronaut is provided with voice communications throughout the entire mission (see Table 12-1). A dual headset and microphone contained within the astronaut's helmet, operate through the audio control circuits to the selected voice communications set (see Figure 12-2). A spacecraft-pad interphone system is available prior to umbilical cable disconnect.

HF reception is available through the HF Voice Communication set during launch and orbit. HF voice transmission may be used only after spacecraft separation by astronaut selection of the HF position of the TRANSMIT switch. The HF set is disabled at antenna fairing separate and re-energized upon landing. The HF Voice Communications set provides reception and transmission, during the post landing phase of the mission, also.

UHF reception is available throughout the entire mission by the UHF Voice Communications set and its UHF Booster Amplifier. Transmissions over this set may be made when the UHF position of the TRANSMIT switch is selected by the astronaut.

The selected transmitter may be energized by operation of a push-to-talk switch, or by a voice operated relay when the VOX switch is in the ON position by speaking into the microphone. The UHF transmitter will automatically be energized upon landing to provide a direction finder signal. This automatic feature may be overridden by the astronaut.

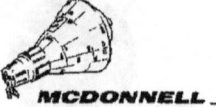

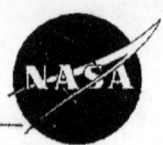

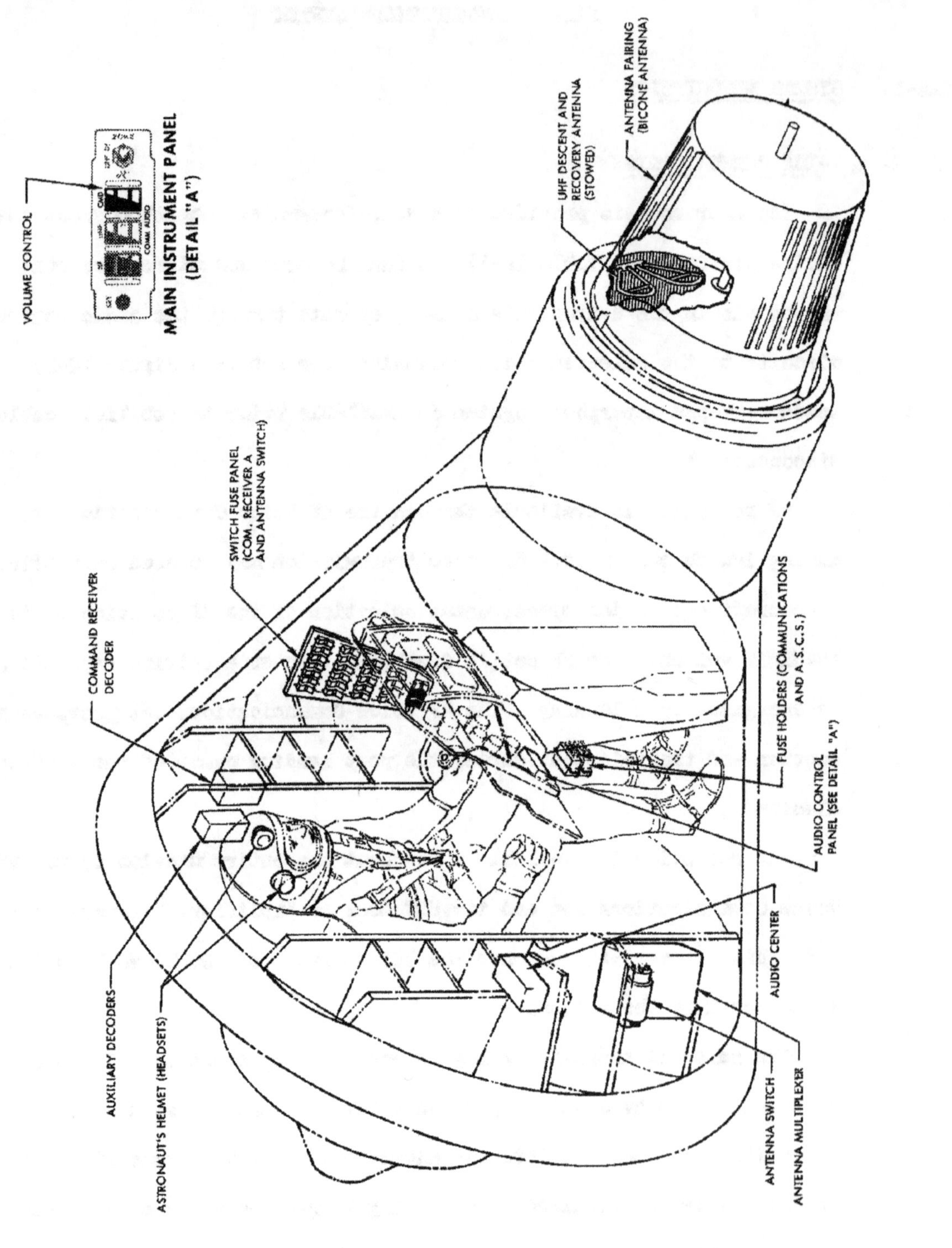

FM18-102A

Figure 12-3 Command Receivers System

A spacecraft to life raft extension cable is provided to allow the astronaut use of spacecraft transmitter-receivers while outside the spacecraft. This is accomplished by placing a push-to-talk switch and an earphone and microphone at the end of a 27 foot extension cable. The spacecraft end of the extension cable is provided with a plug which fits the astronaut's suit disconnect. With the TRANSMIT switch in the HF position, HF and UHF operation is provided. With the TRANSMIT switch in UHF, only UHF operation is provided.

The Command Receiver provides an emergency ground station-to-spacecraft voice communications channel throughout the mission until spacecraft impact. Power for the voice communications systems is supplied through fuses located in the Communications and Communications ASCS Fuse Holders (see Figure 12-2).

12-3. COMMAND RECEIVER

A set of Receiver-Decoder and auxiliary decoder units is used for reception and decoding of ground command signals. These signals are for the purpose of activating various control circuits.

Power for the Command Receivers is supplied through the fuses located in the Communications and the Communications ASCS Fuse Holders (see Figure 12-10).

12-4. TELEMETRY

A Telemetry Transmitter is provided for communicating instrumentation information to the ground stations. Information is picked up throughout the spacecraft in the form of voltages from voltage divider circuits.

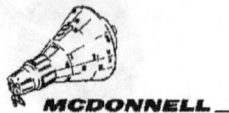

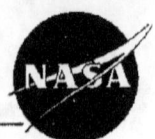

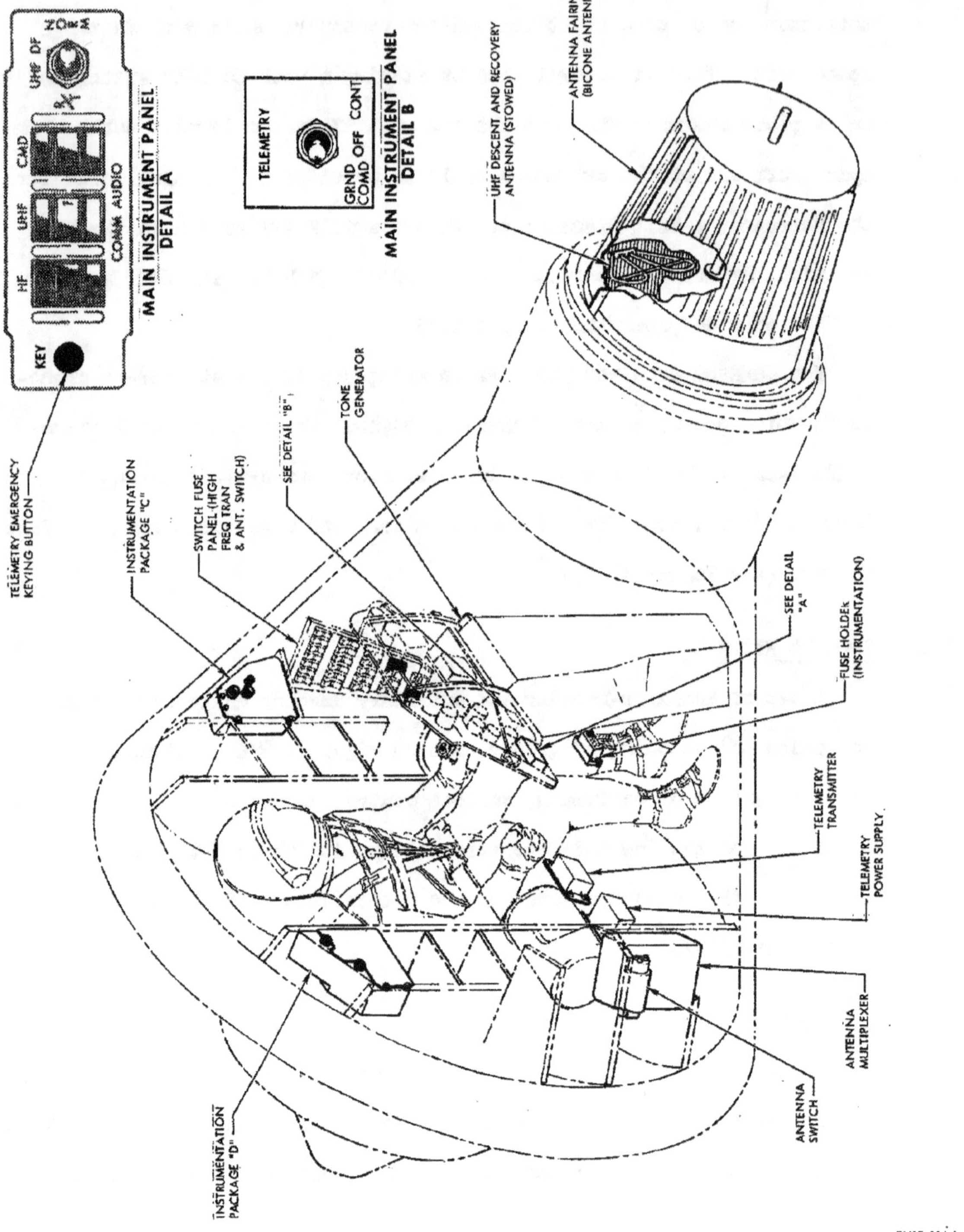

FM18 114 A

Figure 12-4. Telemetry System

These voltages cause VCO frequencies to change to supply suitable inputs to the Telemetry Transmitter. (Refer to the Instrumentation Section XIV of this manual). A transmitter having a power output of 2.0 watts is used for transmission of telemetry information. The operation of the transmitter is controlled by the TELEMETRY switch on the Main Instrument Panel. It's three selections are GRND COMD, OFF, and CONT. The power output of the telemetry transmitter is fed to either the Main or the UHF Descent Recovery Antenna. Power for the system is obtained from fuses located in the Instrumentation Fuse Holders (see Figure 12-4 and Figure 12-11).

12-5. BEACONS

The beacons provided in the spacecraft to aid tracking by ground stations are C-Band and S-Band beacons, a UHF Recovery Beacon and an auxiliary UHF Beacon energized at antenna fairing separation, and an HF Recovery Beacon, energized upon landing. These beacons provide signals compatible with direction finding equipment used by the recovery crews. The UHF Voice Communications transmitter is keyed at antenna fairing separation to provide an additional signal for direction finders. A flashing strobe light is installed for visual location of the spacecraft after landing. (See Section VII of this manual).

Spacecraft power for the beacons system is supplied through fuses located in the Communications and ASCS Fuse Holders. (See Figure 12-5 and Figure 12-12).

12-6. ANTENNAS

The voice communications, telemetry and beacon receivers and transmitters, with their various frequencies and types of outputs

SEDR 104

require an antenna system with wide capabilities. Therefore, five antennas are used to fulfill the entire mission requirements. A main Bicone antenna and a Retro Package HF Dipole Antenna is used for the major portion of the mission. During re-entry, the Bicone antenna must be jettisoned to allow main parachute deployment. The HF Dipole is jettisoned with the retro package. To replace the UHF function, a compact UHF Recovery antenna is automatically placed in operation at bicone separation. Upon landing, an HF Recovery Whip Antenna and an Aux. UHF Beacon Antenna are extended to permit HF and UHF operation. Throughout the entire mission, C- and S-Band antennas are provided for operation of the radar beacons. Antenna switching and multiplexing are performed automatically by the RF circuitry. (See Figures 12-6 and 12-13). Power requirements for antenna switching are supplied through a switch-fuse located on the left console Switch-Fuse Panel. (See Figure 12-13)

12-7. SYSTEM OPERATION

12-8. VOICE COMMUNICATIONS

12-9. AUDIO CONTROL AND GROUND INTERPHONE SYSTEM

HF and UHF receiver outputs are routed to the control panel. This provides one volume control for HF audio and one volume control for UHF audio. The voice output from the command receiver is connected to the Communication Control Panel for mixing and volume control. Separation of command and voice audio signals by a low pass filter, and amplification of resulting voice audio is done in the audio center. (See Figure 12-7).

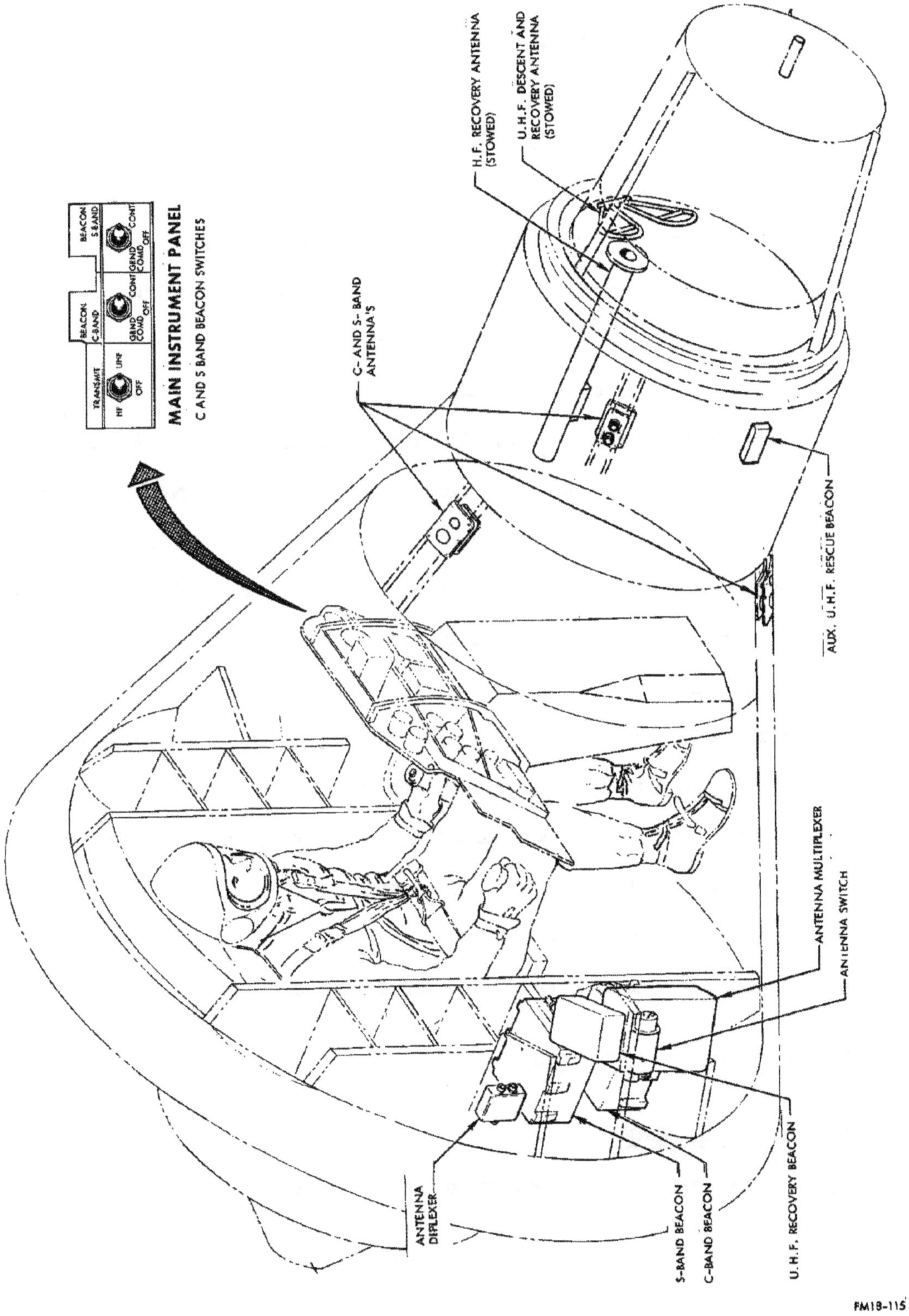

Figure 12-5. Beacon Systems

Communication audio signals from the volume controls, the interphone audio from the pad-to-the-pilot and alarm tones are supplied to the tape recorder relay and the two headset amplifiers in the audio center. The headset amplifiers serve to amplify the audio signals and feed them to the individual earphones in the astronaut's helmet. The de-energized position of the tape recorder relay supplies a path for receiver audio to the main tape recorder.

Audio from the microphones is fed to two separate microphone amplifiers in the audio center. These two amplifiers serve to amplify microphone output to a level sufficient to supply modulation circuits of the voice transmitters. The microphone amplifier output is also fed to the input of the VOX (voice operated relay circuitry). The transmitter is energized by use of the Push-To-Talk switch on the abort handle, or by the VOX circuit when the VOX switch on the instrument panel is in the TRANSMIT position. After impact, the astronaut may disconnect his suit communications connector and attach a microphone-headset-PTT switch assembly to the suit communications connector. This microphone-headset assembly is fitted with a 27 foot waterproof electrical cable and is used after egress to provide the astronaut with two way communications on the spacecraft HF and UHF communications systems.

12-10. HF VOICE COMMUNICATIONS

12-11. HF COMMUNICATIONS

The HF voice communications set is an AM receiver-transmitter unit.

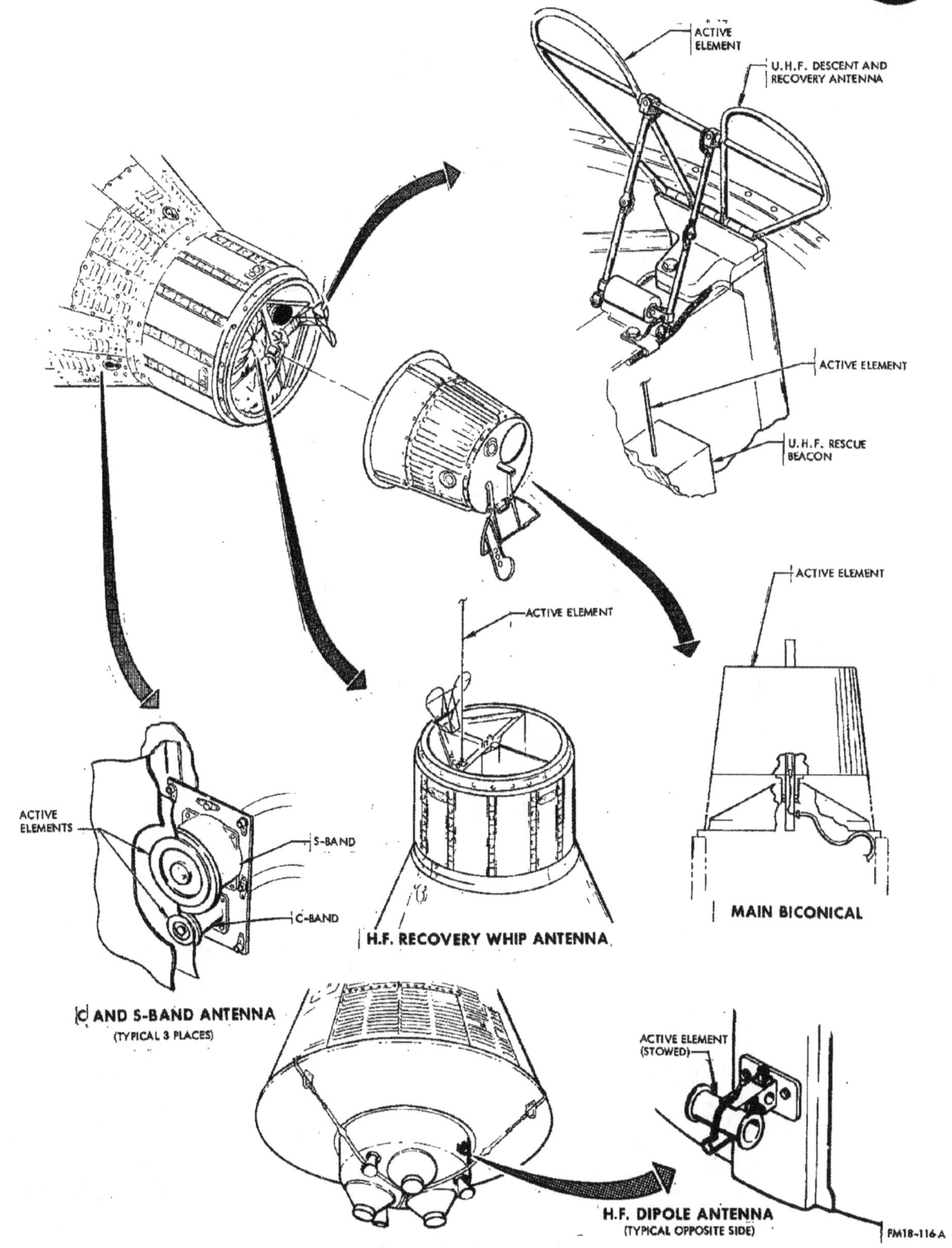

Figure 12-6. Antenna System Utilization

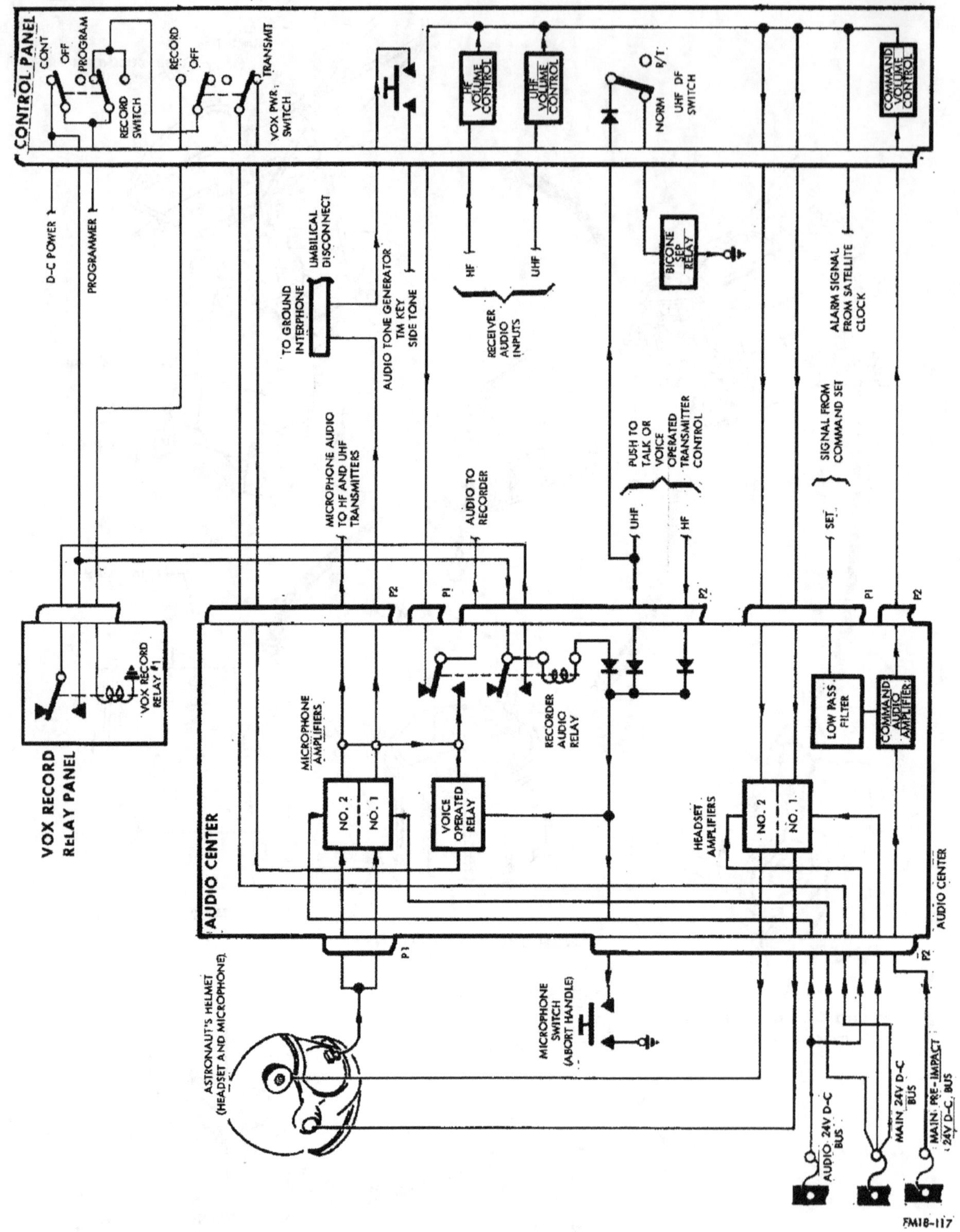

Figure 12-7. Audio Control and Ground Interphone System

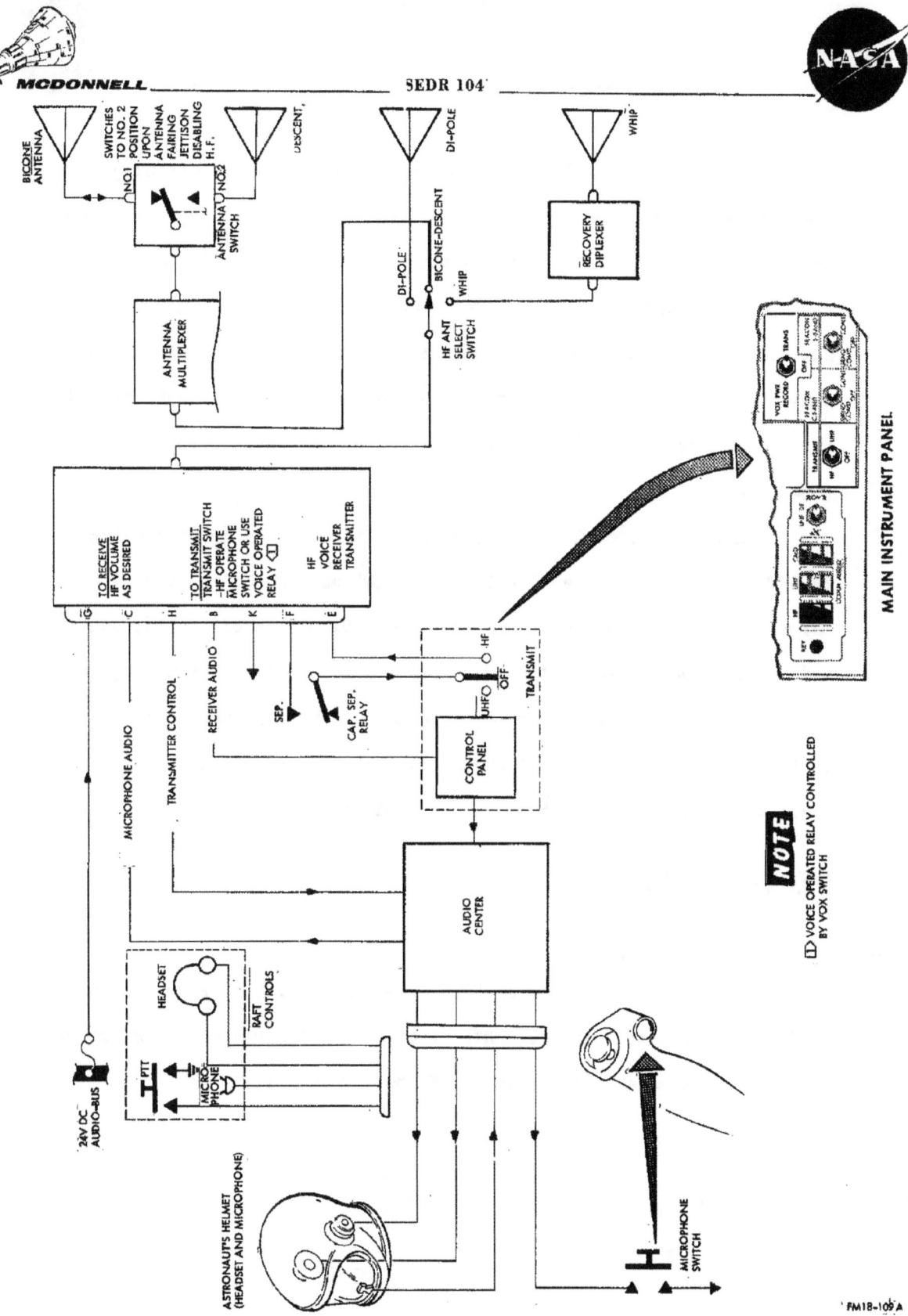

Figure 12-8. H.F. Communications Data Flow

Power from the Main Pre-Impact 24 volt d-c bus is fed directly to the receiver section of the set. The transmitter is fed 24 volts through the HF position of the TRANSMIT switch and the closed contacts of the Tower Separation Relay, after tower separation. Audio input to the transmitter portion of the unit is from the microphone amplifier in the audio center. The transmitter is keyed either automatically through the VOX circuit or manually by the astronaut's use of the push-to-talk switch. (See Figure 12-8).

The antenna connection from the set is through the three position HF ANT SELECT switch located below the right hand console. The HF ANT SELECT switch utilizes three positions, DIPOLE (orbit only), BICONE and WHIP. These three switch positions provide optimum antenna operation. HF operation is preferable through the use of the dipole antenna while in orbit. Audio output from the receiver, including sidetone during transmission is routed to the HF volume control in the control panel.

12-12. UHF VOICE COMMUNICATIONS

12-13. UHF RECEIVER-TRANSMITTER

The UHF voice communications set is an AM receiver-transmitter unit designed to operate on a frequency of approximately 299 MC. The transmitter output is increased by a separate UHF booster amplifier.

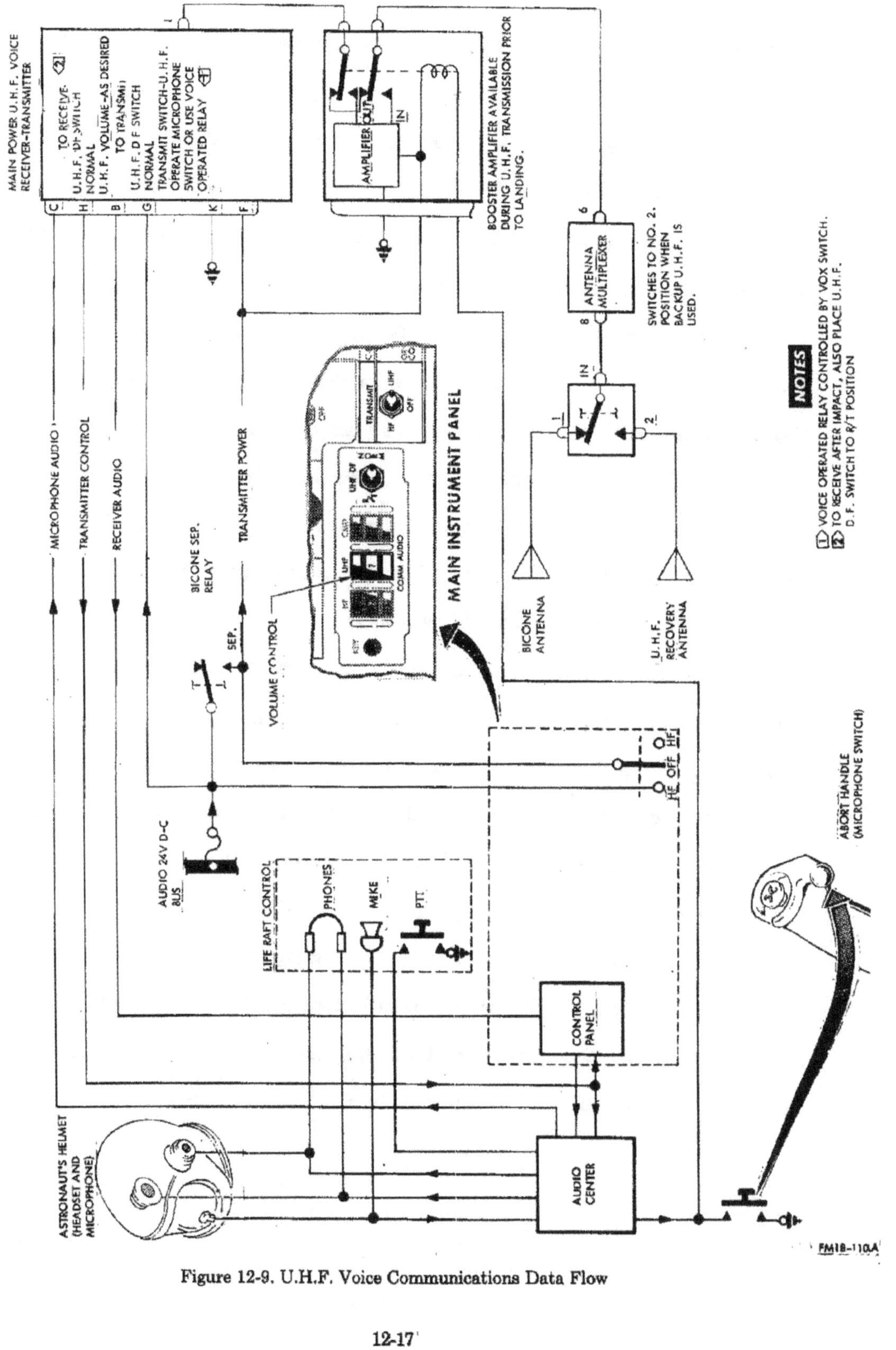

Figure 12-9. U.H.F. Voice Communications Data Flow

Power from the 24 volt d-c Audio Bus is fed through the UHF position of the TRANSMIT switch directly to the receiver section of the set. (See Figure 12-9). Power for the transmitter section of the set is also taken from the UHF position of the TRANSMIT switch. At bicone antenna separation the bicone separation relay contacts assume the same function as the UHF contacts of the TRANSMIT switch thus providing a continuous UHF signal for DF purposes. Audio input to the transmitter portion of the unit is from the microphone amplifiers in the audio center. The transmitter is energized either automatically or manually by the astronaut. It will be energized automatically at bicone separation to provide a UHF signal for direction finding equipment. This feature may be overridden by operation of the UHF-DF switch on the control panel to the R/T position.

Antenna connection from the set is through the UHF booster amplifier, coax switch, antenna multiplexer, and the antenna switch to either the main bicone or UHF rescue antenna. Operation of the PTT switch or energizing the voice operated relay while in the UHF mode causes the booster amplifier to be inserted in series with the coax line. Transmitter output is then boosted by this amplifier to 2 watts. The booster is also available after landing. The multiplexer output is connected through the antenna switch to either the main bicone or the UHF rescue antenna. Audio output from the receiver, including sidetone during transmission, is routed to the UHF volume control in the control panel.

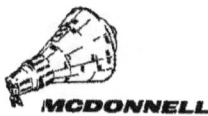

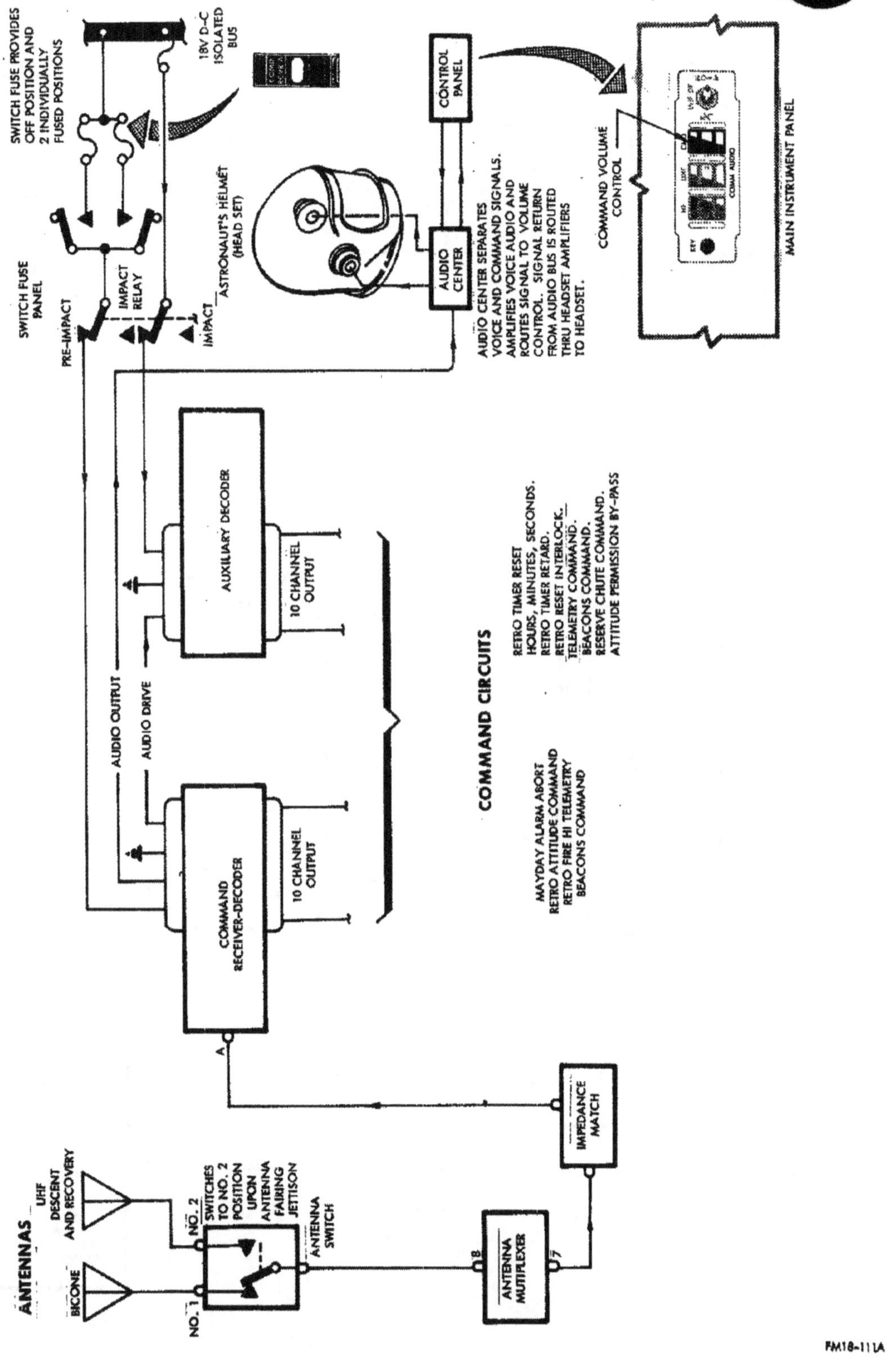

FM18-111A

Figure 12-10. Command Receivers Data Flow

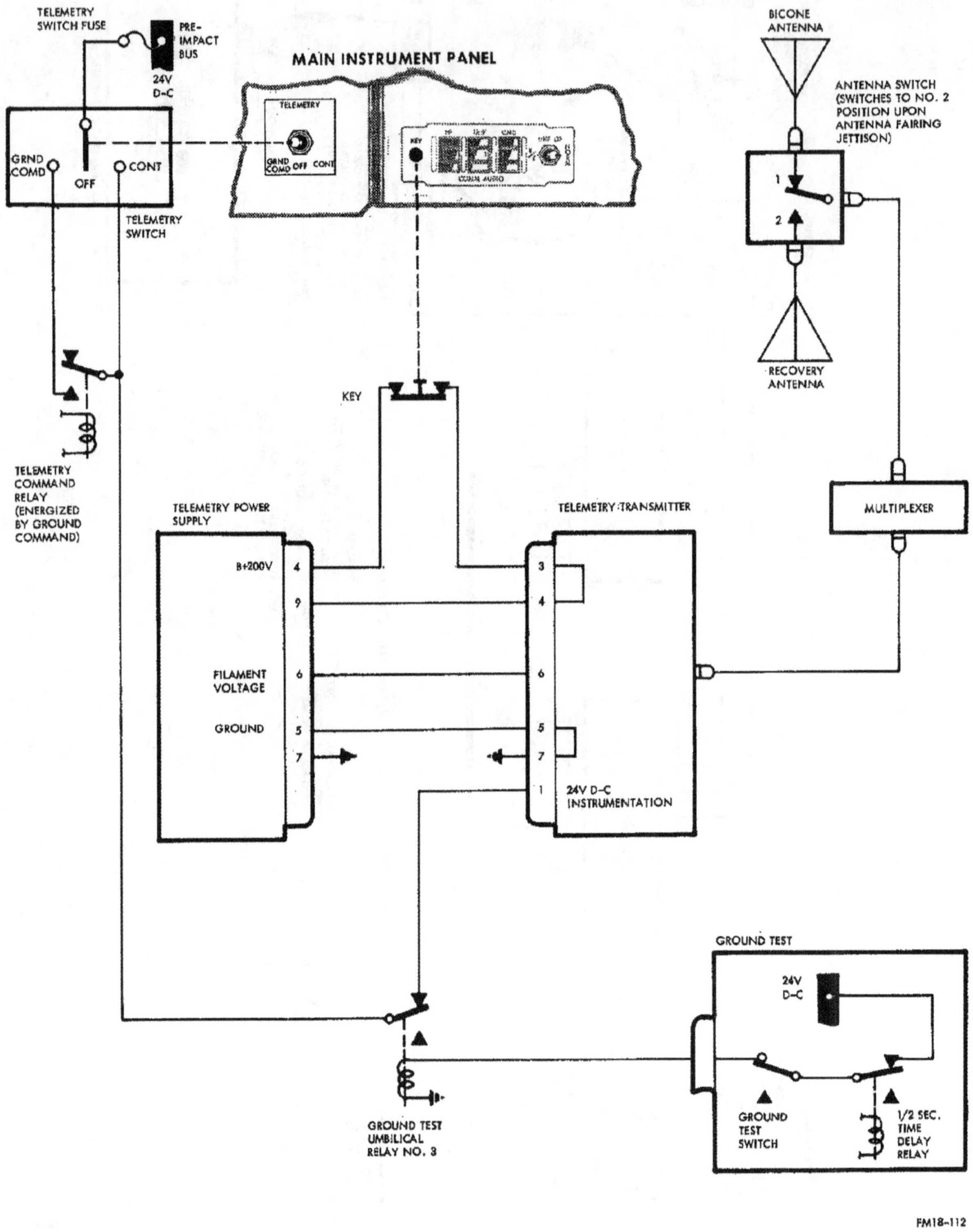

Figure 12-11. Telemetry System Data Flow

SEDR 104

12-14. COMMAND RECEIVER

The receiver-decoder unit consists of an FM receiver. The received signal may be modulated with a maximum of six of a possible twenty audio frequencies. The receiver reduces the input signal to the modulation frequencies which operate individual control relays. (See Figure 12-10).

Each control relay provides contacts from a normally open or a normally closed control channel. Ten channels are provided in the receiver-decoder with an additional ten available in the auxiliary decoder. Command channel assignments are not disclosed for security reasons. Emergency voice communications may be had from the ground station to the spacecraft through the command receiver. Receiver outputs are supplied through a filter and amplifier in the audio center circuits to the astronaut's headset. Power for the command set is supplied from the standby 18 volt d-c bus. Both power circuits are routed through sections of the impact relays in order to de-energize the set upon landing.

Antenna input is from the bicone or UHF rescue antenna through the antenna switch and antenna multiplexer to the receiver.

12-15. TELEMETRY

12-16. TELEMETRY TRANSMITTER

The telemetry transmitter set is an FM transmitter operating on a frequency of approximately 228 MC.

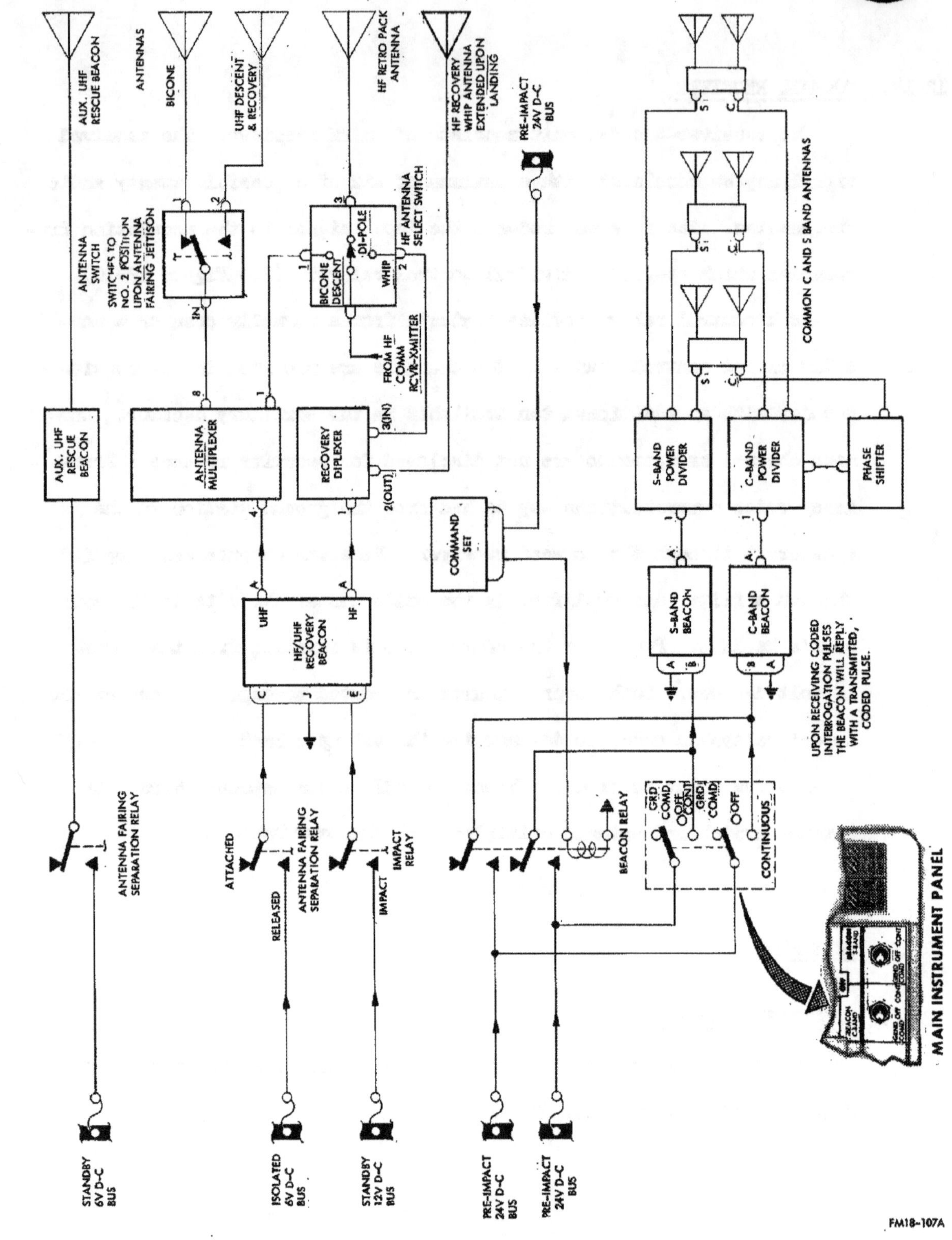

Figure 12-12. Beacon System Data Flow

SEDR 104

Before umbilical drop, the telemetry transmitter and its power supply receive 24 volts d-c from the Main Pre-impact bus through the energized Ground Test Umbilical Relay, normally open contacts. This relay's solenoid is energized through the umbilical until the umbilical is dropped. (See Figure 12-11).

To silence the TM transmitter, the astronaut may place the telemetry switch in the OFF position which will prevent 24 V d-c power from reaching the transmitter. Placing the telemetry switch in the ground command position enables a signal from the ground to energize the telemetry command relay which will apply power to the transmitter. The signal is programmed to last 6 minutes and then it is removed shutting off the telemetry transmitter. The transmitter will remain off until commanded again by ground command. The CONT position of the TELEMETRY switch applies continuous power to the TM transmitter.

Coded instrumentation information is supplied from the Instrumentation Package "D", and used to frequency modulate the transmitter. (See the Instrumentation Section XIV of this manual).

RF power output is fed to the antenna multiplexer where it is routed through the antenna switch to the main bicone or UHF recovery antenna.

12-17. C-BAND BEACON

The C-Band beacon is a transponder unit consisting of a receiver and transmitter operating on a frequency of approximately 5400 to 5900 MC. The beacon is double pulsed and is compatible with modified FPS-16 radar. Upon ground command, through the command receiver, or by astronaut selection of the CONT position of the C-Band beacon switch, the beacon receiver is energized. Interrogation by ground radar will then result in a coded reply from the beacon transmitter. Input power is from the main pre-impact 24 volt d-c bus through the beacon relay controlled by the command receiver, or, for continuous operation, through the C-Band beacon switch. An OFF position of the C-Band beacon switch is also provided. (See Figure 12-12).

The C-Band beacon antenna connection is through the C-Band power divider to the three C- and S-Band Beacon antennas. A phase shifter is used between the C-Band Power Divider and one C-Band Antenna to prevent nulls in the radiation pattern.

12-18. S-BAND BEACON

The S-Band beacon is a transponder unit consisting of a receiver and transmitter. (See Figure 12-12). The unit operates on a frequency of approximately 2700 to 2900 MC and is double pulsed to reduce possibilities of unauthorized interrogation. This unit is compatible with ground based Verlort Radars and operates at a positive acceptance tolerance of $\pm$ 0.5 micro-seconds and a positive rejection tolerance of $\pm$ 1.8 micro-seconds.

Power circuits, interrogation and reply are the same as the C-Band Beacon. (Refer to Paragraph 12-17).

Beacon antenna connection is through the S-Band Power Divider to three "C" and "S" Band Beacon antennas.

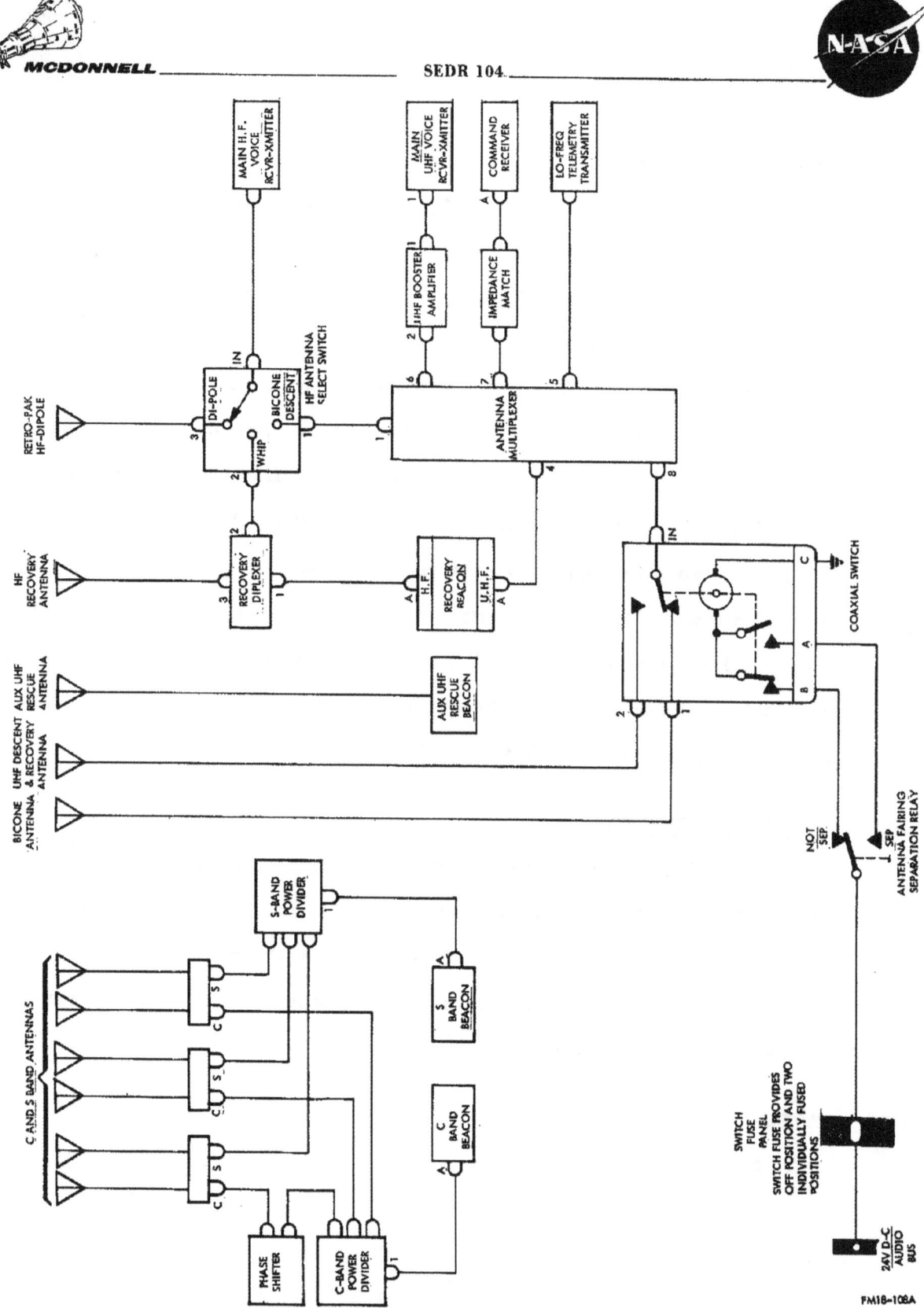

Figure 12-13. Antenna System Schematic

12-19. **HF/UHF RECOVERY BEACON**

Two recovery beacons are combined into one unit. One beacon operates on high frequency, while the other operates on ultra high frequency. Both are energized to provide radio signals for recovery direction finder equipment. (See Figure 12-12).

The HF recovery beacon operates on a frequency of approximately 8 MC with a tone modulated output. It is powered by the 12 volt standby bus through the impact relay and is energized upon landing. The RF power output is fed through the rescue diplexer to the elevated HF Recovery Antenna.

The UHF recovery beacon operates on a frequency of approximately 243 MC with pulse modulation. It is powered by the 6 volt isolated bus through the antenna fairing separation relay. This circuit is energized when the antenna fairing is jettisoned. The RF power output is fed through the antenna multiplexer and the antenna switch to the UHF rescue antenna.

12-20. **AUXILIARY RESCUE BEACON**

The Aux. Rescue Beacon operates on a frequency of approximately 243 MC with pulse modulation. It is powered by the 6 volt standby bus through the Antenna Fairing Separation Relay. These circuits are energized at antenna fairing separation. The RF power is radiated from the Aux. Rescue Beacon Antenna. (See Figure 12-12).

12-21. **ANTENNAS**

12-22. **MAIN BICONE (HF AND UHF)**

A biconical antenna is used for pre-launch, launch, orbit and initial re-entry phases of the mission. This antenna is an integral part of the antenna fairing and is located over the open end of the recovery system compartment of the cylindrical spacecraft afterbody. The biconical antenna serves the HF and UHF voice receiver-transmitters, the command receivers, and the telemetry transmitters. The active element of the biconical antenna forms the upper portion of the antenna fairing while the lower portion of the fairing and the spacecraft body forms the ground plane for the antenna. (See Figure 12-13).

12-23. **UHF RESCUE ANTENNA**

A UHF antenna is used for the final phase of re-entry, landing and rescue. It is a compact antenna located on the open surface of the recovery systems compartment. The antenna is folded when the antenna fairing is installed. Sixteen seconds after the fairing is jettisoned the UHF recovery antenna is erected and serves the UHF voice receiver-transmitters, the UHF portion of the recovery beacon, the command receivers, and the telemetry transmitter. (See Figure 12-6).

12-24. **MAIN BICONE HF DIPOLE AND UHF RECOVERY ANTENNA FEED**

The various radio systems are connected to the bicone antenna, the HF dipole or the UHF recovery antenna in the following manner: (See Figure 12-13).

(1) The HF voice receiver-transmitter antenna leads are connected to the HF ANTENNA SELECT manual coaxial switch which when placed in the following positions feeds the antenna systems as follows:

  (a) WHIP position, HF recovery antenna through the recovery diplexer.

  (b) BICONE position, the bicone antenna through the main multiplexer and the antenna switch.

  (c) DIPOLE (ORBIT ONLY) position, to the HF dipole antenna on the retro package.

(2) The UHF receiver-transmitter antenna lead is connected through the main multiplexer, to the antenna switch whose outputs feeds the bicone antenna until antenna separation. After antenna separation the descent antenna is fed.

(3) The command receiver antenna lead is connected to the antenna multiplexer.

(4) The Low frequency telemetry transmitter feeds direclty to the antenna multiplexer.

12-25. ANTENNA MULTIPLEXER

The antenna multiplexer enables simultaneous or individual operation of the radio systems using one antenna. Effectively this is a radio frequency junction box which allows several receivers and transmitters to operate simultaneously without interference with each other. Final connection to the antenna is through the antenna switch to either the

SEDR 104

biconical antenna, or the UHF rescue antenna. The antenna switch is operated by the antenna fairing separation relay to cause the automatic shift from the main antenna to the UHF rescue antenna upon antenna fairing jettison (see Figure 12-13).

12-26. RECOVERY ANTENNA

An antenna is provided to permit HF radio transmission and reception after landing. The antenna is a telescoping whip antenna which is automatcially extended by a pyrotechnic after impact. Once extended, the antenna is used for the HF voice receiver-transmitter and the HF portion of the recovery beacon.

The HF recovery beacon feeds through the recovery diplexer. The HF voice receiver-transmitter is fed through WHIP position of the manual ANT SWITCH and then into the recovery diplexer to the HF whip antenna. The diplexer allows simultaneous or individual operation over a single lead to the antenna. (See Figure 12-13).

12-27. C AND S BAND ANTENNAS

Three C- and S-Band antenna units are installed in the spacecraft structure for the C- and S-Band beacons. These units are equally spaced about the circumference of the conical section. Each antenna unit consists of one helix as a C-Band antenna and one helix as an S-Band antenna.

Antenna leads from the C-Band and S-Band beacons are routed through individual power dividers to the three associated helix antennas. (See Figure 12-13).

SEDR 104

12-28. HF DIPOLE (RETRO PACKAGE) ANTENNA

An HF Dipole antenna is used while in an orbit condition. The dipole antenna consists of two units, one attached to either side of the retro package. The active elements consist of tubular beryllium copper ribbons. The ribbons are rolled in a flat condition prior to orbit. Upon reaching orbit, a squib is fired on the dipole antennas assemblies, to release the active elements and allow them to unroll and become tubular in an extended condition (See Figure 12-13).

12-29. SYSTEM UNITS

12-30. AUDIO CENTER

The audio center provides transistorized audio amplifiers, a voice operated relay (VOX), an audio filter, tape recorder control circuitry and transmitter control circuitry. (See Figure 12-7). All components are contained in a light weight, foam encapsulated unit.

Two fixed gain headset amplifiers are used to bring audio signals up to headset level and feed the headsets separately. Two fixed gain amplifiers are provided to increase the dynamic microphone outputs to a level suitable to be used with the various transmitters.

A low pass filter, with a cutoff for frequencies above 3000 cps, filters the audio supplied from the command receivers. Outputs from the filter is fed to a variable gain, command audio amplifier.

The "voice operated relay" is a transistorized amplifier with separate adjustable threshold level and release time controls. The amplifier

operates a relay to provide a grounding circuit for transmitter keying. This unit parallels the external PTT switch.

The audio center furnishes a circuit to apply the keying ground potential to the transmitters. Each circuit is protected from the rest by a diode.

A relay is installed in the audio center for supplying power and audio signals to the tape recorder. In the de-energized condition, the relay closes a circuit to the tape recorder input, thus audio received by the spacecraft is recorded whenever instrumentation programs tape recorder operation.

When the microphone switch or VOX is operated, the tape recorder relay is energized if VOX operation of tape recorder is selected. One set of closed relay contacts now completes the recorder power circuit independent of instrumentation programming, while a second set of contacts routes signal from the microphone amplifiers to the recorder input.

The circuits in the audio center operate directly from the spacecraft 24 volt d-c inputs with no further regulation or voltage increase.

12-31. CONTROL PANEL

The audio control panel provides controls and circuits for the audio signals of the various spacecraft receivers (see Figure 12-7).

The HF and UHF circuits are routed through individual T-pads to volume controls. The HF circuit has a single volume control, the same is true of the UHF circuit, while separate volume control is provided

for the command audio circuit. Fixed inputs are used for the alarm tone and ground interphone circuits.

The panel also contains a switch override for the impact keying feature used with the UHF transmitters, and keying button on the panel to interrupt the telemetry B+ supply to provide emergency keying. (Refer to Figure 12-11).

12-32. HF VOICE RECEIVER-TRANSMITTER

The HF voice set is an AM receiver-transmitter designed as a small, lightweight unit. (See Figure 12-8).

The receiver section of the unit is a transistorized superhetrodyne circuit using a crystal filter, crystal diode detector and class B audio amplifier. The final audio amplifier is used for sidetone during transmissions.

The transmitter section of the unit utilises vacuum tube stages for the crystal controlled oscillator, driver and power amplifier. The power amplifier may be modulated up to 90% by a transistorized speech amplifier and modulator. These audio stages are also used for sidetone. Transmitter output is 5 watts. The unit is powered by spacecraft, 24 volts d-c, Power is routed through an external switch and contacts of the spacecraft separation relay which controls transmitter filament power, relay operation and a transistorized power converter. High voltage from this converter is used for the transmitter power amplifier. Antenna feed is through the manual ANT SWITCH to the HF dipole antenna or the bicone antenna or after impact to the HF recovery whip antenna.

SEDR 104

D-c voltage is also removed from the receiver RF stages. Antenna switching is accomplished by a solid state circuit which blocks the receiver during transmission.

12-33. UHF VOICE RECEIVER-TRANSMITTER

UHF voice receiver-transmitter consists of an AM receiver-transmitter designed as a small, lightweight unit operating near 297 MC. Transmitter output is .5 watt. The transmitter output is boosted by a final booster amplifier.

The receiver section of the unit is a transistorized superheterodyne circuit using a crystal controlled local oscillator, crystal filter and crystal diode detector. The audio section of the receiver also serves as the speech amplifier, modulator and provides sidetone for the transmitter. The transmitter section of the unit utilizes a crystal controlled oscillator, tripler and power amplifier. The RF section uses vacuum tubes while the modulation circuits are transistorized.

Spacecraft power, 24 volts d-c, is supplied to the set. This voltage is applied to the receiver, audio circuits and back through an external transmit switch to an internal power converter. This transistorized converter supplies B+ voltage to the transmitter RF sections. Transmitter filament voltage is also applied by the external transmit switch or the bicone separation relay after bicone separation.

Switching from receiver to transmitter operation is accomplished when ground potential is applied to a switching relay and a blocking circuit. The relay provides antenna and power converter switching. The blocking circuit removes receiver voltage.

12-34. UHF BOOSTER AMPLIFIER

A booster amplifier is used prior to landing to increase the .5 watt output of the UHF transmitter to 2.0 watts. The higher power is also available after landing.

Signal input to the booster is routed through a double pole, double throw coaxial relay. When the relay is de-energized, the signal is routed through the contacts to the output jack. Energizing the relay feeds the signal through the amplifier and takes the amplifier output to the output jack.

12-35. COMMAND RECEIVER-DECODER

The command receiver-decoder is a transistorized unit consisting of an FM Receiver and a decoder unit to operate control circuits. (See Figure 12-10).

The receiver section of the unit is a dual conversion superheterodyne circuit. The first local oscillator is crystal controlled and uses two stages of frequency multiplication. Two stages of amplification are used for the first IF, 78 MC signal. The second local oscillator is also crystal controlled, mixing with the first IF and giving a resultant second IF of 10.75 MC. Output from the IF strip is through a limiter to the discriminator. Audio amplifiers boost the discriminator output for the command voice channel and the decoder driver. The driver in turn supplies the ten decoder channels in the set.

The individual decoder channels each provide filters for their specific command frequency and amplifiers to operate a double pole, double throw relay for each channel. The ten relays thus make available normally open and normally closed contacts for external control circuit operation.

Spacecraft power, 18 volts d-c, is used to power the set. A Zener diode circuit, within the unit, is used for voltage regulation.

12-36. **AUXILIARY DECODER**

An auxiliary decoder operates with the receiver-decoder unit, allowing an additional ten channel capability.

The decoder channels in the auxiliary decoder are identical to the decoder channels of the receiver-decoder, with the exception of the command frequencies at which they operate.

The auxiliary decoder operates from spacecraft 18 volt d-c power. No further voltage regulation or increase is required.

12-37. **TELEMETRY POWER SUPPLIES**

The telemetry power supplies generate voltage used in the telemetry transmitter (see Figure 12-11). The unit is transistorized and uses crystal diodes. Spacecraft power, 24 volts d-c, is applied to a transsistor switching circuit operating into the primary of a power transformer.

A full wave, crystal diode rectifier is used on one secondary, with voltage regulation, to provide 200 volts d-c.

12-38. **TELEMETRY TRANSMITTER**

The telemetry transmitter is an all vacuum tube unit with an output of approximately 2.0 watts at approximately 226 MC. The transmitter is an FM unit using VCO inputs from instrumentation circuits. VCO signals are applied to a first stage signal amplifier-isolator circuit. The RF

portions of the transmitter consist of a triode quartz line oscillator into a pentode buffer stage feeding a triode power amplifier. The VCO signals are fed to the oscillator stage. Filament and B+ are obtained from a separate power supply. Spacecraft 24 volts d-c is the telemetry primary power supply.

12-39. <u>C-BAND BEACON</u>

The C-Band transponder is a pressurized superheterodyne receiver and pulse modulated, 400 watt peak output transmitter, operating in the frequency range of 5400 to 5900 MC (See Figure 12-12). With the exception of the magnetron and local oscillator, the unit is transistorized. The receiver consists of a pre-selector, local oscillator, 400 MC IF amplifier strip, pulse detector, pulse amplifier and decoder. Resonant cavities are used for the pre-selector and local oscillator.

The transmitter section accepts decoder outputs and pulse modulates the transmitter output.

The unit contains a power supply for converting spacecraft 24 volts d-c input to filtered 24 volts d-c and regulated, 115 and 150 volts d-c outputs. Antenna sharing is through an internal diplexer.

12-40. <u>S-BAND BEACON</u>

The S-Band transponder is a pressurized superheterodyne receiver and pulse modulated, 1000 watt peak output transmitter operating in the frequency range of 2700 to 2900 MC. (See Figures 12-12).

Receiver and transmitter circuits are the same as those used in the C-Band beacon with the exception of the pre-selector, local oscillator and transmitter which are designed for S-Band frequencies.

12-41. HF/UHF RECOVERY BEACON

The recovery beacon combines an HF, tone modulated, 8.364 MC transmitter and a UHF, pulse modulated, 243 MC transmitter into one small, foam encapsulated unit. (See Figure 12-12). The UHF section of the beacon is a one tube circuit with a pulse coding network. The HF section of the beacon is a transistorized crystal oscillator and two stage power amplifier with tone modulation supplied from a power converter. The beacon utilizes 6 and 12 volts d-c from the spacecraft power system. The UHF section is energized by applying the 6 volt d-c to a transistorized power converter. A full wave, crystal diode circuit is used to rectify the power converter output which is applied to the UHF stage. Applying 12 volts d-c energizes the HF section of the beacon. No power converter is required for the 12 volt input.

Modulation for the HF section is provided by routing the 12 volt supply to the power amplifier stages through a secondary winding of the power converter.

12-42. AUXILIARY UHF RESCUE BEACON AND ANTENNA

The Aux. UHF Rescue Beacon, consists of a pulse modulated transmitter and power supply which is enclosed in a foam encapsulated case. (See Figure 12-12). The unit is connected to the 6 volt standby bus and has an output of 91 watts. The antenna is mounted in the recovery compartment and is self erecting.

12-43. ANTENNA MULTIPLEXER

The antenna multiplexer allows reception and transmission of the many spacecraft frequencies over one line to the bicone or UHF recovery antenna. The unit consists of a number of filters arranged so that all spacecraft frequencies between 15 and 450 MC can be multiplexed on the single feed line. Each input channel is provided 60 db of isolation.

12-44. RECOVERY DIPLEXER

The recovery diplexer unit is used for the HF voice receiver-transmitter and HF section of the recovery beacon. One low pass and one high pass filter is used to diplex approximately 15 MC on one feed line to the HF recovery antenna.

12-45. COAXIAL SWITCHES (ANTENNA SWITCH, MANUAL HF ANTENNA SELECT)

RF switching is accomplished with motor driven SPDT switches. Application of spacecraft 24 volts d-c through external circuits drives the switch to the appropriate RF position and opens the power circuit for that position. The manual HF ANTENNA select switch is hand positioned to the DIPOLE - BICONE or WHIP position.

12-46. BICONE ANTENNA

The spacecraft is electrically divided in two sections. (See Figure 12-6). The antenna fairing structure at the junction of these sections resembles a discone antenna. This junction is center fed by a coaxial cable from the communications sets. At frequencies between 225 and 450 MC the antenna fairing acts like discone antenna. A lower frequency of 15 MC causes the unit to resemble an "off center fed" dipole, between the upper and lower limits.

SEDR 104-

Thus the bicone antenna may serve all spacecraft frequencies, with the exception of C- and S-Bands, allowing reception and transmission within limits of the spacecraft system.

12-47. **BICONE ISOLATOR**

An isolator is provided to shield electrical wires that pass through the bicone antenna fairing structure. The isolator is formed into a tube which is curved to allow mounting beneath the periphery of the antenna fairing.

12-48. **HF DIPOLE (ORBITAL) ANTENNA**

The HF Dipole Antenna is extended to 13 feet, 8 inches after an orbital condition is reached. Selection of this antenna for HF use is achieved by the astronaut through the use of the HF ANT SELECT switch located to the right of the astronaut. (See Figure 12-6)

12-49. **UHF DESCENT AND RECOVERY ANTENNA**

The UHF descent and recovery antenna takes over the UHF functions of the bicone antenna when the antenna fairing is jettisoned. (See Figure 12-6). The UHF descent and recovery antenna is a fan shaped, vertically polarized monopole located on the top of the recovery compartment.

12-50. **HF RECOVERY ANTENNA (WHIP)**

Upon landing, impact circuits initiate a sequence for the HF recovery antenna. (See Figure 12-6). The elevated antenna acts as a vertically polarized monopole for HF frequencies.

12-51. C- AND S-BAND ANTENNAS

Three antenna units serve the C- and S-Band beacons. (see Figure 12-6). Each unit consists of a C- and a S-Band radiator. Each radiator is a cavity mounted helix antenna.

12-52. AUXILIARY UHF RESCUE BEACON ANTENNA

A spring-tape type antenna is vertically mounted in the recovery compartment and is connected by coaxial cable directly to the auxiliary UHF rescue beacon inside the spacecraft. With the bicone antenna installed, the tape antenna is held in a bent position and upon bicone separation the antenna springs into operating position.

SECTION XIII

NAVIGATIONAL AIDS

TABLE OF CONTENTS

TITLE	PAGE
General	13-3
Navigational Aid Kit	13-3
Satellite Clock	13-3
Altimeter	13-4
Longitudinal Accelerometer	13-5
Attitude-Rate Indicator	13-5
Navigational Reticle	13-6

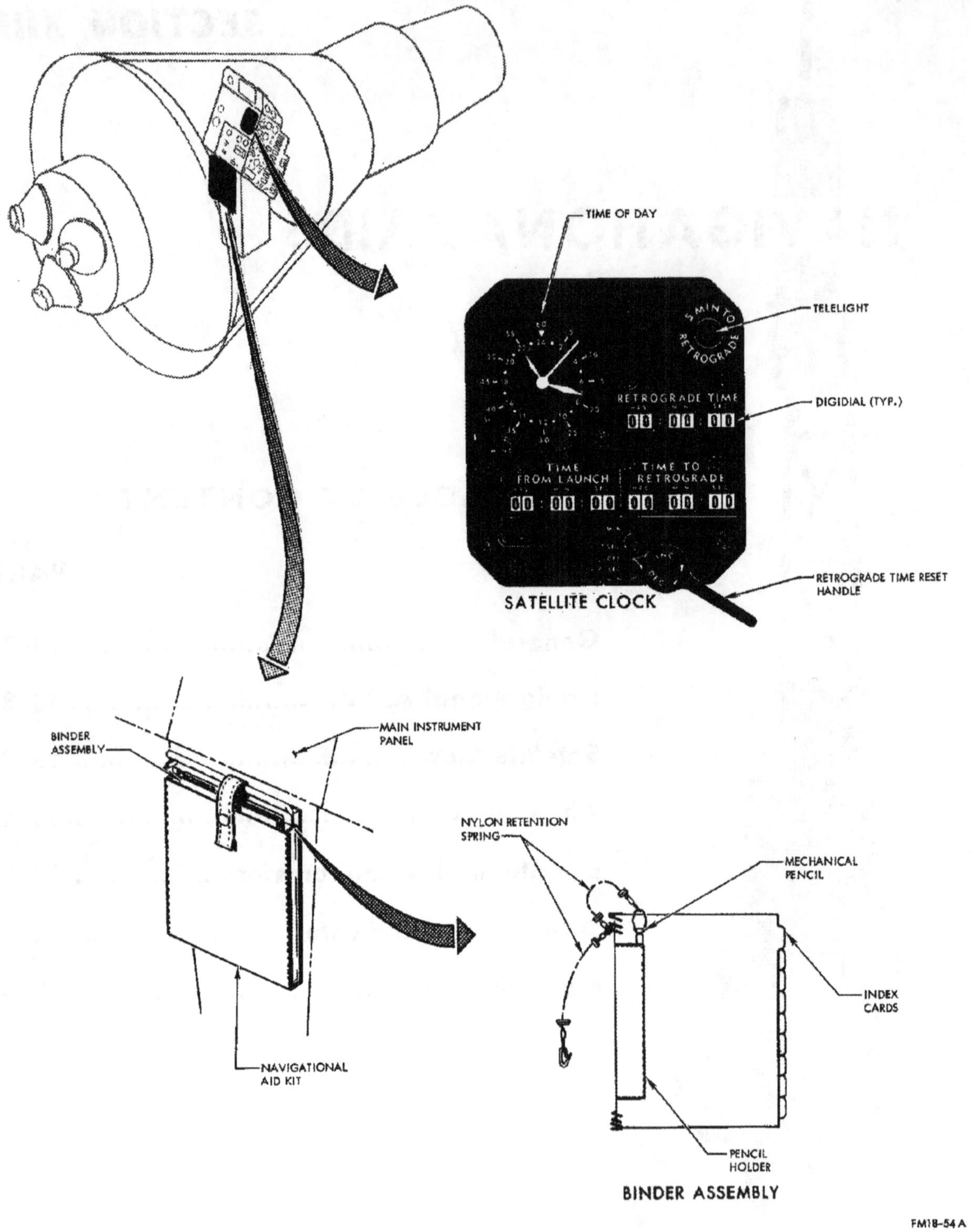

Figure 13-1 Navigational Aid Kit and Satellite Clock

MCDONNELL SEDR 104

XIII. NAVIGATION AIDS

13-1. GENERAL

Navigational aids which are required to obtain altitude, course, velocity and landing data are provided to attain and maintain the proper attitude through each phase of the flight.

13-2. NAVIGATIONAL AID KIT

The navigational aid kit consists of a neoprene coated nylon case, and a binder assembly; and is mounted below the main instrument panel (see Figure 13-1). The binder assembly consists of a number of index cards, pencil holder, mechanical pencil, and two (2) nylon retention springs. The provided index cards are used to file note cards, check lists, trouble cards and navigational charts as required by the spacecraft mission. The pencil holder is fabricated from neoprene coated nylon and is sewn to the case. The mechanical pencil is secured to the binder assembly by means of a nylon retention spring. A second nylon retention spring secures the binder assembly to the neoprene coated nylon case.

13-3. SATELLITE CLOCK

The satellite clock is an electro-mechanical timing device located on the upper right of the main instrument panel. The satellite clock indicates the time of day, TIME FROM LAUNCH, TIME TO RETROGRADE and RETROGRADE TIME (see Figure 13-1). The time of day will be reflected by a manually wound spring driven watch. The manually wound watch is located in the upper left-hand corner of the satellite clock. TIME FROM LAUNCH, TIME TO RETROGRADE and RETROGRADE TIME is displayed on digidial drum counters. The drum counters indicate time in HRS, MIN. and SEC. The time elements move in one step

increments. The TIME TO RETROGRADE digidial is supplemented by a telelight. The telelight is located in the upper right-hand corner of the satellite clock and illuminates five minutes prior to retrograde time; in addition to the telelight, an aural signal to the astronaut's headset is initiated ten seconds prior to retrograde time. The satellite clock is automatically started by 28 V D-C power at lift-off. Should this not occur, a switch labeled TIME ZERO is provided above and adjacent to the clock to allow the astronaut to energize the clock (and maximum altitude sensor). The retrograde time is normally computed and set prior to flight, but the retrograde time can be manually changed by the astronaut with the use of the retrograde time reset handle located on the lower right of the satellite clock or remotely set through the command receivers. Ten minutes prior to retrograde time, the satellite clock transmits a signal to the ASCS to start horizon scanners operating continuously and assures rate gyro operation in preparation for retro sequence. When retrograde time is obtained, a set of contact points within the clock close, initiating the retrograde sequence. TIME FROM LAUNCH and RETROGRADE TIME digidials provide outputs for telemetering. The main instrument panel is lighted externally by the cabin flood lights.

13-4. ALTIMETER

The altimeter is a pressure sensitive device located on the lower left of the main instrument panel (see Section II for exact location). The altimeter indicates the external pressure (in pounds per square inch absolute PSIA) and the altitude (in thousands of feet) above sea level of the spacecraft. The static system provides the atmospheric pressure necessary for the altimeter. It is a single revolution type, calibrated from 0 to 100,000 feet, with a marker at 10,000 feet (MAIN) and 20,000 feet (SNORKEL).

13-5. LONGITUDINAL ACCELEROMETER

The accelerometer which is located on the upper left of the main instrument panel (see Section II for exact location) is a self-contained unit and is housed in a hermetically sealed enclosure. The accelerometer is designed to indicate acceleration in the range -9 to 0 to 21 g units (1 g unit is equal to an acceleration of 32.2 feet per second per second). Attached to the face of the accelerometer are three pointers. One pointer indicates instantaneous acceleration. The remaining two pointers are memory pointers. One memory pointer records positive acceleration and the other memory pointer indicates negative acceleration. The memory pointers incorporate a ratchet device which maintain a deflection until they are reset by means of a reset knob which is located in the lower left hand corner of the accelerometer.

13-6. ATTITUDE-RATE INDICATOR

The Attitude-Rate indicator is a three axis angular rate and attitude indicating system located approximately at the top center of the main instrument panel (see Section II for exact location). The system indicates pitch, roll and yaw angles and angular rates. The unit is a composite arrangement consisting of a rate indicator around which are positioned a roll attitude indicator, a yaw attitude indicator and a pitch attitude

indicator (see Section II). The rate indicator displays three pointers. The rate of roll pointer which is flat white in color is parallel to the pointer of the roll attitude indicator. The rate of yaw pointer which is yellow in color is pointed towards the yaw attitude indicator. The rate of pitch pointer which is pink in color is pointed towards the pitch attitude indicator. The system components are completely interchangeable so that failure of one component shall not necessitate recalibration or replacement of the entire system. The attitude-rate indicator is activated by pitch rate, roll rate and yaw rate transducers. Each transducer is identical and consists of a gyroscope, amplifier and a demodulator. These components function together to produce a d-c output signal proportional to the input rate of change of attitude.

13-7. **NAVIGATION RETICLE**

The navigation reticle is a device which determines when the spacecraft is at the correct angle for retro fire. It is located at the rear and to the left of the astronaut's window (see Figure 13-2). The navigation reticle is mounted on a 180° axis (from left to right), this allows the instrument to be positioned so not to block the view to the astronaut's window but still be accessible. The instrument when it is in its viewable position contains a tinted red lighting system of sufficient brilliance to illuminate four (4) red lines. Three (3) of the four (4) lines are vertical. The fourth line which is horizontal is the line which is required to be tangent to the earth prior to retro fire. The lighting system is automatically turned off when the instrument is in its stowed position. The incoming light may be dimmed by the use of a polar-

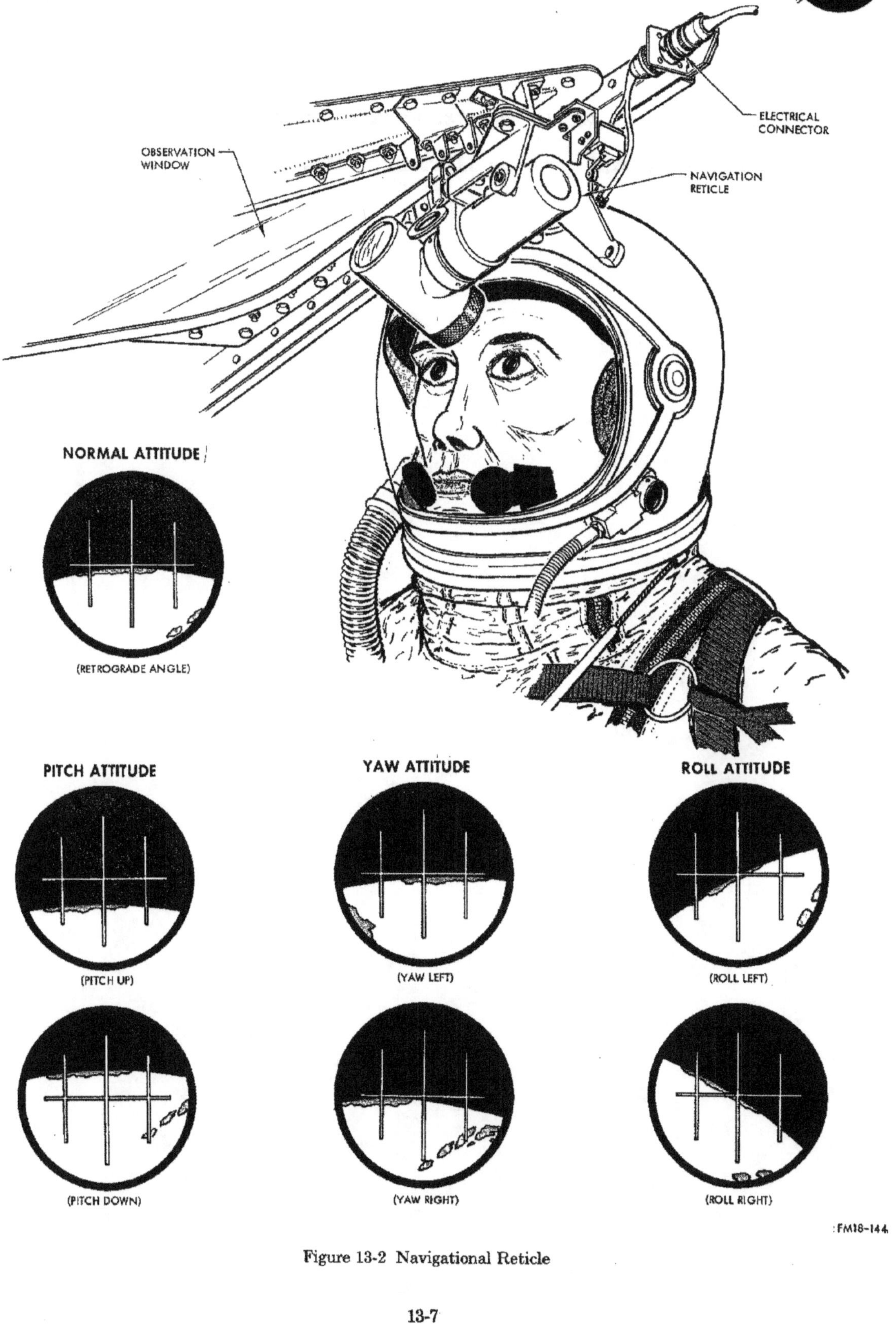

Figure 13-2 Navigational Reticle

old filter. The filter is operated by rotating the outside of the instrument which will vary the light intensity.

SECTION XIV

INSTRUMENTATION SYSTEMS

TABLE OF CONTENTS

TITLE	PAGE
System Description	14-5
System Operation	14-5
System Monitoring	14-6
System Instrumentation Control	14-30
Instrumentation Recording	14-33
System Units	14-36

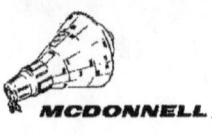

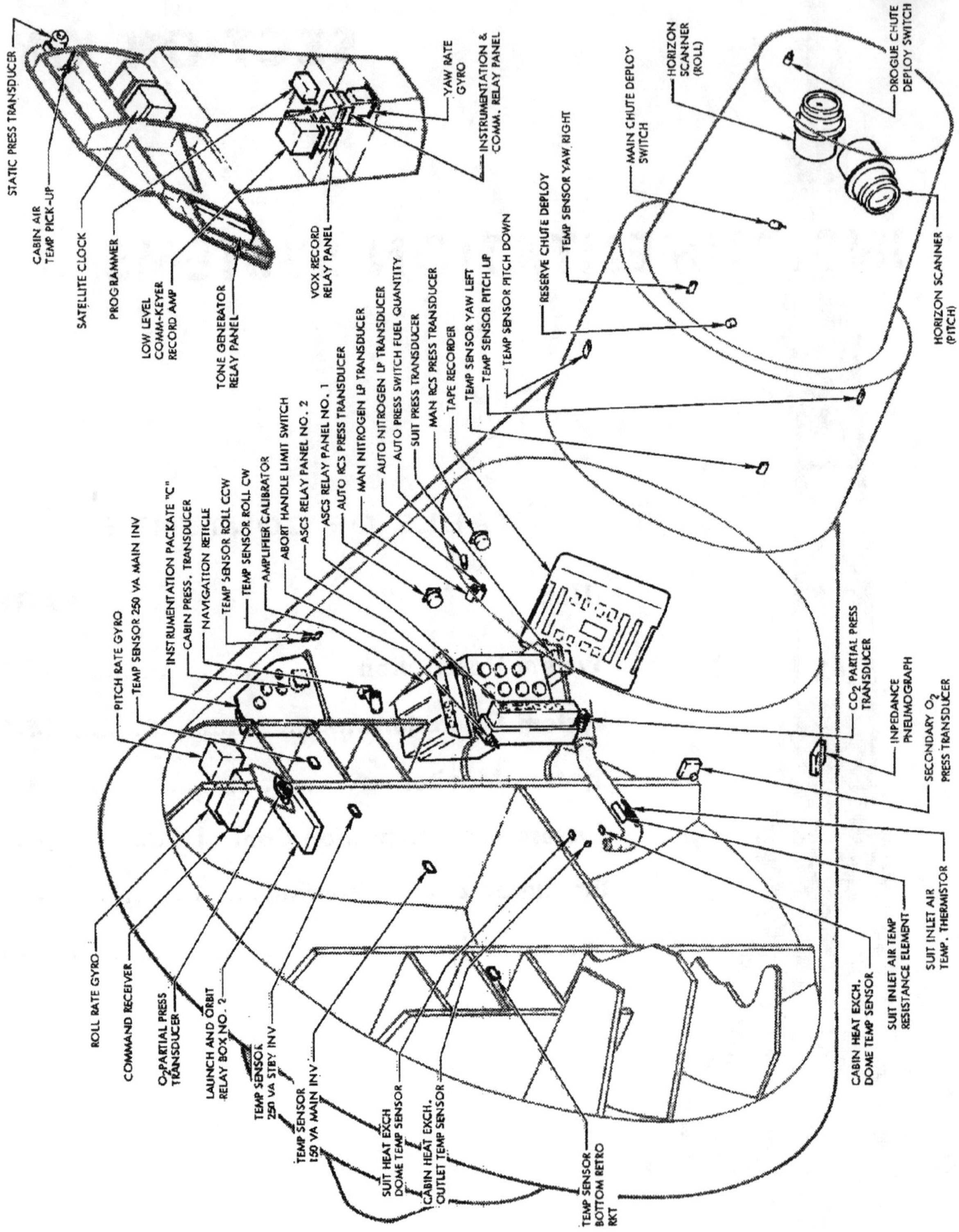

FM18-140-1

Figure 14-1 Instrumentation Component Location-Left Side (Sheet 1 of 3)

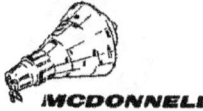

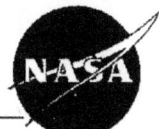

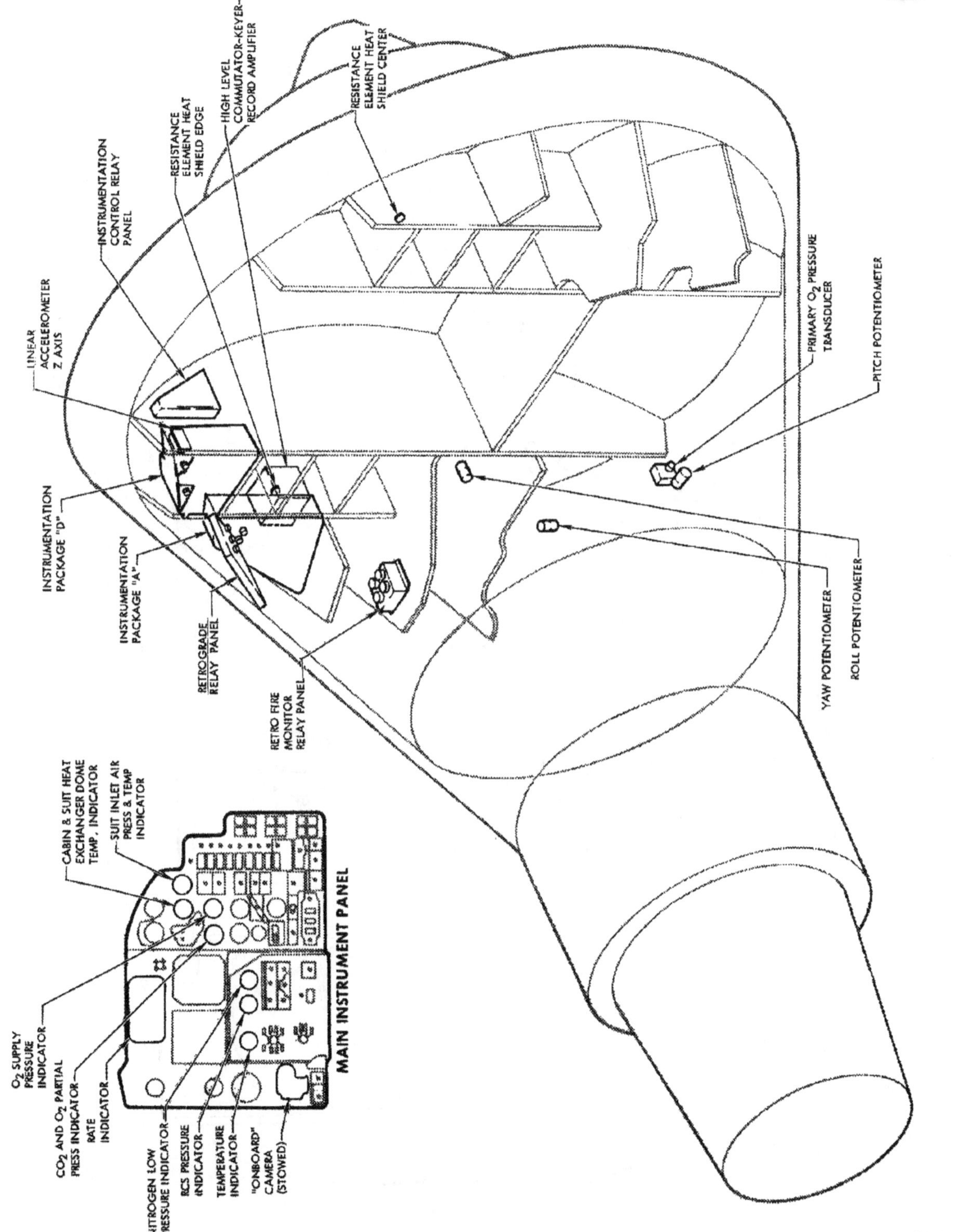

FM18-140-2

Figure 14-1 Instrumentation Component Location-Right Side (Sheet 2 of 3)

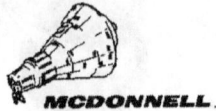

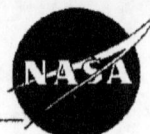

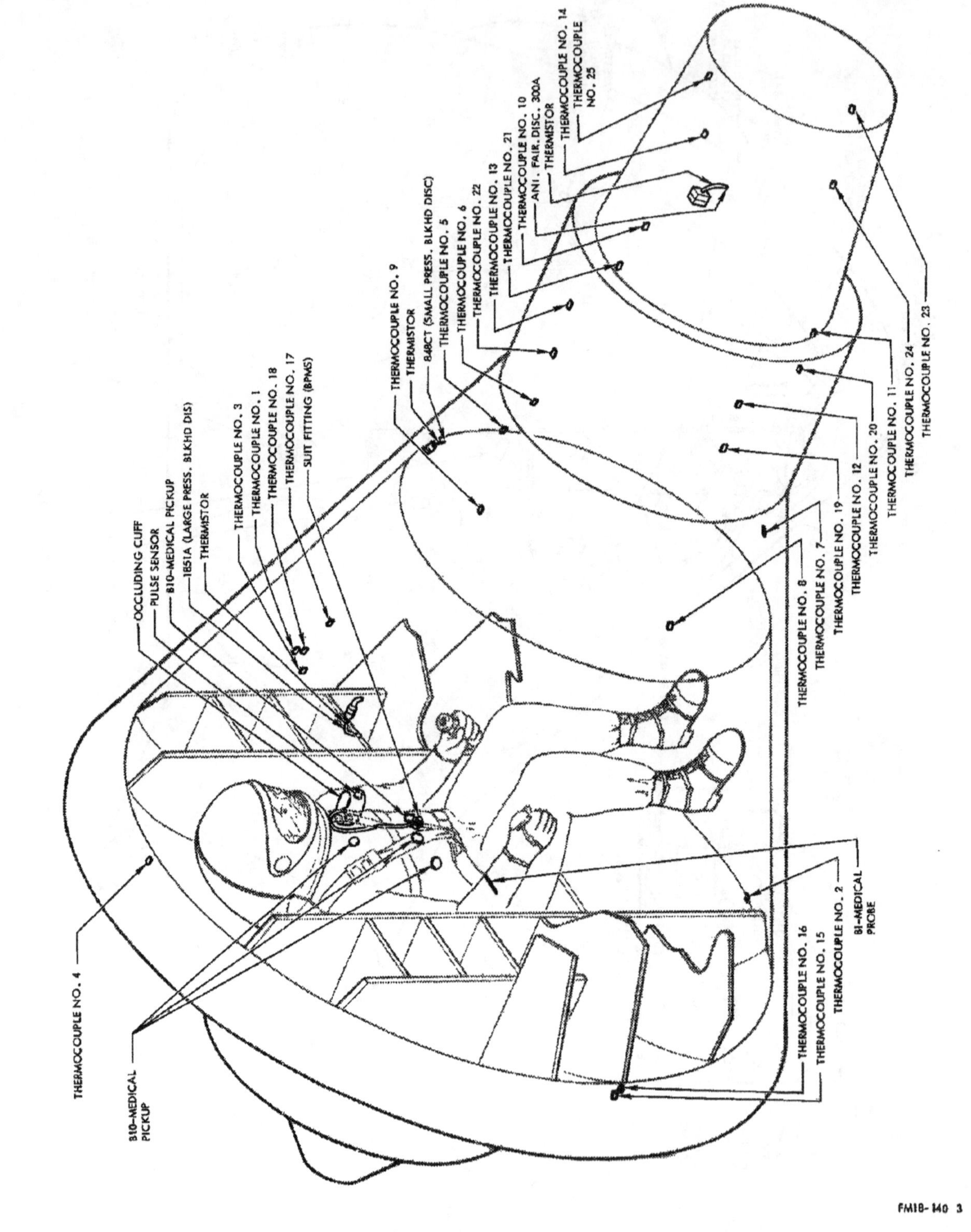

FM18-140 3

Figure 14-1 Instrument Component Location (Sheet 3 of 3)

SEDR 104

XIV. INSTRUMENTATION SYSTEM

14-1. **SYSTEM DESCRIPTION**

The instrumentation system consists of the major components shown on Figure 14-1. These components coupled with various transducers and other pickup devices provide a means of monitoring the physical condition and reactions of the astronaut as well as spacecraft conditions and systems operational performance. This data so obtained is applied to various voltage controlled subcarrier oscillators which modulates the Low-Frequency Telemetry Transmitter and radiates to ground stations for analysis and evaluation; this same data is also recorded on a tape recorder in the spacecraft for subsequent study and interpretation.

A portable 16 millimeter camera is provided for the astronaut to record on film various events of interest during the mission. Provisions are also provided for automatic programmed control over some components not intended for continuous operation.

14-2. **SYSTEM OPERATION**

The instrumentation system is automatic and semi-automatic in operation from the time power is applied to the spacecraft until 10 minutes after landing impact, however, certain components may be controlled or interrogated during flight by either the astronaut or ground command. The instrumentation system is divided into three groups, namely, monitoring, control and recording. These three groups are treated individually in paragraphs 14-3, 14-59 and 14-64.

MCDONNELL SEDR 104

14-3. **MONITORING**

Instrumentation monitoring consists of sampling values of pressures, temperatures, conditions and operations of various units and functions throughout the spacecraft. (See Figure 14-6). These samples are converted into signals composed of voltages proportional to the temperature, pressure and conditions being measured. The proportional voltages are calibrated within common maximum and minimum ranges to provide zero and full scale readings. Instrumentation monitoring is sub-divided into two areas namely, High Level and Low Level. A description of each is provided in the following paragraphs.

14-4. **HIGH LEVEL (0-3V d-c)**

High level input signals are channeled into the commutator (electronic switching device) which is located in instrumentation package A. The commutator continuously samples its input channels, combining the signal voltage pulses into a pulse train from the commutator. This pulse train (PAM) is applied to a 10.5 KC voltage controlled subcarrier oscillator where the changing voltage of the pulse train varies the frequency of the oscillator. The output of the 10.5 KC voltage controlled oscillator and other oscillators are applied to an isolation amplifier which has outputs for the tape recorder, telemetry transmitter and hard line. The High Level Commutator output signals are also converted to pulse duration signals (PDM) and recorded on the spacecraft on-board tape recorder.

14-5. <u>LOW LEVEL (-5 mv to +15 mv)</u>

Low level instrumentation monitoring performs a temperature survey of various structural and spacecraft components. The low level commutator output signals are converted to pulse duration (PDM) signals for recording on the on-board tape recorder, only. The low level PAM may be monitored during ground tests by actuation of the Temperature Survey Switch to TEST. This pulse train is applied to a 10.5 KC voltage controlled subcarrier oscillator where the changing voltage of the pulse train varies the frequency of the oscillator. The output of the 10.5 KC voltage controlled oscillator and other oscillators are applied to an isolation amplifier which provides telemetry transmitter and hard line outputs.

Figure 14-2 is a block diagram showing the parameters of the instrumentation that is monitored with a brief explanation of each parameter given in paragraphs 14-6 through 14-58.

14-6. <u>SPACECRAFT ELECTRICAL POWER</u>

Spacecraft electrical power system instrumentation consists of monitoring the circuit illustrated on Figure 14-2.

14-7. <u>400 cps MONITOR</u>

ASCS bus 115 volt a-c is attenuated, rectified and filtered prior to being applied to the commutator as a zero to three volt d-c signal. The Fan bus 115 volt a-c is applied through a 115 to 6.3 volt stepdown transformer in instrumentation package A. The secondary outputs are attenuated,

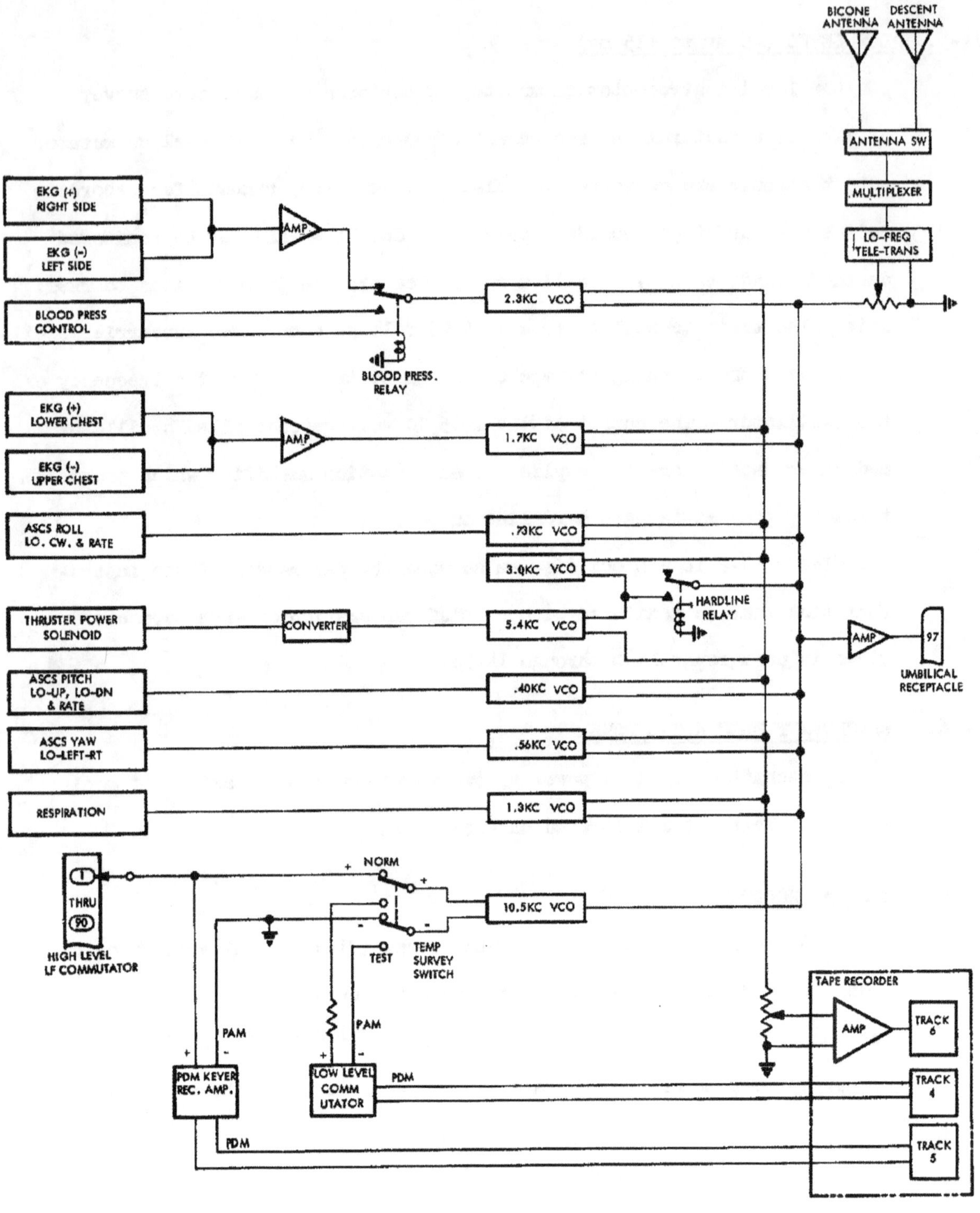

Figure 14-2 Monitoring Instrumentation Block Diagram

14-8

rectified and filtered prior to being applied to the commutator as a zero to three volt d-c signal. A three volt signal (full scale) represents 130 volts for each bus.

14-8. <u>D-C CIRCUIT</u>

D-c current amplitude is sensed by the shunt for the instrument panel ammeter. This shunt is in the negative lead to ground of all spacecraft batteries and senses total battery current. The voltage across the shunt is 50 millivolts when 50 amperes are flowing, and proportionately less for lesser currents. This voltage is applied to a d-c amplifier in package A which amplifies it to a zero to three volt level. A three volt level (full scale) represents 50 amperes battery current. The output of the amplifier is applied to the commutator.

14-9. <u>D-C VOLTAGE</u>

The 24 volt main bus d-c monitor circuit is attenuated prior to being applied to the commutator in the A package. A three volt signal (full scale) represents 30 volts bus voltage.

14-10. <u>INSTRUMENTATION POWER SUPPLIES</u>

Instrumentation power supplies-instrumentation consists of the monitor circuits for the two 3 volt d-c references, zero reference, and 7 V 400 cps power supplies. (See Figure 14-2).

14-11. <u>THREE VOLT REFERENCE</u>

The 3 volt d-c reference power supply furnishes excitation for all potentiometer type instrumentation pickups. There are two 3 volt d-c

SEDR 104

power supplies located in packages A and C respectively. Each supply is zener diode regulated. The 3 volt d-c excitation voltage requirements are divided between the two separate supplies with the power supply in package A furnishing a reference full scale signal.

14-12. ZERO VOLT REFERENCE

The zero reference signal is signal ground and is common to both of the above mentioned power supplies. This signal is also applied to the commutator.

14-13. SEVEN VOLT 400 cps

The 7 volt 400 cps power supply furnishes excitation for the input bridge circuits utilized with the resistance element amplifiers and thermistor signal conditioners. Power supply output is attenuated, rectified and filtered to a zero to three volt level. This zero to three volt signal is applied to the commutator. A three volt signal (full scale) represents a 10 volt output level. The power supply is a transistorized power inverter which operates on 24 volts d-c to provide the 7 volt 400 cps output and is located in package A.

14-14. CALIBRATION ON

Calibration ON instrumentation consists of a circuit which monitors presence of the full scale and zero scale calibration command signals. This signal is present when the CALIBRATION switch in the telemetry trailer is placed to FULL SCALE or ZERO position. When the full scale calibrate command is present, 24 volts d-c is applied to an attenuator in package

C. The output of the attenuator (2.25 volts d-c nominal) is applied to the commutator. When the zero scale calibrate command is present, 24 volts d-c is applied to a different input point on the same attenuator network. The output of the attenuator (.75 volts d-c, nominal) is applied to the commutator. Thus, an upper scale signal indicates presence of the full scale calibrate command and a quarter-scale signal indicates presence of the zero scale calibrate command. These command signals energize relays which apply calibrate signals to numerous other instrumentation channels. The R and Z calibrate function may be initiated by the programmer or from a ground station through the command receivers while in orbit.

14-15. STATIC PRESSURE

Static pressure instrumentation consists of a potentiometer type transducer which is operated by static pressure. The potentiometer is excited with 3 volts d-c from the instrumentation power supply located in the C package. Wiper voltage output is inversely proportional to static pressure. A three volt signal (full scale) is representative of 0 psia.

14-16. ENVIRONMENTAL CONTROL SYSTEM

Environmental control system instrumentation consists of circuitry which monitors primary and secondary oxygen supply pressures, suit inlet air pressure, cabin pressure, temperature, CO_2 partial pressure, O_2 partial pressure and emergency O_2 rate.

14-17. CABIN O_2 PARTIAL PRESSURE

O_2 partial pressure is sensed by a transducer in the cabin. The signals are amplified and transmitted to a pressure gage on the instru-

SEDR 104

ment panel. The signals are also applied to the commutator in package A where they are converted to PAM and PDM signals. The PAM signals are telemetered to ground and the PDM signals are recorded on the spacecraft tape recorder.

14-18. SUIT CO_2 PARTIAL PRESSURE

CO_2 partial pressure is sensed by a transducer in the environmental control system's suit circuit. The signals are amplified and transmitted to a pressure gage on the instrument panel. The signals are also applied to the commutator in package A where they are converted to PAM and PDM signals. The PAM signals are telemetered to ground and the PDM signals are recorded on the spacecraft tape recorder.

14-19. OXYGEN SUPPLIES

Primary and secondary oxygen supply pressures are sensed by pressure actuated dual potentiometers in the environmental area. One potentiometer operates a panel indicator while the other wiper picks off a value for instrumentation. Wiper voltage output is linearly proportional to pressure. Excitation is applied from the 3 volt d-c instrumentation power supply. The zero to 3 volts wiper output represents a pressure range of zero to 9,375 psi with 0 to 125% meter indications. Outputs from the primary and secondary oxygen supply pressure transducers are applied to the commutator.

14-20. SUIT INLET PRESSURE

Suit inlet air pressure instrumentation consists of a potentiometer type transducer which is pressure actuated. The potentiometer is excited

14-12

with 3 volts d-c from the instrumentation power supply in package A. Wiper voltage output is linearly proportional to pressure. The zero to three volt (full scale) output represents a folded transducer output with a pressure range of 15 psi = 0V, 7.5 psi = 3V and 0 psi = 0V applied to the commutator.

14-21. CABIN PRESSURE

Cabin pressure instrumentation consists of a potentiometer type pressure transducer installed in package C. The potentiometer is excited with 3 volts d-c from the instrumentation power supply in package C. Wiper voltage output is linearly proportional to cabin pressure. The zero to three volt (full scale) output from the wiper represents a folded transducer output with a pressure range of 15 psi = 0V, 7.5 psi = 3V, and 0 psi = 0V applied to the commutator.

14-22. CABIN TEMPERATURE

Cabin temperature instrumentation consists of a temperature sensing probe mounted on the back of the main instrument panel. Transducer resistance varies proportionally with temperature and is part of a bridge input circuit to an amplifier in package A. The zero to three volt (full scale) output from the amplifier is representative of a temperature range of 40 to 200°F. The output from the amplifier associated with the transducer is applied to the commutator.

14-23. EMERGENCY O_2 RATE

Emergency O_2 rate instrumentation consists of a switch actuated by the oxygen emergency rate valve. Upon closure of this switch to emergency, 24V d-c is provided to an attenuator which steps down this voltage

SEDR 104

to 3 volts. This output is then applied to the commutator.

14-24. CABIN HEAT EXCHANGER GAS TEMPERATURE

Cabin heat exchanger gas temperature instrumentation consists of a thermistor type transducer. Thermistor resistance varies inversely proportional with temperature and is part of a bridge input circuit to an amplifier in package C. The zero to three volt (full scale) output from the amplifier is representative of a temperature range of 30 to 100°F. The output from the amplifier associated with the transducer is applied to the commutator and the temperature indicator on the main instrument panel.

14-25. CABIN HEAT EXCHANGER DOME TEMPERATURE

Cabin heat exchanger dome temperature instrumentation consists of a thermistor type transducer. Thermistor resistance varies inversely proportional with temperature and is part of a bridge input circuit to an amplifier in package C. The zero to three volt (full scale) output from the amplifier is representative of a temperature range of 40 to 100°F. The output from the amplifier associated with the transducer is applied to the commutator and a cockpit meter. If excessive dome temperature is present, an audio tone will be heard by the astronaut along with a warning light.

14-26. SUIT HEAT EXCHANGER DOME TEMPERATURE

Suit heat exchanger dome temperature instrumentation consists of a thermistor type transducer. Thermistor resistance varies inversely proportional with temperature and is part of a bridge input circuit to an amplifier in package C. The zero to three volt (full scale) output from the amplifier is representative of a temperature range of 40 to 100°F. The output from the amplifier associated with the transducer is applied to the commutator and a cockpit meter. If excessive dome temperature is present, an audio tone will be heard by the astronaut along with a warning light.

14-27. **REACTION CONTROL SYSTEM**

Reaction control system instrumentation consists of monitors for automatic and manual reaction control supply pressure and astronaut hand control position. High pressure instrumentation is monitored on the high level commutator while the low pressure is monitored on the indicator located on the main instrument panel.

14-28. **HORIZON SCANNER**

Horizon scanner instrumentation monitors for the pitch and roll horizon scanner outputs and ignore signals for each of these outputs.

The horizon scanner system utilized two identical infrared scanning units to provide pitch and roll reference signals. The horizon scanners are on continuously, from launch until re-entry at which time the scanners are de-activated by the 0.05g relay, but during the orbital phase the reference signals are applied to the ASCS attitude gyros only upon command from the programmer. (Refer to Table 14-1). The signals that are applied to the gyros are monitored by instrumentation. The pitch and roll signals range between $\pm$ 10 volts d-c. These signals are applied to a biased attenuator card to provide a zero to 2.68 volt out-

put which is coupled to separate channels of the commutator. The signals represent an output range of ± 35°.

Occasionally a scanner sweeps across the sun. Since the scanners are infrared devices, sweeping of the sun introduces error voltage. To prevent utilization of this voltage, the scanner supplies an ignore signal to the ASCS. This "ignore" signal is monitored as an on-off type of signal by instrumentation. Pitch ignore and roll ignore signals are applied to the commutator also. A full scale signal represents presence of the pitch ignore signal and full scale level indicates presence of the roll ignore signal. Each ignore signal is monitored on separate channels.

14-29. ATTITUDE

Attitude instrumentation consists of telemetry channels which monitor spacecraft pitch, roll and yaw attitudes. Each attitude is read out of a synchro actuated potentiometer. The synchros are driven by the automatic stabilization control system. Excitation for the potentiometers is furnished by the three volt d-c instrumentation power supply in the C package. Signal voltage varies along a multiple slope function with spacecraft attitude. Pitch and roll signals cover a range of +130° to -190°. Yaw signals cover a range of +70° to -250°. Each of the attitude signals is applied to channels of the commutator. After retrograde assembly jettison and energizing of a 0.05g relay, the potentiometer-positioning synchros become inoperative.

14-30. ATTITUDE RATE

Attitude rate instrumentation utilized signals from rate gyros. The gyros are part of the attitude rate indicating system. A zero to three

SEDR 104

volt signal level represents a rate level of decreasing 40° per second to increasing 40° per second. Pitch, roll and yaw rates are assigned to separate voltage controlled oscillators (Pitch Rate and Pitch low solenoids .40 KC VCO, Roll Rate and Roll low solenoids .73 VCO and Yaw Rate and Yaw low solenoids .56 KC VCO)

14-31. REACTION CONTROL SYSTEM SOLENOIDS

The reaction control system solenoids control the thrust jets used for spacecraft stabilization in flight. These solenoids can be energized manually or automatically. When a solenoid is energized (high solenoids only), 24 volts d-c is applied through an attenuator in package C to the commutator. This on-off signal is presented to instrumentation circuitry from the ASCS system.

14-32. SUPPLY PRESSURES - H_2O_2

High pressure monitor circuits for reaction control supply pressures are basically identical in operation. A helium source of approximately 2,770 psi for automatic and approximately 2,830 psi for manual is utilized to expel hydrogen peroxide from a bladder. As hydrogen peroxide is expelled, the confined volume of the helium increases and helium pressure decreases. A pressure potentiometer senses this change in pressure. The potentiometer is excited with three volts from the instrumentation power supply located in both the A and C packages. Wiper output voltage is applied to the commutator and through an attenuator to an indicator. Transducer range is 600 to 3,400 psi. A pressure of 2,770 psi for automatic and 2,830 psi for manual provides a reading of

SEDR 104

100% on the panel indicator. Hydrogen peroxide is exhausted at approximately 1,580 psi for automatic and 1,960 psi for manual helium pressure. Indicator reading at this pressure is approximately 0%. Low pressure is monitored on the low side of the regulator with a transducer range of 400 to 700 psi.

14-33. HAND CONTROL

Astronaut hand control position is monitored by three potentiometers. The wipers of these potentiometers are driven by linkage to the hand control. Three volts from the instrumentation power supply located in the C package is utilized to excite the potentiometers. Zero to three volt signal level represents $\pm 13°$ hand control movement in the roll and pitch planes and $\pm 10°$ movement in the yaw plane. Wiper output is applied to the commutator.

14-34. SPACECRAFT ACCELERATION

An accelerometer installed in package C provide zero to three volt d-c outputs proportional to acceleration along the longitudinal (Nz) axis of the spacecraft. The accelerometer output is linear with a zero acceleration providing a 1.5 volt d-c signal. The longitudinal axis accelerometer covers a range of $\pm 30g$ to provide zero to three volt output signals. These zero to three volt signals are applied to the commutator.

14-35. HIGH LEVEL COMMUTATED STRUCTURAL TEMPERATURES

High level commutated structural temperature instrumentation consists of monitor circuits for the ablation shield.

14-36. ABLATION SHIELD TEMPERATURES

The ablation shield temperatures are monitored through the telemetry system. Two transducers are embedded in the inner face of the shield. The transducers have a temperature range from -55 to 2,000°F. with a nominal resistance of 100 ohms at 70°F. Input power of 7 V d-c 400 cps applied to the transducers is attenuated to a value dependent upon the transducer resistance. Transducer resistance varies proportionally with temperature and is part of a bridge input circuit to an amplifier in package A where the voltage is converted and amplified to a zero to 3 V d-c signal and applied to the telemetry commutator.

14-37. LOW LEVEL COMMUTATED STRUCTURAL TEMPERATURES

Temperatures monitored on the low level commutator use thermocouples with reference junction thermistors in the first available disconnect. Attenuator circuits if needed are located in the low level commutator. Refer to Figures 14-1 and 14-3.

14-38. METERED STRUCTURAL TEMPERATURES

Various structural temperatures are monitored by the astronaut through the use of temperature selector switches and a temperature indicator located on the main instrument panel. Thermistor transducer resistance varies inversely proportional with temperature and is part of a bridge input circuit to an amplifier in package A where the voltage is converted and amplified to a zero to 100 micro-amp signal and applied to the temperature indicator. (Refer to Figure 14-4).

SEDR 104

Description	Location	TC#	Part No.		Ch	Range
FULL SCALE REFERENCE (+ 15 MV INPUT)				FROM INTERNAL SOURCE	1	
ZERO REFERENCE (-5 MV INPUT)					2	
CONICAL SECTION SHINGLE AT HEAT SHIELD	Z111.5, LX, BY 4.3	TC#1	(1) 45-79012-123		3	1800°F (DIFF.)
CONICAL SECTION SHINGLE AT HEAT SHIELD	Z109.1, RX, TY 2	TC#3	(1) 45-79012-123			1800°F
CONICAL SECTION SHINGLE AT HEAT SHIELD	Z108, RX2, BY	TC#2	(1) 45-79012-123		4	1800°F (DIFF)
CONICAL SECTION SHINGLE AT HEAT SHIELD	Z111.3, LX2, TY	TC#4	(1) 45-79012-123			1800°F
THERMISTOR REFERENCE AT PLUG 1851A	(LARGE PRESS.BLK'HD. DISC.)		GB34P91	BRIDGE	5	50-150°F
THERMISTOR REFERENCE AT PLUG 848 CT	(SMALL PRESS. BLK'HD. DISC.)		GB34P91	BRIDGE	6	50-150°F
			SPARE		7	
			SPARE		8	
CONICAL SECTION SHINGLE	Z148, LX, BY2	TC#5	(2) 45-79012-123		9	1800°F (DIFF.)
CONICAL SECTION SHINGLE	Z154, LX, BY2	TC#6	(2) 45-79012-123			1800°F
CONICAL SECTION SHINGLE	Z154, RX2, BY	TC#7	(2) 45-79012-123		10	1800°F (DIFF)
CONICAL SECTION SHINGLE	Z157, RX2, TY	TC#9	(2) 45-79012-123			1800°F
CYLINDRICAL SECTION SHINGLE	Z178, LX, BY2	TC#10	(2) 45-79012-87		11	600°F (DIFF)
CYLINDRICAL SECTION SHINGLE	Z178, RX, TY2	TC#12	(2) 45-79012-87			600°F
CYLINDRICAL SECTION SHINGLE	Z178, RX2, BY	TC#11	(2) 45-79012-87		12	600°F (DIFF)
CYLINDRICAL SECTION SHINGLE	Z178, LX2, TY	TC#13	(2) 45-79012-87			600°F
INLET B NUT, AUTO. YAW LEFT SOLENOID	Z173, RX, TY1.5	TC#19	(2) 45-79012-65		13	250°F MAXIMUM
INLET B NUT, AUTO. PITCH UP SOLENOID	Z173, RX1.5, BY	TC#20	(2) 45-79012-65		14	250°F MAXIMUM
INLET B NUT, AUTO. YAW RIGHT SOLENOID	Z173, LX, BY1.5	TC#21	(2) 45-79012-65		15	250°F MAXIMUM
INLET B NUT, AUTO. PITCH DOWN SOLENOID	Z173, LX1.5 TY	TC#22	(2) 45-79012-65		16	250°F MAXIMUM
			SPARE		17	
			SPARE		18	
THERMISTOR REFERENCE AT PLUG 300A	(ANTENNA FAIRING DISC.)		GB34P91	BRIDGE	19	20-150°F
CONICAL SECTION SHINGLE	Z154.5, RX, TY2	TC#8	(2) 45-79012-121	ATTEN.	20	1800°F MAXIMUM
MAN. LOW ROLL CCW THRUSTER	Z115, RX, BY1.1	TC#15	(1) 45-79012-123	ATTEN.	21	1500°F MAXIMUM
SHINGLE OVER MAN. LOW ROLL CCW THRUSTER	Z110, RX, BY2	TC#16	(1) 45-79012-121	ATTEN.	22	1800°F MAXIMUM
AUTO. LOW ROLL CW THRUSTER	Z114, LX, BY2.5	TC#17	(1) 45-79012-123	ATTEN.	23	1500°F MAXIMUM
SHINGLE OVER AUTO LOW ROLL CW THRUSTER	Z111.5, LX, BY2	TC#18	(1) 45-79012-121	ATTEN.	24	1800°F MAXIMUM
ANTENNA CANISTER	Z202, LX2, TY	TC#14	(3) 45-79012-15	ATTEN.	25	2000°F MAXIMUM
ANTENNA CANISTER	Z202, RX2, BY	TC#23	(3) 45-79012-15	ATTEN.	26	2000°F MAXIMUM
ANTENNA CANISTER	Z202, LX, BY2	TC#24	(3) 45-79012-15	ATTEN.	27	2000°F MAXIMUM
ANTENNA CANISTER	Z200, RX, TY2	TC#25	(3) 45-79012-15	ATTEN.	28	2000°F MAXIMUM

TEMP. SURVEY TEST SWITCH — NORM — PAM HIGH LEVEL
TO 45-88243-19 VCO — TEST — PAM
TO TRACK 4, TAPE RECORDER — PDM DIFF

NOTES

(1) REFERENCE JUNCTION, CH. 5 (1851A)
(2) REFERENCE JUNCTION, CH. 6 (848 CT)
(3) REFERENCE JUNCTION, CH. 19 (300A)
CR-CN ARE 45-79012-65, -87
CR-AL ARE 45-79012-15, -121, -123
ONLY END T.C. ASS'YS SHOWN

LOW LEVEL COMMUTATOR-
KEYER-RECORD AMPLIFIER
45-88128-301

FM18-141

Figure 14-3 Low Level Commutated Structural Temperatures

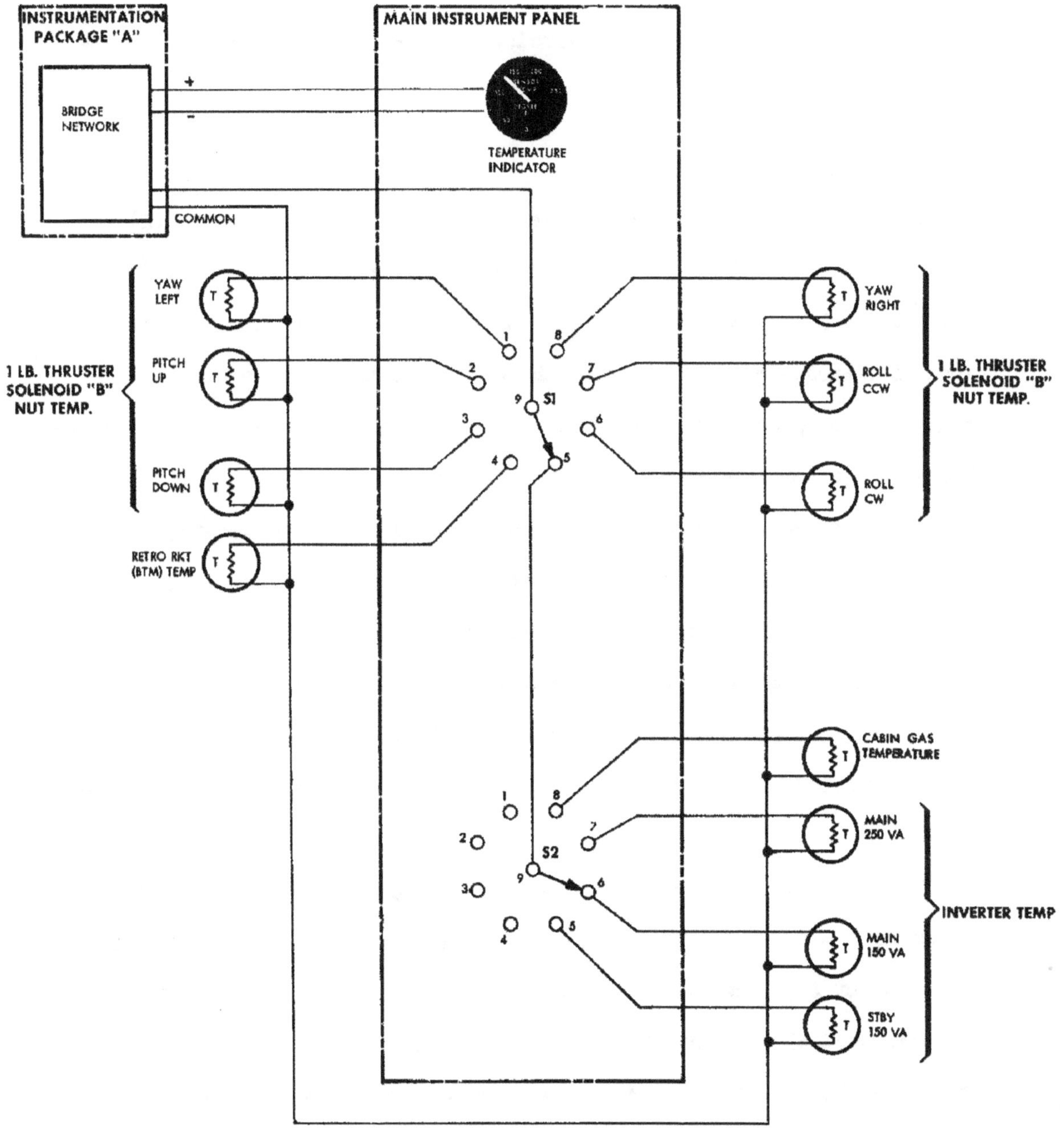

Figure 14-4 Metered Structural Temperatures

14-39. <u>AEROMEDICAL</u>

Aeromedical instrumentation consists of monitor circuits for electrocardiograph (EKG), respiration signals, body temperature and blood pressure.

14-40. <u>ASTRONAUT BLOOD PRESSURE</u>

The blood pressure system consists of (1) an occluding cuff, (2) a pulse sensor, (3) differential transducer, (4) pressure source, and (5) a controller system. The occluding cuff is attached to the astronaut's arm. A transducer which measures the differential pressure between the cuff and the spacecraft cabin pressure is located in the controller system. The pulse sensor is a small transducer (microphone) attached inside the suit. The pressure source is a separate oxygen bottle containing sufficient oxygen to provide the desired aeromedical information during the mission. The system measures the astronaut's blood pressure, converts the pressure to a corresponding electrical signal which is then applied through a blood pressure relay to the 2.3 KC voltage controlled oscillator and transmitted by the telemetry transmitter. This signal is also recorded on the tape recorder.

The blood pressure system may be put into operation by the astronaut actuating a START switch on the instrument panel. Upon actuation, a 24 V d-c pulse of five seconds duration, causes the system to pressurize to 4.4 psi differential pressure from the pressure source. After pressurizing, the system bleeds off at a linear rate to 0.75 psi in approximately 22 seconds. The output signal from the pulse sensor is routed through the pressure

suit disconnect and mixed with the differential pressure signal in a superimposing manner. This combined signal is routed through a relay and relay contacts to a 2.3 KC VCO located in D package, and then to the telemetry transmitter. Directing the signal through the relay is necessary in order to share the 2.3 KC VCO with the EKG signals.

The first appearance of the blip indicates systolic pressure with a minimum peak amplitude of 150 mv while the last occurence of the blip signal indicates diastolic pressure with a minimum peak amplitude of 150 mv. The maximum pulse pressure is 1 volt peak.

Upon completion of the cycle, the system will remain at rest (below 3/4 psi pressure). Each measurement is manually initiated by the astronaut, if he does not use the STOP button, a 110 second timer will run out and will then automatically commence to measure EKG. A light on the panel indicates when the system is operating.

14-41. ASTRONAUT EKG

Electrocardiograph signals are obtained from four transducers attached to the astronaut's right and left side, and on the upper and lower chest. The outputs from the transducers are applied to two amplifiers in D package (left and right side paired to one amplifier and upper and lower chest paired to the other). Signals from the amplifiers are then directed to a 2.3 KC and 1.7 KC voltage controlled oscillator, which in turn apply their outputs to the telemetry transmitter. The 2.3 KC VCO input signals are divided between the astronaut's EKG and blood pressure outputs.

14-42. <u>RESPIRATION RATE</u>

The astronauts breathing rate is monitored through the telemetry system. Two EKG type probes are attached to the sides of the astronauts chest. An impedance pneumograph senses a capacitance change between the two probes as the astronaut breathes. This signal modulates a 50 KC carrier wave. The changing output voltage of the pneumograph varies the bias of a transistor in an in-line amplifier. The amplifier signal is then applied to a 1.3 KC VCO. A potentiometer in the pneumograph provides an adjustment for sensitivity.

14-43. <u>ASTRONAUT BODY TEMPERATURE</u>

Body temperature is sensed by a rectal temperature probe. The probe is a thermistor element which is utilized as one leg of a bridge circuit, which forms the input to a d-c amplifier. The output of the amplifier is applied to the telemetry commutator. The zero to 3 V d-c output represents a temperature range of 95° to 108°F. The amplifier is R and Z calibrated by either ground command or the programmer.

14-44. <u>SEQUENCE SYSTEM NORMAL LAUNCH</u>

Normal launch sequence instrumentation consists of monitor circuits for tower release, spacecraft separation, retrograde attitude command, retrograde fire and retrograde rocket assembly jettison. These signals are all on-off type functions and each is applied to the commutator.

14-45. <u>SATELLITE CLOCK</u>

The satellite clock utilizes potentiometers to provide electrical signals representative of elapsed time from launch and retrograde time. These potentiometers are excited with three volts from the instrumentation

power supply located within the A package. The outputs for each type of time are divided in signals representative of 0 to 10 seconds, 0 to 1 minute, 0 to 10 minutes, 0 to 1 hour, 0 to 10 hours and 0 to 60 hours. Wiper output is linearly proportional from zero to three volts for each time span. Wiper outputs are applied to the commutator. Instrumentation monitors ELAPSED TIME from LAUNCH and also EVENT TIME of retrograde. Elapsed time from launch is the length of time spacecraft has been in motion. Prior to liftoff, elapsed time will be zero. Instrumentation recording devices also will indicate zero time. Output signals for elapsed time therefore are directly proportional to time. As time increases, so will output voltage; for example, elapsed time recorded by clock is 10 hours, 5 minutes and 10 seconds. Output signals will then be as shown below:

SATELLITE CLOCK OUTPUTS FOR 10 HOURS, 5 MINUTES, 10 SECONDS, AFTER LAUNCH

TIME POTENTIOMETERS	POTENTIOMETERS WIPER TRAVEL IN %	SIGNAL VOLTAGE
0 - 60 Hours	21.26%	.638 Volts
0 - 10 Hours	0%	0 Volts
0 - 1 Hour	0%	0 Volts
0 - 10 Minutes	55.3%	1.66 Volts
0 - 1 Minute	21.3%	.638 Volts
0 - 10 Seconds	0%	0 Volts

Event time of retrograde is preset prior to lift off. After retrograde time has been set, instrumentation will receive this time signal continuously throughout the mission. Event time of retrograde can be

changed at any time during the mission by either the astronaut or by ground command. When retrograde time is changed during the mission, instrumentation will receive this change also. Signal output voltage is proportional to retrograde time. For example, if retrograde is set to commence at 20 hours, 10 minutes and 10 seconds, instrumentation will be receiving the signal voltage outputs as shown below:

SATELLITE CLOCK OUTPUTS FOR RETRO-FIRE AT 20 HOURS, 10 MINUTES AND 10 SECONDS

TIME POTENTIOMETERS	POTENTIOMETERS WIPER TRAVEL IN %	SIGNAL VOLTAGE
0 - 60 Hours	40.4%	1.213 Volts
0 - 10 Hours	0%	0 Volts
0 - 1 Hour	21.3%	.638 Volts
0 - 10 Minutes	0%	0 Volts
0 - 1 Minute	21.3%	.638 Volts
0 - 10 Seconds	0%	0 Volts

14-46. TOWER SEPARATION

When the tower separates from the spacecraft, the No. 3 tower separate sensor relay de-energizes and applies 2.4 ± 0.3 volts d-c to the commutator. This signal is present for the remainder of the mission.

14-47. SPACECRAFT SEPARATION

When the spacecraft separates from the booster, a limit switch closes and causes the No. 1 spacecraft separation sensor relay to energize. While this relay is energized a 2.4 ± 0.3 volt d-c signal is applied to the commutator. This relay remains energized for the remainder of the mission.

14-48. RETROGRADE ATTITUDE

The retrograde attitude command signal normally occurs when the retrograde clock runs out. It may also be caused by ground command or by operation of a bypass switch on the instrument panel. This signal remains present until the retrograde rocket assembly is jettisoned (approximately 90 seconds). Signal level is approximately 2.4 ± 0.3 volts. Normally open contacts of the retrograde attitude command relay in retrograde relay box No. 2 closes to route the signal to the commutator.

14-49. ELAPSED TIME SINCE #2 RETRO FIRE

During a mission, the spacecraft may be out of range of a monitoring ground station during retrograde rocket fire. To provide the ground station with some idea when retro-rocket fire took place, a solid state timing device is provided. Upon closure of the No. 2 retro-rocket monitor relay, a start signal is applied to the timing device. Outputs for each type of time are divided into signals representative of 0 to 10 seconds, 0 to 1 minute, 0 to 10 minutes, and 0 to 1 hour. Instrumentation will continuously monitor these time signals until 10 minutes after impact.

14-50. RETROGRADE ROCKET ASSEMBLY JETTISON

The retrograde rocket fire occurs at five second intervals. The first fire occurs thirty seconds after reception of retrograde clock runout if the retrograde interlock is closed in the ASCS.

The retrograde rocket assembly jettison signal occurs 60 seconds after the initiation of the retrograde fire signal. The signal is routed through normally open contacts of the retrograde rocket assembly separation sensor relay in retrograde relay box No. 1. This relay is energized by limit switches which close when the retrograde assembly is blasted away from the spacecraft. The relay remains energized until the 0.05g relay drops out. (The 0.05g relay de-energizes at 10,000 feet.) A d-c signal of approximately 2.4 ± 0.3 volts is applied through normally open contacts of this relay to the commutator.

14-51. EMERGENCY ESCAPE SEQUENCE

Emergency escape sequence instrumentation consists of mayday abort and tower escape rocket fire signal monitors.

14-52. MAYDAY

The Mayday signal is produced by the mayday alarm relay. This relay is energized by any abort signal. With the relay energized, 2.4 ± 0.3 volts d-c is applied to the commutator. Once initiated, this signal is present for the remainder of the mission. The mayday alarm relay is in launch and orbit relay box No. 4.

14-53. ASTRONAUT'S ABORT SWITCH

Instrumentation is provided to monitor an abort signal originating from the astronaut's ABORT Handle; this signal is applied to the commutator.

14-54. **TOWER ESCAPE ROCKET**

The tower escape rocket signal is obtained from the emergency escape rocket fire relay in launch and orbit relay box No. 2. This relay remains energized until orbit attitude is attained.

14-55. **LANDING SYSTEM SEQUENCE**

Landing system instrumentation consists of monitor circuits for chute deploy and jettison and release of the antenna fairing. These signals are approximately 2.4 ± 0.3 volts and are applied to the commutator. Main and reserve chute deploy signals are obtained from toggle switches in the chute compartment. Lanyards from the chutes operate these switches when the chutes deploy. The main chute jettison signal is obtained through a limit switch in the chute compartment. The antenna fairing release signal comes from the antenna fairing separation relay in the communications relay box. This relay is energized through a limit switch. All landing system signals remain on until impact.

14-56. **0.05G RELAY**

Instrumentation of 0.05g relay operation consists of an on-off type signal which indicates whether the relay is energized or de-energized. The relay may be energized by operation of the 0.05g sensor or by the command receiver. When the relay is energized, a d-c signal is applied to the commutator.

14-57. DROGUE CHUTE

Drogue chute deployment is monitored by a 2.4 ± 0.3 volt signal controlled by the drogue chute sensor, through a set of contacts on the antenna separation relay and is applied to the commutator.

14-58. LANDING BAG

The landing bag operation is monitored by a voltage signal applied to the commutator through two sets of unlock signal limit switches.

14-59. INSTRUMENTATION CONTROL SYSTEM

14-60. HIGH LEVEL CODING

The signals applied to the commutator are sampled once every 0.80 seconds. Commutator outputs are square wave pulses with amplitude between -1 and +3 volts. These pulses are applied to voltage controlled oscillators and pulse duration modulation converters.

(a) The frequencies of the voltage controlled oscillator is varied between 10.5 KC ± 6-3/4% by the commutated pulse amplitude signals. This frequency band corresponds to IRIG Channel 12. The frequency modulated outputs of the 10.5 KC voltage controlled oscillator is applied to an isolation amplifier.

(b) Commutator pulse amplitude modulation signals are also applied to a pulse duration modulation converter. The converter reshapes the pulse amplitude wave-shapes to obtain pulse duration wave trains. These wave trains are then applied to the tape recorder.

(c) Amplifier aeromedical signals are coupled to a 1.3 KC, 1.7 KC and 2.3 KC voltage controlled oscillators. A zero to full scale signal causes a deviation in center frequency of $\pm$ 6-3/4%. The frequency bands of the oscillators correspond to IRIG Channels 5, 6 and 7. The oscillator outputs are applied to an isolation amplifier.

(d) Attenuated ASCS rate signals are coupled to a .40 KC, .56 KC and .73 KC voltage controlled oscillator. A zero to full scale signal causes a deviation in center frequency of $\pm$ 6-3/4%. The frequency bands of the oscillators correspond to IRIG Channels 1, 2 and 3. The oscillator outputs are applied to the isolation amplifier.

(e) The commutated outputs, aeromedical signals and ASCS rate signals are combined. The isolation amplifier also accepts a signal from the compensating oscillator which serves as a reference during data evaluation to indicate fluctuations in tape speed. The composite signal from the isolation amplifier is applied to the tape recorder, the ground test umbilical and telemetry transmitter.

14-61. **LOW LEVEL CODING**

The signals applied to the commutator are sampled once every 0.30 seconds. Commutator outputs are square wave pulses with amplitude between -5 mv and 15 mv. These PAM pulses are applied to a 10.5 KC voltage controlled oscillator upon closure of the Temperature Survey Switch to

TEST, and also to a pulse duration modulation converter for tape recording. The converter reshapes the pulse amplitude wave to obtain a pulse duration wave train. While the instrumentation system is in the low level mode (Temperature Survey Switch to TEST), all high level commutated inputs are removed from the 10.5 KC VCO.

14-62. **TRANSMISSION**

Ground testing and control of the instrumentation system is provided through the umbilical receptacle. Nonradiating checks can be performed to evaluate system operation. Radiating checks are performed through the telemetry link. Refer to Section XII for further information regarding telemetry.

14-63. **INSTRUMENTATION CONTROL**

The instrumentation system controls and programs (see Table 14-1) power to its own and other systems equipment by means of mode relays and programmer. (See Figure 14-5).

(a) The water extractor in the environmental control system is also programmed at regular intervals during the mission.

(b) Calibration voltages, R-calibrate for maximum readings and Z-calibrate for minimum readings, are supplied periodically to the monitoring instrumentation circuits. This is done prior to launch and by ground command or programmer at intervals during orbit.

SEDR 104

14-64. **INSTRUMENTATION RECORDING** (See Figure 14-5)

Recording instrumentation consists of a tape recorder and a portable 16 millimeter camera.

14-65. **TAPE RECORDER**

A low power, lightweight tape recorder provides seven channels for data recording. Voice communications, high level commutator pulse duration modulation signals and VCO's are applied to tracks 3, 5 and 6, respectively. Low level commutator pulse duration signals are applied to track 4. Stick position VCO's are recorded on track 2. The tape recorder operates continuously from umbilical separation to spacecraft separation plus five minutes and from TR-30 seconds to impact plus 10 minutes. During orbit, the tape recorder is on 1 minute every 10 minutes with a total of 30 seconds every 1 hour allowed for zero and full scale calibration. The astronaut is provided with a CONTINUOUS - OFF - PROGRAMMED switch on the main instrument panel which allows manual as well as automatic control of the tape recorder. A VOX switch located on the main instrument panel when placed to RECORD will allow the astronauts conversation to be recorded. A recording indicator light is located on the center of the main instrument panel when the recorder is operating so that the astronaut may record voice at these times if he desires.

14-66. **ASTRONAUT PORTABLE CAMERA**

A light weight portable 16 millimeter camera is provided for the astronauts use during the mission. The camera operates at a single speed of 6 frames per second. During non use, the camera is stowed along with

TABLE 14-1

INSTRUMENTATION PROGRAMMING

Equipment	Before Umb Separation	Umb Ej. To Cap Sep +60 Sec.	Cap Sep Plus 60 Sec to Tr-30 Jettison	Tr-30 to Retro	Retro Jett to .05G	.05G To Ht Shld Deploy	Ht Shld Onward	Mayday Onward
ECS Water Extractor	30 Sec/ 30 Min.	──►						
Tape Recorder	Continuous	Continuous	1 Min/ 10 Min.	─── Continuous ──────────────────────────► (Impact +10 min)				
Telemetry	6 Minutes on Command	────────────►	15 Sec/60 Min and 15 Sec/TIM Command 15 Sec/60 Min 15 Sec/TIM Command					
R-Cal Z-Cal	15 Seconds each on Command							

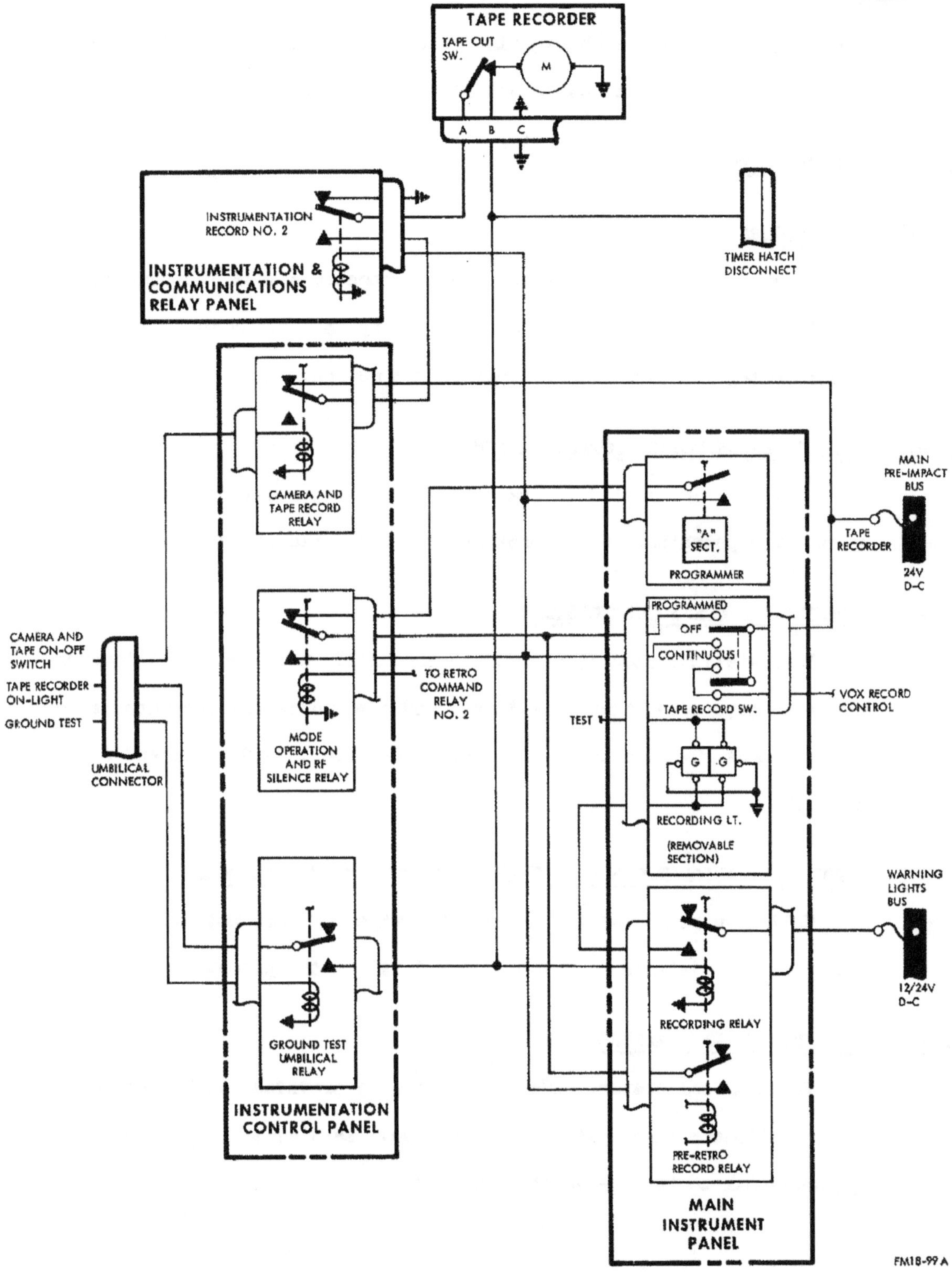

Figure 14-5 Tape Recorder Control Circuit

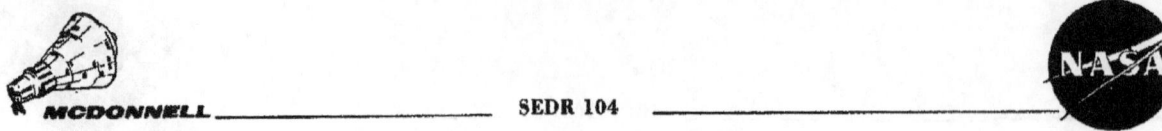

6 magazines of film (each magazine contains approximately 120 ft. of film) in the personal effects container on the main instrument panel.

14-67. SYSTEM UNITS

14-68. TRANSDUCERS

Potentiometer type transducers are connected across instrumentation 3 volt d-c power. The wiper is activated by the action to be measured. Wiper voltage is then proportional to the action.

14-69. CONTROL STICK MOTION POTENTIOMETERS

Control stick motion is translated into rotary potentiometer movement. One potentiometer is provided for each axis of motion.

14-70. SATELLITE CLOCK POTENTIOMETERS

The satellite clock (refer to Section XIII) utilized potentiometers to indicate elapsed time from launch and time to retrograde outputs for 0-10 seconds, 0-1 minute, 0-10 minutes, 0-1 hour, 0-10 hours and 0-60 hours.

14-71. MANUAL AND AUTOMATIC SUPPLY HELIUM PRESSURE POTENTIOMETERS

Helium supply pressures actuate the wiper of each potentiometer transducer to a resistance position proportional to the pressure.

14-72. ATTITUDE POTENTIOMETERS

The ASCS calibrator (refer to Section V) provides synchro actuation of potentiometers for pitch, roll and yaw. Each wiper output is then proportional to the spacecraft attitude for that axis.

14-73. **MAIN AND RESERVE OXYGEN PRESSURE POTENTIOMETERS**

Each oxygen bottle pressure actuates a dual potentiometer transducer. A low resistance linear element is used to operate a panel indicator while a higher resistance linear element is used for instrumentation. Wiper voltage outputs are proportional to applied oxygen pressure.

14-74. **STATIC PRESSURE AND SUIT PRESSURE POTENTIOMETERS**

Each pressure transducer is used to provide a folded linear output proportional to the applied pressure. Pressure ranges are 0-15 psia for static pressure and 0-15 psia for suit pressure.

14-75. **RESPIRATION RATE PICKUPS**

Respiration rate is sensed by two, one-half inch square stainless steel wire screens. These electrodes are attached with silver metalized adhesive to the astronauts body. Small connecting wires leading to the astronauts suit disconnect complete the circuit.

14-76. **RESISTANCE ELEMENT TRANSDUCERS**

Resistance elements are used to measure temperatures. The resistance of the element varies proportionally to its temperature. Mounting of the element depends on the application. Stick-on surface temperature elements are small, lightweight units. Other elements are mounted as an integral part of the spacecraft structure. The following list indicates the purpose and approximate temperature and resistance ranges for each transducer.

(a) Ablation Heat Shield Temperature: -55° to 2000°F
(b) Suit Inlet Temperature: 35° to 100°F -249 to 300 ohms
(c) Cabin Air Temperature: 40° to 200°F -232 to 316 ohms

14-77. **BODY TEMPERATURE TRANSDUCER**

The body temperature transducer is a rectal temperature pickup which consists of a thermistor imbedded in sealing compound at the end of a flexible wire lead.

14-78. **D-C CURRENT SHUNT**

The shunt resistance used for the instrument panel ammeter also supplies voltage for instrumentation. This shunt is discussed in Section XI of this manual.

14-79. **ELECTROCARDIOGRAPH PICKUPS**

Cardiac activity is sensed by four, one-half inch square, stainless steel wire screens. These electrodes are attached with adhesive to the astronaut's body. Small connecting wires leading to the astronaut's suit disconnect, complete the circuit.

14-80. **OXYGEN PARTIAL PRESSURE TRANSDUCER**

The O_2 Partial Pressure transducers are used to convert O_2 partial pressures to a signal compatible with high level telemetry. The voltage range of 0-3 V d-c output is representative of 0-6 psi oxygen partial pressure.

14-81. **TAPE RECORDER**

A low power, lightweight tape recorder is used in the spacecraft to make available 7 channels of recorded data. At the present time channel

assignments are as follows: Track 3, voice communications; Track 5, commutator PDM; Track 6, VCO's, Track 4, low level commutator PDM, Track 2, stick position VCO's. Tape speed is 1-7/8 ips. Tape capacity is 6,250 feet of ½ inch wide mylar base tape. The tape transport consists of a capstan drive, supply reel and take-up reel mechanism. A d-c motor is used, through reduction gearing, for capstan drive. A limit switch is provided to interrupt recorder power should the tape break. Record amplifiers are incorporated in the unit for Channels 2, 3 and 6. Channel 5 is fed by an amplifier incorporated in the commutator located in instrumentation package A. Channel 4 is fed by an amplifier incorporated in the low level commutator.

14-82. INSTRUMENTATION PACKAGE C

The C package incorporates units of various functions into one compact panel allowing convenience of mounting and of making electrical connections. These various sub-units are discussed in the following paragraphs.

14-83. CABIN PRESSURE TRANSDUCER

Cabin pressure actuates the wiper of a 10,000 ohm potentiometer located in the C package. Three volts d-c from the C package is applied across the potentiometer. Wiper output voltage is then proportional to the cabin pressure.

14-84. CONVERTER CARDS

The instrumentation package utilizes a unique method of construction and mounting of the transistorized amplifiers, power supplies, attenuators and monitors. Each unit consists of the necessary component parts mounted

SEDR 104

on a printed circuit, dielectric card with printed connector contacts at the bottom for plug-in insertion. The card is then covered, with the exception of base connector contacts and side mounting edges, with a thin layer of epoxy resin. This coating is used to provide moisture protection, to insure operation in a 100% oxygen atmosphere and to improve mechanical rigidity of components. These Converter Cards are package mounted in boxes providing side rails, base contact receptacles and printed circuit interconnections.

14-85. D-C CURRENT AMPLIFIER CARD

An amplifier located in the A package is used to bring the d-c current shunt voltage up to a maximum of 3 volts d-c. The amplifier feeds the commutator. The amplifier is transistorized and card mounted.

14-86. RESPIRATION RATE CALIBRATION AND ATTENUATOR CARD

Resistors are mounted on the card for attenuation of 24 volt d-c voltages to proportional voltages compatible with the commutator.

14-87. VOLTAGE MONITOR CARDS

Resistors, capacitors and circuit isolating crystal diodes are used to attenuate reaction control solenoid valve energizing voltages, command receiver signal strength and emergency O_2 rate prior to application to the commutator. Each attenuator circuit output, a maximum of 3 volts d-c, is applied to the commutator.

14-88. HORIZON SCANNER CARD

The horizon scanner amplifier card provides circuitry for processing the scanner roll and pitch signals, roll and pitch ignore signals prior to being applied to the commutator. During launch and orbit, the scanners are operating continuously; however, during orbit the signals applied to the ASCS and the commutator are timed by the programmer located on the pedestal.

14-89. PROGRAMMER

The programmer contains switch contacts which operate control circuits for specific intervals. The programmer is mounted forward near the center of the main instrument panel.

14-90. A - SECTION

The programmer used for orbital missions consists of two sections. When power is applied to the programmer, electronic controlled timers continuously operate the following contacts:

CONTACTS CLOSED

WAFER SECTION	DURATION	RATE
1. Water extraction	30 Seconds	1 per 30 minutes
2. Full Scale Calibrate	15 Seconds	1 per 60 minutes
3. Zero Calib.	15 Seconds	1 per 60 minutes
4. Tape	1 Minute	1 per 10 minutes
5. Horizon Scanner	8.5 Minutes	1 per 30 minutes

SEDR 104

14-91. B - SECTION

The B Section of the programmer is energized through the command receiver-decoders and the electronic timers to provide full scale and zero calibrate signal, as follows:

WAFER SECTION	CONTACTS CLOSED DURATION	RATE
1. Full Scale Calibrate	15 Seconds	On Command
2. Zero Calibrate	15 Seconds	On Command
3. Telemetry	6 Minutes	On Command

14-92. INSTRUMENTATION PACKAGE A

The A package also incorporates units of various functions into one panel. These sub-units are discussed in the following paragraphs.

14-93. CABIN AIR TEMPERATURE TRANSDUCERS

A platinum resistance wire is used to measure cabin air temperature. Temperature changes from $40°$ to $200°F$ cause the element to change resistance from 236 to 316 ohms. The resistance element forms a part of an amplifier circuit. A transducer is used in conjunction with an amplifier to supply signal to the commutator.

14-94. FILAMENT TRANSFORMER

A filament transformer is used to step down 115 volts 400 cps spacecraft power to 6.3 volts for use in package A

14-95. CONVERTER CARDS

The instrumentation package A also utilizes the Converter Card principle for amplifiers and power supplies.

14-96. **RESISTANCE ELEMENT AMPLIFIER CARDS**

The same type amplifier is used for heat shield, suit inlet air and cabin air temperature transducer signals. Each amplifier is of dual channel design in order to accomodate two transducers. Seven volts, a-c is supplied from the resistance element power supply. This voltage is applied across a bridge circuit in each amplifier. The transducer associated with each bridge circuit causes the voltage in the circuit to vary proportionally to the transducer temperature. This voltage change appears across a transformer and is rectified, using crystal diodes, to a maximum output of 3 volts d-c. The output from the amplifier is supplied to the commutator. Two relays on each card amplifier allow full scale and zero calibration of each channel. Calibration resistors are also provided for each channel.

14-97. **RESISTANCE ELEMENT A-C POWER SUPPLY CARD**

Resistance element amplifier circuits require 7 volts, a-c. Spacecraft power, 24 volts d-c, is applied to a transistorized power inverter. The inverter, using zener diode-transistor voltage regulation and transistor switching, supplies a 7 volt a-c output which is monitored by an attenuator, rectifier circuit. The monitor output, a maximum 3 volts d-c, is applied to the commutator.

14-98. **BODY TEMPERATURE AMPLIFIER CARD**

A transistorized d-c amplifier is used to increase the output of the temperature transducer to a maximum 3 volt d-c level prior to application to the commutator. This amplifier is the same type as that used for the d-c current signal amplification.

SEDR 104

14-99. SIGNAL CONDITION AND D-C SUPPLY CARD

This Converter Card provides four functions in the instrumentation system. Filament transformer output is applied to a monitor circuit. This circuit attenuates and rectifies to provide a maximum 3 volt d-c signal indicating transformer operation. Spacecraft power, 24 volts d-c, is applied to an attenuator circuit which provides a 3 volt d-c output for the monitoring circuit. This 3 volt output is then applied to the commutator as a monitor of the main 24 volt d-c bus voltage. The signal condition and d-c power supply card also provides meter attenuator resistors which limit current flow in the panel indicator circuits. These circuits involve the primary and secondary oxygen supply pressures, and the manual fuel supply pressures.

14-100. COMMUTATOR-KEYER-RECORD AMPLIFIERS

A unit is provided in the A package for commutating transducer data and supplying PDM and PAM outputs. The commutator portion of the unit is a 90 x $1\frac{1}{4}$ solid state device which samples sequentially, 88 channels of signal input information. The output produced is a pulse amplitude modulated signal wave train. Each 0 to 3 volts d-c input to the commutator is sampled $1\frac{1}{4}$ times per second per IRIG standards. The PAM wave train output is fed through a buffer stage to a PAM/PDM converter. The PDM output is then applied to a record amplifier which produces a signal capable of directly driving the recorder head in the spacecraft tape recorder. The PAM output is fed through a gating circuit which introduces a master pulse and negative pedestal pulses to operate automatic decommutation

MCDONNELL SEDR 104

equipment in the ground station. A power supply is incorporated in the unit to provide the positive and negative voltages required in the circuits.

14-101. **INSTRUMENTATION PACKAGE D**

The primary function of the D package is to convert spacecraft information to signals capable of modulating the telemetry transmitter. Transducers and amplifiers are also contained in the package to complete spacecraft information circuits.

14-102. **ACCELEROMETER**

The Z axis accelerometer is mounted in the D package and used, to determine the static longitudinal, acceleration of the spacecraft. The accelerometer unit gives a d-c output which is applied to 3 channels of the commutator.

14-103. **ELECTROCARDIOGRAM AMPLIFIER CARDS**

Two amplifiers are used for the EKG transducer inputs. Each amplifier increases the transducer output to a 3 volt peak to peak signal.

14-104. **OSCILLATORS**

The D package supplies sub-carrier oscillators to allow one channel of instrumentation data. The following paragraphs describe the sub-carrier oscillators.

14-105. **COMPENSATING OSCILLATOR CARD**

During playback of a tape, the recorded signal from this oscillator is monitored to detect changes in tape recorder speed. A frequency

SEDR 104

shift indicates a change in speed. The oscillator is of module construction and operates at 3000 cps. Output level is adjustable.

14-106. **VOLTAGE CONTROLLED OSCILLATOR (V.C.O.) CARDS**

Instrumentation data voltages are applied to the sub-carrier oscillators causing oscillator frequency shift proportional to the input amplitude. The transistorized oscillator consists of a free running multivibrator and filter. The oscillator functions and frequencies are given below:

1. High and (Low Level commutator test only) - 10.5 KC VCO
2. Left Side (Comm) EKG, Right Side (+) EKG and Blood Pressure - 2.3 KC VCO
3. Upper Chest (Comm) EKG and Lower Chest (+) EKG - 1.7 KC VCO
4. Respiration Rate - 1.3 KC VCO
5. Solenoid Malfunction Detector - 5.4 KC VCO
6. ASCS Roll Rate and Low Solenoid - .73 KC VCO
7. ASCS Pitch Rate and Low Solenoid - .40 KC VCO
8. ASCS Yaw Rate and Low Solenoid - .56 KC VCO

14-107. **MIXER AMPLIFIER CARD POWER SUPPLY**

Spacecraft power, 24 volts d-c, is converted to 6 volt d-c for use by the sub-carrier oscillators. A mixer circuit combines the sub-carrier oscillator outputs. Solid state components for these circuits are combined on one Converter Card.

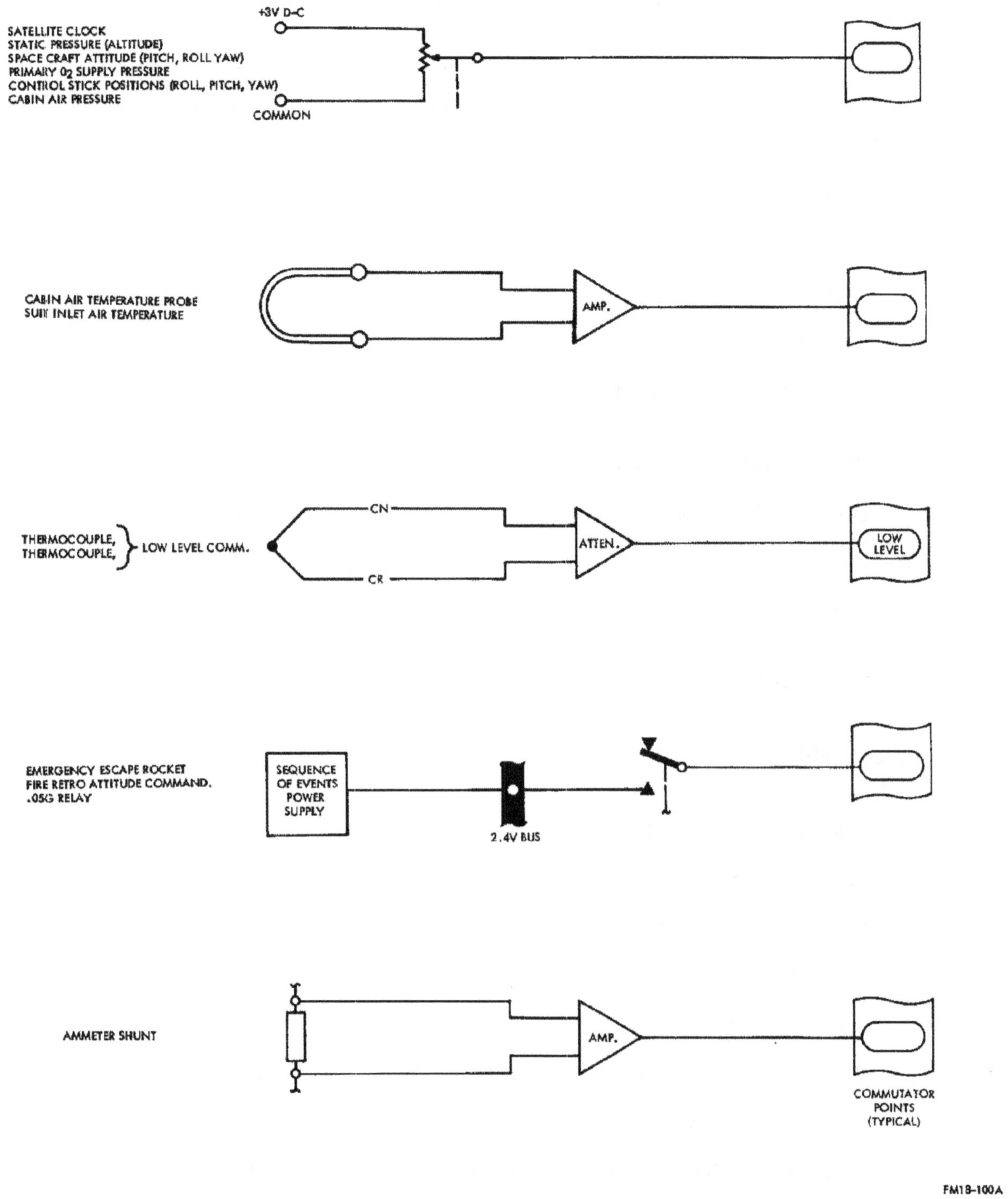

Figure 14-6 Typical Commutator Signal Sources

TABLE 14-2 INSTRUMENTATION COMMUTATOR POINT
ASSIGNMENT (HIGH LEVEL 0-3 VOLTS.)

PARAMETER	CHANNEL
ELECTRICAL POWER SYSTEM	
Main 24 V D.C. Bus	26
D.C. Current	34
ASCS A.C. Bus	77
Fans A.C. Bus	33
Secondary Power Supply	19
INSTRUMENTATION POWER SUPPLIES	
3 V D.C. Reference	1
0 V (Zero) Reference	2, 72
7 V, 400 CPS	3
STATIC PRESSURE	22
CALIBRATION ON	64
ENVIRONMENTAL CONTROL SYSTEM	
Suit Inlet Pressure	8
Primary O_2 Pressure	9
Cabin Heat Exchanger Gas Temperature	74
Cabin Temperature	10
Cabin Heat Exchanger Dome Temperature	20
Suit Inlet Air Temperature	11
Suit Heat Exchanger Dome Temperature	21
Secondary O_2 Supply Pressure	12

SEDR 104

PARAMETER	CHANNEL
Cabin Pressure	82
Cabin O2 Partial Pressure	6
Emergency O2 Rate	68
Suit CO2 Partial Pressure	27
REACTION CONTROL SYSTEM	
Reaction Control Supply Pressure (Auto.)	39
Reaction Control Supply Pressure (Man.)	40
Stick Control Positions	
Roll	23, 58, 80
Yaw	25, 50, 83
Pitch	24, 81
ACCELERATION	
Z Axis (longitudinal)	16, 38, 61
STRUCTURAL TEMPERATURE	
Heat Shield (Center)	75
Heat Shield (Edge)	76
AEROMEDICAL DATA	
Body Temperature	4
EKG	1.7 & 2.3 KC VCO
Respiration	1.3 KC VCO

SEDR 104

PARAMETER	CHANNEL
0.05G RELAY	87
HORIZON SCANNER	
Roll Ignore	84
Pitch Ignore	85
Roll Output	86
Pitch Output	88
ATTITUDE	
Pitch (ASCS)	15
Roll (ASCS)	17
Yaw (ASCS)	18
ATTITUDE RATE	
Pitch	.40 KC VCO
Roll	.73 KC VCO
Yaw	.56 KC VCO
RCS CONTROL SOLENOIDS	
H.P. Pitch Up	65
H.P. Pitch Down	66
H.P. Roll +	69
H.P. Roll -	70
H.P. Yaw -	78
H.P. Yaw +	79

PARAMETER	CHANNEL
SATELLITE CLOCK (ELAPSED TIME)	
10 Seconds	28, 59
1 Minute	29
10 Minutes	30
1 Hour	31
10 Hours	32
60 Hours	67
SATELLITE CLOCK (RETROGRADE TIME)	
10 Seconds	41
1 Minute	42
10 Minutes	43
1 Hour	44
10 Hours	45
60 Hours	63
NORMAL LAUNCH SEQUENCE	
Tower Separation	46
Spacecraft Separation	47
Retro Attitude Comd.	48
Retro Rkt. Assy. Jett.	53
ELAPSED TIME SINCE #2 RETRO FIRE	
10 Seconds	13
1 Minute	14
10 Minutes	52
1 Hour	62

PARAMETER	CHANNEL
Ref. Voltage Elapsed Time Since #2 Retro Rocket Fire	57
EMERGENCY ESCAPE SEQUENCE	
Pilot Abort	71
May Day	60
Escape Rocket Fire	49
LANDING SYSTEM SEQUENCE	
Drogue Chute Deploy	54
Antenna Release	55
Main Chute Deploy	56
Reserve Chute Deploy	73
Landing Bag Deploy	51
Retro Fire No. 1	35
Retro Fire No. 2	36
Retro Fire No. 3	37
COMMAND RECEIVER	
Signal Strength	7
All Channel Signal	5

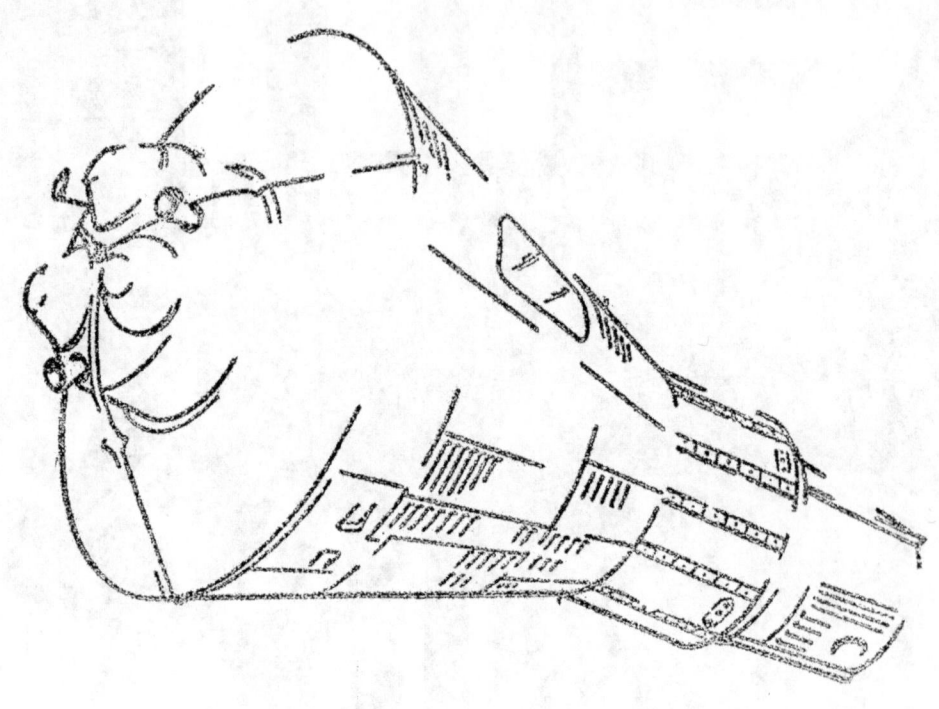

©2011 Periscope Film LLC
All Rights Reserved
ISBN #978-1-935700-68-5

www.PeriscopeFilm.com

www.ingramcontent.com/pod-product-compliance
Lightning Source LLC
Chambersburg PA
CBHW082025300426
44117CB00015B/2359